W9-AYX-603

STOREY'S GUIDE TO RAISING HORSES

2ND EDITION

Storey's Guide to

RAISING HORSES

Breeding ▪ Care ▪ Facilities

Heather Smith Thomas

Storey Publishing

*The mission of Storey Publishing is to serve our customers by
publishing practical information that encourages
personal independence in harmony with the environment.*

Edited by Rebekah Boyd-Owens, Sarah Guare, and Deborah Burns
Art direction and book design by Cynthia N. McFarland
Cover design by Kent Lew
Text production by Erin Dawson

Front cover photograph by © 2009 Mark J. Barrett www.markjbarrett.com
Author photo by Lynn Thomas
Interior photography by the author, except for pages 12, 14, 15, 18, 25, 26, 30, and
    77 by © Dusty Perin www.dustyperin.com
Illustrations by Elayne Sears, except for page 50 by James E. Dyekman

Expert read by Rhonda Massingham Hart
Indexed by Andrea Chesman

Storey Publishing
210 MASS MoCA Way
North Adams, MA 01247
www.storey.com

Printed in the United States by Versa Press
10 9 8 7 6 5 4 3 2 1

Library of Congress Cataloging-in-Publication Data

Thomas, Heather Smith, 1944–
  Storey's guide to raising horses / by Heather Smith Thomas. — 2nd ed.
    p. cm.
  Includes index.
  ISBN 978-1-60342-471-4 (pbk. : alk. paper)
  ISBN 978-1-60342-472-1 (hardcover : alk. paper)
  1. Horses. I. Title. II. Title: Guide to raising horses.
SF285.3.T496 2009
636.1—dc22
                    2009045289

To Velma Ravndal — my 4-H leader and mentor — who shared her love and knowledge of horses, and her high standards of horsemanship, with many eager young riders.

We learned from her that good horsemanship always has a purpose: to promote unity of horse and rider, always with the best interests of the horse at heart. My commitment to horses and to sharing that commitment with others in my writing is a direct result of her inspiration and guidance during my youth.

Contents

Preface

FOR THOUSANDS OF YEARS, the horse has served humankind; our history is inextricably linked with his. Until recently, he has been our chief means of transportation and source of power for farming. After the invention of tractors and cars, horses' numbers dwindled, but, since the 1950s, there's been renewed interest in horses — for sport and pleasure and for working small farms.

The horse of the twenty-first century has a distinct advantage over his forebears. In the early days, nearly everyone owned a horse, the principal means of conveyance. Many people did not have the aptitude, understanding, or patience to be good horse handlers or caretakers. Today, horses are no longer needed for transportation, and people who own horses enjoy and appreciate them. Now, perhaps as in no other time in history, the horse in the United States is owned by folks who respect him, seek to understand him, and want to do their best to take good care of him.

Our biggest stumbling block as horse owners comes from our Computer Age way of thinking: wanting to accomplish everything quickly, whether training a horse or learning to ride. Contemporary conveniences and time-saving devices have spoiled us; we want shortcuts. There is no shortcut to becoming a good horseman or horsewoman, a knowledgeable horse owner, and a good horse breeder. There's no substitute for time invested and the desire to learn all we can. It can take a lifetime.

Dedication makes a horse person. Only one who seeks to understand a horse can truly enjoy him. The casual horse enthusiast may admire a horse from a distance or ride one on weekends, but that person is missing a great deal. One who lives with horses, cares for them daily, and works with them, sincerely trying to improve horsemanship and understanding while continually thinking about them, their problems and personalities — that person will get the most satisfaction from horses.

Fully enjoying horses involves more than admiring them in pictures or watching one gallop across a pasture. The key to really enjoying horses is doing — feeding, grooming, breeding, showing, moving cattle, racing, jumping, or riding for pleasure. We all need a sense of identity, of belonging — a realm in which we feel at home and with which we identify. Working with horses and accomplishing something with them can give us this sense of purpose.

Understanding and experience make a good horseman. An intuitive "feel" for horses — a sense of what a horse needs at a certain time and how best to care for him — makes a responsible horseman. Part of fully enjoying a horse is caring for him as best we can: being able to tell when he's fit and healthy and when he isn't, and, if he isn't, being able to tell what he needs and whether he needs veterinary attention.

Your horse needs proper care — the right amount and right kind of feed, and enough of the right kind of exercise to keep him fit and in condition for the work he is to do. The good horse person cares for his or her horses as regularly and responsibly as possible, as one would care for and love a child or friend. When it comes to horses, the dedicated horse person is a perfectionist and will do anything and everything to take proper care of them and learn what that proper care entails.

Horses are unique creatures. The horseman must come to understand each horse in his care, learning what's best for each individual in a given situation. There is no substitute for the "eye of the master." The person who understands a horse will give him the proper amount of work and rest, reward and discipline. Horse owners have an enormous responsibility, for these animals are totally dependent on us for their existence and well-being.

Caring for a horse is a challenge, and there is much to learn. The serious student of horses today is fortunate because there are many sources of help and advice. This book does not attempt to cover all aspects of horse care and breeding but will be useful to the reader who wants to learn about the basics of good horsekeeping, raising horses, and their health and soundness.

My aim in this book is to provide useful information to the horse owner with special emphasis on horse care and "intelligent horsekeeping." This is not a book on training (although every time you handle a horse you are training him, for better or worse) but on horse husbandry — the art of keeping a horse fit and healthy in body and spirit. Learn from your horses and they will learn from you.

In this updated edition, I have included cutting-edge information and have deleted a few passages that were based on outdated ideas or theories. Modern technology and medical research, for example, have given us new feed products for insulin-resistant horses and horses with muscle problems, new treatments for EPM, new vaccines for influenza, and a vaccine for West

Nile Virus. West Nile Virus was unheard of in North America when this book was originally written.

Research has also given us better strategies for parasite control. Genetic research has shed more light on several devastating inherited defects in certain breeds and family lines of horses, and new DNA tests have given us ways to check for these when selecting breeding stock. Several new sections have been added to various chapters to bring the reader up to date on current horsecare practices.

For specific information regarding training horses, see *Storey's Guide to Training Horses.*

Author's Note: Throughout the text, except when referring to mares or fillies, I use the masculine pronoun to refer to horses. This is done for simplicity and reader ease.

Acknowledgments

I owe special thanks to many people who helped make this book possible. I thank the many horsemen and horsewomen who have given me good advice and encouragement. This is a wonderful aspect of the world of horses — other horse people are always willing and happy to help you. We share a common bond; our love and respect for horses drives and unites us in a special way.

I thank my husband, Lynn, for his support, patience, help, and encouragement in all my endeavors with horses. His partnership and contributions have been indispensable. He's helped me with everything. He's built foaling pens and hay sheds, held sick horses for treatment, accompanied me on a "babysitter horse" while I started a youngster under saddle on first rides into the mountains surrounding our cattle ranch, and so much more.

Lynn and I raise beef cattle on a ranch that depends on four months of summer grazing in the mountains, and we need our horses to help us take care of the cattle. We appreciate athletic and sure-footed horses with the stamina to travel day after day in rough country, working cattle.

Our horses are an integral part of our ranch; we could not raise cattle without them. Yet they are a joy and pleasure as well as a necessity for the job. We delight in good horses who enjoy their work, and we respect the amazing partnership that we have with them. We travel many miles with our horses and have, in past years, enjoyed endurance riding with other horse people, relishing their camaraderie and benefiting from the fact that our horses were already fit for this demanding sport just from doing the range riding.

Our life with horses has given me much of the knowledge for writing this book. Therefore I must also thank several generations of horses who have grown up here on Withington Creek over the past 50 years. My firstborn foal, Khamette (1959), was raised as a 4-H project when I was in high school, and we still have grandsons of hers on the ranch. My father's old Thoroughbred mare Nell blessed us with many half-Arab offspring and their descendants; some have been the best cowhorses I've been privileged to know.

Over the years, we've had numerous horses of various breeds — Arabian, Thoroughbred, Quarter Horse, Appaloosa, Morgan — and appreciated the talents of all. Working with them has been one of my greatest satisfactions and accomplishments. They have been my teachers as well as my pupils, and I owe a large debt of gratitude to them. I cannot imagine my life without horses. They are the reason for my writing and the inspiration for this book.

PART I

Basic Horsekeeping

1

Facilities

GOOD HORSEKEEPING INVOLVES two important, interrelated components: dedicated and conscientious personal care of each animal and the use of good facilities. Conscientious care entails regular daily feeding and maintenance if horses are confined and daily inspection if they're at pasture. But even with the best of care, if you don't have good facilities for your horses, they may become ill or injured. For example, a poorly ventilated barn may cause respiratory problems, a stall that's too small may get a foaling mare into trouble, a poorly designed or worn-out fence may invite injury, and a weedy pasture containing toxic plants may cause colic or fatal illness. If you strive to make sure that your facilities are adequate, safe, and in good repair, you can greatly reduce the risk of injury to your horses. And you make your own job of caring for each animal much easier.

The more natural the conditions in which your horse lives, the healthier and happier he will likely be. This usually means outdoors rather than indoors, at pasture rather than confined in a corral, living with a buddy instead of in isolation, and so forth. But we can't always keep our horses in an ideal situation and must make do with artificial conditions, striving to make the facilities as safe and horse-friendly as possible.

Pasture

If you have adequate acreage to provide pasture for your horse or horses, the pasture will be your most important "facility." All too often a pasture area is used as just a turnout space, with little thought as to keeping the plants healthy and protected from overgrazing. Horses tend to use some parts of a pasture more heavily than others. They graze the same plants over and over. They also trample out the plants in areas where they spend time resting or swatting flies,

SMALL-PASTURE WOES

In a small pasture, horses may beat out the grass in certain areas near fences or watering places, under shade trees, and so on. They'll overgraze their favorite types of grass, leaving others. Horses also don't like to graze grass that grows where they have defecated.

or interacting with neighbor horses across the fence. To keep a pasture viable, you need a management plan to control or alternate grazing use.

Good pasture is the best feed for mares and foals and for growing young horses. Mature idle horses may not need all the nutrients green pasture provides and often get too fat (unless their time on pasture is limited), but having pasture gives you some options for feeding and for allowing horses room to exercise.

The size of pasture needed to support a horse varies greatly depending on climate (a wet climate grows more grass than a dry one, unless you provide extra water with irrigation), soil fertility, and types of grasses growing there. It takes a lot more dry-land acreage to support a horse than well-watered green pasture.

A pasture should be large enough for the number of horses intended to graze it and should carry several types of palatable grasses. A mix of clover or some other legume with grass makes good pasture; legumes are high in protein. A pasture with too much clover, however, can be detrimental to horses — if they eat too much, they may founder. Some types of clover, such as alsike, may also contain dicoumarin, which is toxic.

If you plan to have pasture supply most or all of your horses' feed, assess soil fertility and grass types. Your county agricultural Extension agent can help with this. Then you'll know whether the pasture can provide proper nutrients for horses or if it needs fertilizer, renovations, or reseeding to produce better types of forage plants. The county Extension agent can advise you on the best type of fertilizer for your soil or plants, as well as what kind of plants that are nutritious for horses do well in your climate and soil type.

Weed Control and Pasture Management

Check pastures for weeds. Some are harmless (or nutritious, like dandelions and lamb's-quarters), others take up space that could be better used by good forage plants, and others are poisonous to horses. Certain poisonous plants are common to the Southwest, the East, the plains states or mountain West, or coastal areas, so check with your veterinarian or county Extension agent about what to look for in your area. (Some of the most common poisonous plants are listed in chapter 11.)

Assess the weeds you have. Large patches of thistles or other undesirable plants should be eliminated. Patches of foxtail or cheatgrass (downy brome) are undesirable because the sharp awns of their seeds can become embedded in a horse's mouth and cause sores or abscesses.

The best defense against undesirable plants is a good stand of vigorous forage plants to keep weeds from becoming established on bare or disturbed ground. Control patches of weeds by spraying (check with your county Extension agent for proper types of herbicides and how to use them) and seeding the area back to grasses, or by mowing diligently. If you mow weed patches before they mature, you keep them from going to seed and spreading. You also reduce their competition with grass plants, which come in thicker when given a chance. If weeds regrow, you may have to mow several times during the growing season but, eventually, this will weaken and eliminate them.

To keep pasture healthy and prevent it from going to weeds, don't have more horses than it can support, and don't leave them on it too long. Horses are selective grazers who eat favorite plants down to the roots, grazing them over and over, letting others go to waste. They like new, green regrowth, and they keep eating the short grasses, leaving mature plants untouched. In wet seasons, they tromp part of the pasture into mud and eventually destroy the plants.

Pasture Rotation

For best pasture health, let it grow 4 or 5 inches (10 or 12 cm) tall before you turn horses onto it, and then remove them after they've eaten their favorite plants down to about a 2-inch (5 cm) stubble. If you leave them too long, they'll graze the best grasses down to the dirt or even paw them up to eat the roots. The best solution is to divide your pasture into several sections so you can rotate the grazing, moving horses to new pasture periodically so the grazed area can regrow. If you don't want to use permanent fencing for dividing a pasture, you can use temporary electric fencing to facilitate your rotational grazing. If you don't have enough pasture area to make rotation feasible, lock the horses in a pen part of the time; limit their time on pasture to what the plants can withstand. If they graze down some areas and leave others, mow the pasture periodically to clip off tall, mature plants. When the plants regrow, they'll be more palatable and the horses may eat them.

Barns

What you need will depend on the climate where you live and your situation. Many horses are at pasture year round, being fed hay if snow gets deep. They never see a barn and are none the worse for it, but they appreciate a windbreak, natural or human-made. In such cases, a grove of trees or brushy draw can provide a lot of shelter from wind or driving rain and snow.

If you work with your horses during winter or have mares who might foal during bad weather, it's practical to have a good barn. In a cold or rainy climate, horses in a pen or pasture with no windbreak should have a run-in shed as a place to get out of the weather. Shelter for a sick or injured horse is also helpful.

If there is already a barn on your property, it may just need some repair and maintenance to be hazard-free and safe for horses, or it may need a little remodeling to suit your purposes. If you are building a new barn, consider visiting other horsekeeping facilities to get ideas for design and layout that will work well for you and your horses. There are many good designs, and some that are not so good. A barn is expensive and should last your lifetime, so make sure it is well designed before you begin construction.

Building a Barn

In a cold climate, your horses need a barn that is not drafty and retains heat. In an area where summers are hot and humid, they benefit from a barn that stays cool. Choose the site carefully, considering climate as well as present and possible future needs.

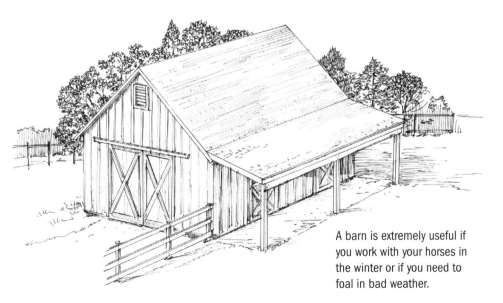

A barn is extremely useful if you work with your horses in the winter or if you need to foal in bad weather.

Location

Choose a well-drained site. The barn should be on higher ground than surrounding pasture, not in a swale or floodplain that gets wet with every major rain or snowmelt. You don't want a sea of mud encircling the barn, moisture seeping through stall floors, or spring runoff coming in the doors. If your ideal barn location is not in a good spot for drainage, arrange for a contractor to haul in enough fill to build up the area for good drainage.

Convenience is key. The barn should be within reach of pens and pastures and your house. It should be far enough away so flies and odors aren't a problem for your household but close enough that you can get to it quickly for chores and emergencies.

The barn should be located where there's access to electricity and water. After all, you'll want lights, outlets for electrical equipment and a heater, and a source of water for horses confined in stalls. Consider also whether there's room for additional barn area if, in the future, you decide to add more stalls or an indoor work area.

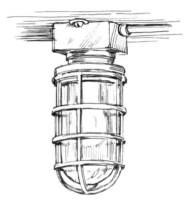

Lights in the barn (in the aisle, stalls, and feed areas) should be at least 8 feet (2.5 m) high and covered with a wire cage to protect them from breakage.

A safe light fixture has a wire framework to protect it from breakage, and a glass jar inside the wire cage to contain bulb fragments if the light shatters.

Permits

You may need a permit to build a barn if you live on the edge of town or in a subdivision. Check the local zoning regulations before you start building. Your barn must also be constructed in accordance with local building codes.

Building Site and Related Issues

Prepare a good building site before you start construction. The ground underneath should be solid with some rock so nothing will shift, sink, or sag. If the existing soil will not support a solid building with good drainage, the area should be excavated to a depth of at least 4 feet (1.5 m), and large coarse rocks put into the hole, with crushed rock on top of that layer. Fine gravel can be spread on top of the crushed rock. This will provide good drainage. One foot

CONVERSION CAVEATS

If you're converting an existing building (chicken house, garage, or some other type of shed) into a horse barn, make sure it is high enough inside — at least 8 or 10 feet (2.5 or 3 m) for large horses — so that your horse will never hit the ceiling with his head, even if he rears. And make sure the building does not have a concrete floor, which has poor drainage and offers poor footing. A horse may be injured from slipping and falling. Wood has some give but can be slippery; dirt makes the best floor.

(0.3 m) of soil or tamped clay (or a mixture of clay and sand — usually about three parts clay to one part sand) can be put on top of the gravel for the floor base, unless you choose to have a wooden or synthetic floor.

The barn should have a foundation of concrete, cinder blocks, or pressure-treated wood. An inexpensive pole structure can be built using tall, treated posts set deeply into firm ground as the main roof and wall supports.

Dirt floors are often used for stalls because they are easy to keep dry (they provide some drainage) and are not too firm or unforgiving if a horse lies down a lot or crashes down hard on the floor. You may want rubber mats on stall floors in some situations — such as for a foaling mare, young horse, or any horse who might spend time lying down.

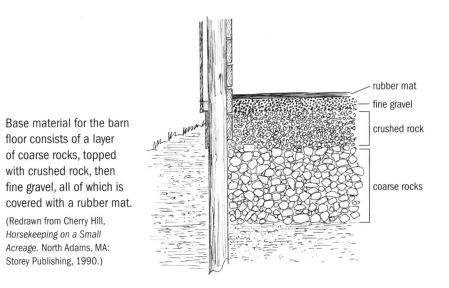

Base material for the barn floor consists of a layer of coarse rocks, topped with crushed rock, then fine gravel, all of which is covered with a rubber mat.

(Redrawn from Cherry Hill, *Horsekeeping on a Small Acreage*. North Adams, MA: Storey Publishing, 1990.)

Cover stall windows with bars so they aren't accidentally broken.

Windows should be covered with bars so they cannot be accidentally broken. They should let in light and sunshine but keep out wind and moisture, as should doors. Openings should face east and south for solar warmth in winter. The barn should be designed to stay warm in winter but cool in summer; take summer sun into consideration when placing windows.

Ventilation

You do not want an airtight barn that holds in humidity or stall dust. Dust from hay and bedding can cause respiratory problems in horses, as can an environment that stays too warm and humid. The barn shouldn't be drafty, but it must have adequate ventilation to continuously move out any contaminants in the air such as dust, mold spores, and ammonia fumes. When designing your barn to suit your site in terms of ventilation requirements, consider the following guidelines. In warm climates, a row of covered stalls that open to outside paddocks will do. In cold climates, two rows of enclosed stalls separated by a center aisle work well. The aisle should be at least 8 to 10 feet (2.5–3 m) wide; 12 feet (4 m) is even better if you have the space.

The main reason a horse is better off outside than in a barn — even in cold or stormy weather — is that it's unnaturally damp and dusty indoors. If your horse must be stalled in a barn, good ventilation can minimize the problems of dampness and dust.

Whenever an animal exhales, moisture accompanies air expelled from the lungs. This is why when you breathe in an enclosed space (such as a car on a cold day), condensation is produced — droplets of water "steam up" the windows. With a large animal, an even greater volume of moisture is exhaled with every breath. The average 1,000-pound (454 kg) horse puts out about 2 gallons (8 L) of moisture daily with air expelled from his lungs. A barn with six horses can have 12 gallons (45 L) of water vapor put into the air daily, as well as moisture created by the evaporation of urine.

Damp air (along with irritation of air passages from ammonia fumes produced by urine and manure) can contribute to respiratory problems and encourage growth of fungus, molds, and certain bacteria. Also, when weather is cold, moisture condenses on the underside of a metal or uninsulated roof, dripping down like gentle rain or freezing and making icicles that drip when they melt.

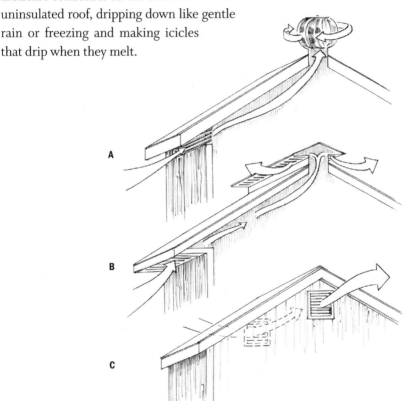

Three methods for improving barn ventilation: **A.** Air enters the upper wall louvers and is pulled out the top of the roof by a turbine vent. **B.** Air enters through soffit louvers under the eaves and exits through a continuous ridge vent. **C.** Air enters and exits through gable louvers. (Redrawn from Cherry Hill, *Horsekeeping on a Small Acreage*. North Adams, MA: Storey Books, 1990.)

KEEP THE AIR FLOWING

Ammonia fumes near the stall floor are a prime cause of lung irritations in foals (they spend a lot of time lying down), so make sure stagnant air from the lower part of the stall is moved out. In addition to the ventilation methods listed below, you may need louvers in the sides of solid stall walls to help clear stagnant air.

Air movement and ventilation are important to remove damp air and replace it with dry, fresh air. Condensation does not occur as much if the air is slowly moving, taking the moisture with it. Ventilation can be obtained by:

- Opening windows, barn doors, and tops of stall doors
- Making sure stall walls don't go clear to the ceiling
- Creating louvers or air spaces in the bottoms of stall doors so air flow can clear out the dust and ammonia fumes in the lower portion of the stall to improve air quality
- Installing vents, such as louvers under the eaves or roof ridge
- Installing a spinning ventilator on the roof to pull air out of the barn to supplement the ridge vents
- Using a fan to force air through the barn, making sure this does not create a draft

Heat

You probably don't need a heated barn; it requires good insulation and a source of heat, and it is very expensive. Moreover, a heated barn is an unnatural environment for horses. An unheated barn is usually much more healthful for horses; they are less prone to develop colds and respiratory problems. You won't need a source of heat except in unusual situations, such as for a

WET AND WINTRY CLIMATES

Install rain gutters and downspouts to handle water flow off the roof. If you get snow, build overhangs to shelter the doorways and keep entryways less muddy from snowmelt. If exercise paddocks are attached to the barn, locate them so that rain and snow coming off the roof won't make mud holes.

seriously sick horse or a new foal during very cold weather. At these times, you can usually supply enough warmth with a space heater. Consider cost, durability, maintenance needed, and fire resistance.

ADVANTAGES AND DISADVANTAGES OF COMMON BUILDING MATERIALS

When choosing building materials, consider cost, durability, and insulating quality. Different materials have various advantages and disadvantages.

Material	Advantages	Disadvantages
CONSTRUCTION MATERIALS		
Metal	■ Quick to build ■ Less expensive than wood ■ Fireproof ■ Minimal maintenance	■ Noisy during rain, hail, wind ■ Cold in winter, hot in summer (low insulating ability) ■ Can be damaged by kicking and rubbing
Wood	■ More forgiving (less likely to cause injury) than metal or stone	■ Materials and labor are expensive ■ Wood must be treated to prevent chewing ■ Not fireproof
Brick, cinder block, concrete, stone masonry	■ Cool in warm climate ■ Fireproof	■ Damp and cold in winter ■ Require good ventilation ■ Expensive
ROOFING MATERIALS		
Wooden shakes or shingles	■ Warm in winter, cool in summer (good insulating ability) ■ Do not cause condensation in barn	■ Materials and labor are expensive ■ Fire hazard ■ Cannot be used on flat roofs (water won't run off well) ■ Prone to weather and rot
Asphalt shingles, rolled roofing	■ Durable (20 years or more if properly installed) ■ Less expensive than wooden shingles ■ Easy to install ■ Less noisy than metal ■ Better insulating quality than metal	■ Prone to damage by heat, wind, or ice buildup ■ Become brittle in temperatures below freezing
Metal	■ Quick and easy to install ■ Relatively inexpensive and maintenance free ■ Translucent fiberglass sections can be installed to let in light	■ Hot in summer ■ Noisy during storms ■ Condensation forms inside and drips down ■ Moisture leaks around screws ■ Snow may slide off and build up in front of doorways and walls

Stalls

Most horses can be comfortable in a stall 12 × 12 feet (3.7 × 3.7 m), but for adequate room, foaling stalls should be at least 14 × 14 or 14 × 16 feet (4.3 × 4.3 m or 4.3 × 4.9 m). A narrow tie stall may be fine for working with a horse but should never be used for keeping a horse more than a few hours. Many horses are reluctant to lie down when tied and cannot get any exercise. A box stall is much more comfortable.

The stall should be large enough so the horse has a clean space in which to lie down and to eat. Many horses accustomed to living in a stall urinate and defecate in the same area, leaving the rest of the stall somewhat clean. Because no one size is perfect for every horse, it's wise to have several stall sizes in your barn. A 10- × 10-foot (3.1 × 3.1 m) stall may be adequate for a pony or a horse who must be strictly confined after injury. A large stall is needed for a large horse, a foaling mare, a horse who rolls frequently and gets cast against the wall in a small stall, or a recuperating horse who cannot be turned out but needs room to move around. It's handy to have a double stall divided by a mov-

able partition so you can use it as two separate stalls or as one large stall for a foaling mare or a mare with a young foal.

Stalls should be sturdy and durable to withstand rubbing, kicking, chewing, and the corrosive effects of urine and manure. Their walls should be smooth and free of hazardous projections that might injure a horse (wood splinters, protruding nails, bolt ends, and sharp edges). Stall doors should be at least 4 feet (1.2 m) wide and 8 feet (2.4 m) high so that a horse can't run into the edge or hit his head on the top of the opening.

The top portion of the front wall can be pipe or metal bars so the horse can see out and

The top of the stall wall or stall door should be made of pipe or metal bars so the horse can see out. This door top has a drop panel that can be closed if needed.

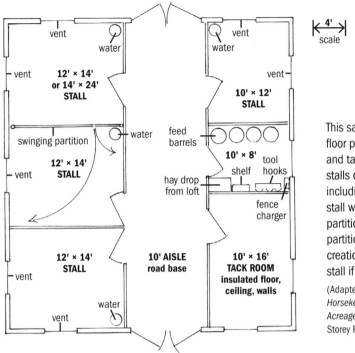

This sample barn floor plan shows feed and tack rooms and stalls of various sizes, including a double stall with a movable partition. A movable partition facilitates creation of one large stall if needed.

(Adapted from Cherry Hill, *Horsekeeping on a Small Acreage*. North Adams, MA: Storey Publishing, 1990.)

have proper ventilation. Spaces between bars should be small, no more than 3 to 4 inches (7.6–10.2 cm), so that the horse can never get a foot or nose through, or teeth or lower jaw caught when trying to nibble and chew. Dividers between stalls usually must be solid so horses in adjacent stalls cannot smell each other and start an argument. In some cases, however, horses who get along well should be allowed to socialize because they will be happier and less bored in their confinement.

Flooring

Flooring should be a material easily kept clean, because the average horse produces up to 50 pounds (23 kg) of manure and more than 10 gallons (38 L) of urine daily (the actual amount depends on whether he is eating hay or pellets and whether the weather is warm or cold — there's less water consumption and urine production in cold weather). A clay floor has good traction, some cushion, and a moderate degree of warmth, but moisture does not drain through it as readily as through a mix of clay and sand or crushed rock. Sand or crushed rock, however, may get mixed with the feed or bedding if the horse likes to paw.

Wooden floors are fine for tie stalls but not for box stalls; they are too slippery for good traction. Stall mats can be used over wood or dirt floors. Such mats are usually a combination of rubber, clay, and synthetic materials. If they have a textured surface, they afford good traction and aren't slippery. They keep horses from pawing holes in the floor or eating dirt, and they have enough cushioning to be comfortable. Stall mats also can be easily cleaned.

Bedding

What you use for bedding in a barn stall or shed will depend on the materials available in your area, their cost, and your budget. Straw and wood shavings or chips are often the two best options but, sometimes, alternative materials such as shredded paper or peat moss, which are less dusty, may be feasible.

In a farming region, **straw** may be the cheapest and most plentiful material. Wheat straw is often the bedding of choice because it is less palatable than oat straw (horses won't eat the bedding as much) and less abrasive than barley straw. The sharp awns that protrude from the grain heads of barley — if there is any grain left in the straw — can become embedded in a horse's mouth.

If you live where wood products are processed, **shavings** or **wood chips** may be economical. Soft woods like pine and fir make good bedding, but hardwoods have less absorbency and may even be toxic to horses (black walnut is especially toxic and should never be used). Sawdust is usually too fine and dusty for bedding, getting into the eyes and nose. Wood products should not be used as bedding for foaling mares or newborns because shavings or chips can harbor *Klebsiella* bacteria (see chapter 22). Make sure bedding is not dusty or moldy. If it is, your horses may develop respiratory problems.

Manure Pile

Plan stalls and barn with ease of cleaning in mind, and have a place nearby to put soiled bedding and

Clean stalls at least once a day so that horses have fresh, dry bedding.

manure. The manure pile should be far enough away that flies aren't a prob-lem, but close enough for easy access, and at least 150 feet (46 m) away from any water source. If horses spend much time in the barn, stall cleaning must be a daily chore. You don't want your horses lying in wet bedding, at increased risk for health problems.

Compost

Composting manure and old bedding is a good way to reduce the bulk of this material and make it safer to spread over your fields, yard, or garden as fertilizer. Properly composted, the break-down process creates heat that kills most of the parasites, pathogenic bacteria, and weed seeds in the manure and bedding, and reduces odor.

Hay Feeders

Horses in stalls can be fed in mangers or hayracks. If stalls are kept clean, hay can be fed on the floor — the most natural position for a horse to eat. But if the floor is sandy or gravelly, a horse may pick up sand when cleaning up the last wisps and leaves of hay, which can lead to sand colic, unless you put a rubber mat on the floor or under the feeder. And, unless you keep the horse regularly dewormed, there is also some risk of parasitism if hay gets mixed with manure.

An alternative is a walled-off feed bunk where horses eat at floor level (kept clean!) or a good hayrack or wall feeder in a corner, with no protrusions that might cause injury. The rack should be high enough that a horse can't get a foot caught in it but low enough that he does not have to reach up for feed because hay leaves and dust may fall into his eyes as he eats. As the horse tugs at the hay, falling particles and dust may be breathed in, causing irritation to respiratory passages and inviting lung problems.

A combination hay and grain feeder

Tack Room

You might want a special room for storing horse gear, medications, and first-aid equipment. Bridles, halters, and other equipment can be hung neatly on walls, and saddles stored on saddle racks, to keep them in good condition. The door should be wide enough to allow access by a person carrying a large saddle. It is wise to have a strong door that can be locked to prevent theft. A wide area in the aisle or entryway of the barn also can be used for tack storage, with tack hangers and saddle racks mounted on the wall.

Feed Room

Many barns include a place where feed can be stored. Most of your hay should be stored somewhere else (in a hay shed or on well-drained ground with a good tarp over it), but it's handy to have a few bales in the barn along with any grain you are feeding. Be sure to have a horse-proof door and latch for the feed room and to have lids on all grain storage containers. Do not store sacks of feed in a barn; not only will mice get into them, but there is always risk of a horse foundering (developing laminitis from eating too much grain) if he happens to get into the feed room.

Feed Storage

Large quantities of hay should not be stored in a horse barn; it takes up too much space and is a fire hazard. It's best if you have a hay shed with a tall roof to protect hay from rain and melting snow, preventing weather damage and spoilage. An inexpensive hay shed can be built using tall posts to support a metal roof. The shed should be on high ground that never floods, and the base should be of crushed gravel so bottom bales do not draw moisture from the ground. Build the hay shed as large as needed — for storing several tons of hay, or a year's worth of hay for many horses.

An open-sided pole shed for hay storage protects the hay from moisture while allowing good ventilation and easy access.

Hay stored outside should be kept on pallets (so that bottom bales do not draw moisture, which can cause mold or rot) and covered with a good tarp. It keeps the top bales from spoiling, but you may get occasional leakage and some ruined hay anyway.

One risk in storing hay in or near a barn is spontaneous combustion from the natural heating of hay that has been baled too "green"—not dried enough at harvest or wet from rain or heavy dew. In some instances, fermentation and chemical changes can cause the hay to heat so much that it ignites. If you must store hay in the barn, make sure it is well cured. Open a few bales and check that there is no dampness, heat, or mold.

FIRE PREVENTION AND SAFETY

Fire is the greatest danger for a stabled horse, especially if the fire starts at night and you do not discover it in time to save the barn and horses. Always build barns with this in mind, using the most fireproof materials you can afford. Have electric wiring installed (or repaired) by a professional, make sure there are no fire hazards in the barn, and insist that everyone who comes in or uses electrical equipment is careful to adhere to the following basic safety rules:

- No smoking in the barn
- Store a minimal amount of carefully checked hay in the barn
- Be sure all light fixtures, switches, and wires are out of reach of horses and all lights are covered with protective cages to prevent breakage
- Run all wires through conduit and outfit switches with safety covers
- Unplug all electrical appliances when not in use
- Check all cords frequently and keep them in good repair
- Do not leave a heater running when you are not there to watch it
- Have fire extinguishers in the barn in handy locations
- Keep the barn free of dust, cobwebs, and feed spillage
- Hang a halter and rope by every stall door (but out of reach of the horse) and always replace it after use
- Post the phone number of the fire department beside all telephones, along with easy directions to tell them how to get to your place and the official name of your road or street

This run-in shed provides protection from the elements for several horses.

Run-In Shed

An outdoor shelter for horses at pasture or in paddocks can be used instead of or in addition to a barn. A run-in (three-sided) shed gives adequate protection from cold wind, driving rain, or blizzards. It also gives some relief from flies (horseflies do not like to go into shady areas) and hot sun.

The open side should face south (away from cold winds coming from the north) or away from prevailing winds. Horses often use the shelter more as a windbreak than to get out of wet weather. The shed should be built on a high spot, so horses have dry footing instead of mud, and with the slope of the roof away from the opening so moisture can't accumulate in front of it. There should be good drainage away from the shed. It should be large enough (about 140 square feet [13.0 m²] per horse) to accommodate all horses in that pen or pasture to avoid injuries when horses jostle for space.

If there is a natural windbreak of trees or brush, you won't need a shed. This saves the expense of building one and eliminates the risk of horses being injured when trying to get into the shed all at once or chasing one another around or through it.

Fences and Paddocks

Fences must be secure enough to hold horses and safe enough to prevent injury. Choice of fencing material will be influenced by what you can afford, the type of horse you are enclosing (for example, a paddock for a stallion must

Avoid sharp corners in a pen, pasture, or paddock where a horse might be injured while running along the fence line or being chased by another horse.

be more substantial than a pasture fence for a group of geldings), and the time and money you can afford for fence maintenance.

Remember: A fence that won't hold a horse is no fence at all! You don't want your horse to get out on a highway to be hit by a vehicle or cause a fatal accident, escape into a neighboring pasture, or be injured during fence-crashing. You also don't want your horse in the neighbor's garden or flower-beds. Fencing for horses should be 5 feet (1.5 m) high — even higher if you have a horse who likes to jump.

All pens, paddocks, and corrals should have secure fences because any confined horse will test them diligently. At pasture, he has more to keep his mind occupied, such as room for grazing and exercise. In a small area, he can become bored and unhappy, spending more time checking out the fence or trying to find a way out.

Be aware that a poorly maintained fence can become dangerous and unsightly or make you vulnerable to a lawsuit if your horse gets out and dam-ages property or causes an accident. Check fences often. Tighten loose wires, replace splintered boards or poles, pound in loose nails or staples, restretch sagging net wire, and test electric fences periodically to be sure they are work-ing. Keeping fences well maintained not only lengthens their life but also helps ensure that horses won't be injured during encounters with their boundaries.

Types of Fencing

Many types of fencing can be used to confine a horse, but not all are safe. The materials you choose will depend on the situation — whether a roomy pad-dock for mares and foals, a pasture for a group of horses, a stallion pen, a yard for playful yearlings, or a boundary between other horses next door.

- Paddocks for mares and foals should be very safe and durable because foals are inquisitive and can get into trouble. You don't want them injured trying to put a head or foot through the fence. **Best bet:** Wood or net wire, possibly reinforced with electric wire, depending on the size of the paddock

- Any enclosure used for weaning should be completely hazard-proof. **Best bet:** Diamond mesh wire
- A stallion pen should be at least 5 feet (1.5 m) high and the materials stout and durable. **Best bet:** Wood or pipe fence, possibly reinforced with electric wire
- Fencing around small paddocks or pens should be chew-proof, paw-proof, kick-proof, and lean-over-proof because horses confined in small areas think up all sorts of fence-terrorizing activities to entertain themselves. If turned out in the paddock only for short periods, they may gallop and buck around the small enclosure, kicking up their heels — so the fence must be strong. **Best bet:** Diamond mesh wire
- You may get by with less complicated fencing around large pastures where horses have room to roam and lots of grass to graze, but if part of the pasture borders another, you'll need safety measures to prevent neighboring equines from rowdying each other at the fence. **Best bet:** Wood or wire, synthetic materials — possibly reinforced with electric wire

Barbed Wire

When it is tight, barbed wire makes a good barrier for cattle (which have thick, tough hides) or even mature ranch horses accustomed to this kind of fence. However, it can be a disaster for young horses learning about fences or for any horse not used to barbed wire. Even horses who respect barbed wire can be ruined by it — if pushed into it by another horse, by pawing at it, or by attempting to jump or lean over it.

With modifications, barbed wire can be safely used for horses. If your place already has a barbed-wire fence and you cannot afford to replace it yet, make sure the fence is in good repair, with wires stretched tight. Sagging wires are more apt to get a horse in trouble because he may try to reach through to grab a bite of grass or get a foot through it. To ensure that horses won't be injured by a barbed-wire fence, put a strand of electric fence along the inside to teach the horses to respect the fence and stay well away from it. This eliminates the leaning, reaching, rubbing, or pawing that gets a horse into trouble.

Wood Fencing

Under some conditions, wood poles or boards make good fences for horses. Strong and durable, wood is more forgiving and resilient than metal if a horse runs into it. Wood works best where horses have a lot of room. Confined, bored horses like to chew wood and can demolish a pole or board fence in a short time. As a general rule, horses who grow up in large areas and big pastures don't become wood chewers (unless they have mineral deficiencies or dietary imbalances). The chewing habit is learned early, however, by confined youngsters. Once a wood chewer, always a wood chewer; that is, unless a dietary deficiency (such as not enough fiber, lack of salt or certain minerals, and so on) that is later corrected has provoked the behavior.

Nail boards to the inside of the posts to create a smoother fence; horses running along it won't run into the posts.

Nail poles to the inside of the posts so they can't be knocked off by a horse kicking, rubbing, or leaning on the fence.

MAINTENANCE IS CRUCIAL

Wood fences require a great deal of maintenance to be safe for horses. Board fences should be checked for splinters. All types of wood fence should be thoroughly checked at least once a year to see if more paint or preservative is needed. Broken, cracked, splintered, chewed, or rotten poles or boards must be replaced, and nails redriven if they start to work out. Nails that creep out are a serious hazard to any horse who runs into them.

Make sure poles or boards are nailed to the inside of posts so that a horse rubbing, leaning on, or kicking the fence can't knock them off. This makes a smoother fence line; horses running along it will not run into the posts. Wood posts should always be thoroughly treated before being set in the ground, or they will rot.

In many instances, the only way you can get by with wood fences is to add an electrified wire or two, or electrified tape, to keep horses away from the fence or prevent their chewing it. An electric wire can also keep a horse from rubbing on the fence or trying to lean over it — habits that take their toll on fences.

Wood fencing with an electric wire along the top and one out to the side to keep horses back from it will last a long time. In a dry climate, untreated poles or boards may last 20 to 30 years — even longer if treated. Wood preservatives are necessary in a wet climate and should be reapplied periodically. Most preservatives today are nontoxic, but it pays to read labels.

PVC Tubing

Polyvinyl chloride (PVC) tubing doesn't rot or splinter, and horses won't chew it. This material, which was originally developed for plumbing pipes and drain tiles, was first used for fencing in the early 1980s. The PVC tubing used for fencing today has been treated to resist color change caused by sunlight and has been modified so it won't crack in extremely cold weather.

Most companies that sell PVC fence give a 15- to 20-year warranty. PVC tubing is quite expensive (about twice as much as treated wood) but requires little maintenance and usually lasts longer. It never needs painting or preservatives, won't rot, and can't be damaged by horses or termites. The material is somewhat flexible and won't splinter if a horse hits it. PVC tubing will occasionally break if an impact is great enough, but it is safer than metal or wood.

Polyethylene-Coated Wood

Planks and square posts covered with a white polymer resin combine some of the advantages of wood and synthetic materials. The coating makes a nice white fence and protects the wood from moisture; the ends of boards and the tops of posts have protective caps to prevent moisture damage. This type of fencing requires less maintenance than wood and is highly visible, durable, and safe. Also, horses don't chew it as readily as wood.

Rubber and Nylon Strips

This fencing, made from reprocessed tires and conveyer belts, is stretched tightly between wood posts. A fence of this sort is durable, although strips may occasionally have to be tightened between posts. It has a lot of give if a horse runs into it. Newer versions don't fray or deteriorate as badly as earlier materials. If a horse chews on this type of fence, however, he can get indigestible fibers in his stomach that could lead to impaction and colic. If you have a horse who chews, add an electric wire or tape to keep him from nibbling at the fence. (See chapter 15 for a more detailed discussion of impaction.)

Net or Mesh Wire

Fencing made of net or mesh wire is much safer than that of barbed wire and more foolproof than that of smooth wire, although it's possible for a horse to get a foot caught if spaces are hoof size or larger. "No-climb" netting (2- × 4-inch [5 × 10 cm] rectangular spaces) is fairly safe for adult horses, but a foal can get a foot through. Larger mesh, such as that used for cattle, pigs, and sheep, is not safe for horses. The best type of net wire for horses is diamond mesh because even a small hoof rarely gets through it, and a shoe is less likely to be caught.

Wire mesh will stay tighter longer if properly stretched and braced and topped by a pole, board, or pipe to keep horses from leaning over it. A pole at the bottom and a diagonal brace between posts help reinforce it.

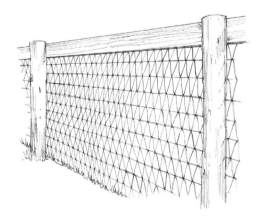

Diamond mesh is the safest type of net-wire fence for horses.

Smooth Wire

Five to seven strands of smooth, 12- to 13-gauge barbless wire (two wires twisted together without the barbs) on wooden posts may work as boundary fencing for large pastures, but in smaller areas, horses may lean over or reach through it unless the fence is topped with a pole, pipe, or board, or used in conjunction with an electric fence.

Electric Fencing

By itself, electric fencing is not adequate for horses — it's more a psychological barrier than a physical one. Electric fencing is not very visible, and a horse may run into it before seeing it, unless you use brightly colored electric-fence tapes. Electric fencing may break if a horse hits it at full speed or tries to jump it. Even the newer plastic tapes cannot hold a determined horse. Flagging an electric fence with bright cloth or using highly visible tapes will not stop a horse being chased by a pasturemate and seeking an out.

Horses unaccustomed to electric fencing must learn about it; otherwise, they may crash through. The horse can best learn about the fence during a situation when he is not upset and trying to go through it; if he crashes through, he will get shocked but will not necessarily learn to respect it. When you first turn him out in a pasture with electric fence, watch to make sure he checks it (most horses try to smell the fence and get a shock on the nose). If he does not check it, tempt him to touch it by offering a bucket of grain or a

In this solar-powered fence charger, sunlight provides the power to charge the wire. When a horse touches the wire, his body shorts it out, with the electricity taking a shortcut through him and back into the ground (completing the circuit), thereby giving the horse a shock.

bite of green grass from the other side of the fence. This may seem like a mean trick, but it will alert him to the hot wire and make him stay back from it after that; it may even prevent serious injury later because he will respect the fence.

Some horses know when the electricity is off or the fence is disconnected or shorting out in the rain. If the power goes off, the battery wears out, or the fence is broken, your horse may walk through it or over it, or crawl under it. If you live in a sunny climate, you can use a solar-powered fence charger, but you still need to check the fence periodically to make sure it is not broken or shorting out. You also need to keep the charger area free of weeds or tree branches that might block the sun.

Lightweight push-in post with a cap on top protects horses from injury; clips hold electric tapes or wire.

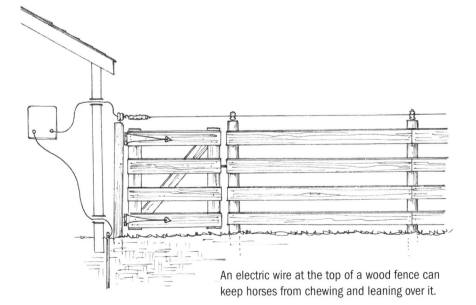

An electric wire at the top of a wood fence can keep horses from chewing and leaning over it.

MAINTAINING ELECTRIC FENCING

Follow a program of regular checking and maintenance to make sure electric fencing is always working properly. Wet grass, weeds, or leaves may short it out, or deer jumping over it may push the hot wire or tape into the solid fence next to it. If the hot wire gets caught on a wire fence, nail, or staple, it will short out. Periodically trim grass and branches away from wires during summer when plants and bushes are growing rapidly; otherwise, wires may become choked with green foliage and short out.

Many fence chargers have a blinking light to show the fence is working. If it's not plugged in or is shorting out, the light will be off. You can use an inexpensive fence tester to see if there's an electrical charge in the fence. The tester has a small rod that you stick in the ground and a wire that you touch to the fence. If the fence is working when you do this, the light on the tester goes on.

Check the electrical current in the fence by simultaneously touching the fence wire with the tester and poking the ground rod into the soil.

Electric fencing is best used in conjunction with a solid barrier. The hot wire teaches the horse to respect the fence; even a bold one won't keep trying to lean over, jump, or push on it after having been zapped by the electric wire a time or two. Used in conjunction with other fencing, an electric wire can do wonders in preventing over-the-fence battles and keeping maintenance costs of the primary fence to a minimum.

Steel Pipe

Steel-pipe fencing was originally created from surplus or discarded drill-rig pipe from the oil industry, but now some companies make steel pipe specifically for horse fencing. Pipe rails are generally secured to metal (pipe) posts set in the ground in concrete. Often 2-inch (5 cm) drill-stem pipe is used for top and bottom rails, with three or four strands of cable or sucker rod in between. Because pipe has more structural strength than wood, pipe rails may be 10 to 20 feet (3–6 m) long, requiring that fewer posts be installed.

Pipe fence is durable and lasts forever, especially if coated with antirust paint. Because of its strength and durability, pipe makes good fence for stallion pens, road boundaries, or other places where you want a fence that never fails. However, it can cause injury if a horse runs into it at full speed or if the spacing of pipes enables a horse to get a leg or head caught. A handful of horses have died when they put their heads through the pipes and tried to chew on the next pipe down — getting mouth and jaw caught on the lower pipe while the upper pipe pressed into the windpipe.

Fence Posts

Make sure posts are set properly; they are the foundation of your fence. If posts are not well set, the entire fence will be less durable or even unsafe because wires will sag or sections of pole or board panels will get pushed over. Wood posts should be treated with a preservative, set at least 2.5 to 3 feet (0.8–0.9 m) deep, and well tamped so they won't give.

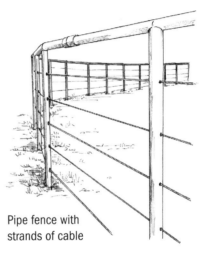

Pipe fence with strands of cable

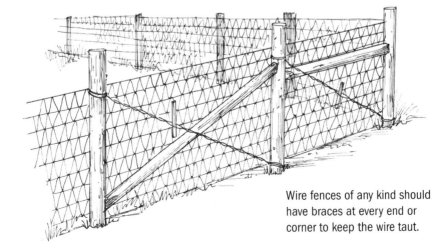

Wire fences of any kind should have braces at every end or corner to keep the wire taut.

Posts for any kind of wire fence must be braced to keep the wire tight, and the wire itself (or netting) must be stretched tightly when installed. There must be a very solid brace at each corner or end, as at a gate, and every so often along a lengthy stretch of fencing; exact spacing will depend on the terrain and the pressure placed on the fence. If a fence is not braced properly, the pull of the wire will loosen some of the posts or pull the corner posts out of the ground.

Gates

Gates should be as durable and safe as the fence to which they are attached. A horse may spend a lot of time waiting by a gate or leaning on it because he knows it is the place to go in and out. The gate should be at least as tall as the fence and have a horse-proof latch. A chain with a snap will work if horses fiddle with a latch. Because an impatient horse may paw at a gate, never use wire or netting for a gate.

A gate gets more wear than the fence, so it should be sturdy (but easy to open and close), well hung, and well braced so it won't sag. The post on which the gate is hung should be stout enough to hold its weight and set deep enough to never move or lean. The longer and heavier the gate, the more likely it is to pull the post out of line unless the post is well set and well braced. The gate should be smooth so a horse pressing against it won't be injured, and there should not be much gap between gate and post so a horse can't get his head or foot caught if he tries to wedge through or paw at it.

For safety, electric wire can be routed over the gate using tall posts.

Gates should be wide enough (at least 8 feet [2.5 m]; 12 feet [4 m] is better) for horses to go through easily without running into the gate or post. Gates between pastures should be situated in good locations — not in a corner or next to an obstacle — and wide enough that horses do not crowd one another and cause injury while passing through. Don't leave gates open between pastures or pens; many horses have been injured by running a shoulder or stifle into a gate or gatepost.

A gate in an electric fence should have an insulated handle so you can unhook it without getting shocked. Design the gate in such a way that the handle disconnects from the side toward the charger so the gate wire is "dead" when unhooked. Then, it won't shock you or a horse when open or spark and snap (or start a fire in dry grass or weeds) if you put it on the ground or throw it over a fence, out of the way. You can bypass the gate by using tall posts on both sides to run the electric wire over the top.

Water

Horses need water available at all times and should be provided with fresh, clean water frequently. The average horse drinks 8 to 12 gallons (30–45 L) a day and even more if he is eating dry feed such as hay, if the weather is hot, or if he's working and sweating. A horse drinks less than 10 gallons (38 L) a day if the weather is cold or if he is eating lush, green pasture that has high moisture content.

Water can be provided by a stream or pond, water tank, or tubs. A paddock or stall should have a tub, bucket, or some other

Provide water in a rubber tub that can easily be rinsed out daily and filled with a garden hose.

type of waterer. For filling tubs or buckets, a water faucet or hydrant in the barn or barnyard is handy for hooking up a garden hose. All water containers should be checked daily and cleaned often to make sure they are free of debris and feed particles, which can ferment and lower the water quality.

Automatic waterers are nice, but they have disadvantages. They can malfunction and flood a stall or pen, or they can quit working. Many people neglect to check them daily, and a horse may go without water for too long if they stop working. It is also impossible to know how much water a horse is drinking when he uses an automatic waterer. A sick horse may stop drinking, but you won't necessarily see this early warning signal. If a horse goes without water for very long, he may suffer impaction. You need to know your horse's drinking habits and be aware of any change.

Cold-Weather Precautions

If you live in a cold climate, be sure to bury water lines to hydrants well below the frost line so they never freeze in winter. This usually means at least 3 to 5 feet (1–1.5 m) under the ground. Water lines under driveways or areas where there is heavy vehicle or horse traffic have to be even deeper because this type of impact drives frost down; the ground freezes deeper and stays frozen longer in spring.

Freezeproof, self-draining water hydrant

In cold weather, drain water hoses completely after each use. Water buckets, tubs, and troughs should be insulated. Buckets with insulation, acting like a thermos bottle, can keep water from freezing for many hours; consider using these in stalls or paddocks when watering single horses using small containers (a small amount of water freezes more quickly than a large volume in tank or trough). If water is warmer, horses drink more in cold weather and are less at risk for impaction. Electric water warmers can be used in buckets or tanks but are dangerous if they malfunction. Remember that all electrical equipment is a potential fire hazard in a barn.

An insulated water container works like a thermos bottle: the water is kept from freezing without the risks of using a bucket heater.

Even with insulation and depending on the container's size, you may have to break ice out of a tub, bucket, or tank daily. Rubber tubs and buckets are better than plastic ones, which crack at low temperatures. Plastic tubs are also not as durable if you have to pound on them to get the ice out.

Water hydrants in a cold climate should be freezeproof and self-draining. This means the water drains out of the upright portion, going deep into the ground, each time you turn the hydrant off, so there is no water left in the upright pipe to freeze. They should be located where horses don't have access to them — outside the pen or pasture, or in a barnyard area well out of the way. Otherwise, a horse may be injured by running into the hydrant, or a playful horse may nibble the handle and accidentally turn on the hydrant.

Make Sure All Facilities Are Hazard-Free

Before putting horses into a new pen or pasture, check it carefully for anything that might cause injury. Make a habit of looking closely at all facilities every time you enter a pen, pasture, or barn stall. Maintaining a watchful eye can prevent serious problems.

This is a hazardous fence: sagging wire and netting could catch a horse's foot.

COMMON HAZARDS AND SOLUTIONS

Hazard	Examples	Danger
Sharp projections	Exposed nails, wire hooks, projections from broken hardware or fencing	Can cause eye injuries or tear an eyelid if a horse rubs on them or bumps them.
Fences in disrepair	Sagging wire fences	Strangulation; cuts on legs if pawed or on head and neck if horse reaches through
Halter left on		A horse may injure himself, catching a foot in it or hooking the halter on a projection of stall hardware or fence; a horse rubbing on a fence or sticking his head through poles or around a post may catch it on a nail or staple, wood stub, or bolt and pull back and hang himself
Unfilled holes in pastures		If a horse steps in a hole, he can pull a tendon or break a leg
Wood posts sawed off at ground level that have rotted		If a horse steps in it, he can pull a tendon or break a leg
Wood or metal posts broken off above ground level		May puncture a foot or cause lower-leg injuries
Badger holes, ground-squirrel holes, or other burrows made by rodents		If a horse steps in a hole, he can pull a tendon or break a leg
Sharp edges	Feeders, waterers, water tanks, mangers	Horse may suffer cuts and wounds
Protruding corners or edges	Feed mangers, gates, metal sheds, roofs	Horse may suffer injury
Narrow spaces	Doorways; gateways; space between feeders, waterers, wall, or fence	Horse might get caught
Low overhangs	Roofs, guywires from power pole or telephone pole	Horse may run into them
Junk, old machinery, garbage	Old batteries; sharp edges; rusty bolts	Batteries contain poisonous substances such as lead; sharp edges and rusty bolts can cause injury or snag the horse
Unkempt road borders	Broken beer bottles, cans, paper, other litter	Cans and broken glass can cause injury
Baling wire or twine		Young horses may chew or eat

Solution	Prevention
Remove or cut off flush with surrounding surface	Use bolts with rounded ends when building fences, gates, or stalls, or cut them off flush with surrounding surfaces; remove broken or unused gate hinges
Repair or replace	Maintain fences at all times
	Never leave a halter on a horse at pasture or a horse in stall
	Fill all holes
	Remove and fill hole
	Remove and fill hole
	Fill all holes
Repair or cover sharp edges	
Repair, cover, or buffer sharp areas	
Rehang gates, fill spaces, redo gateposts	
Make them more visible by installing post and horizontal pole	Install post and horizontal pole
Remove or fence off junk pile; remove or fence off the old machinery	
Patrol and pick up litter regularly	
Make a habit of never leaving twine or wire where horses can get to them; dispose of these hazards in a safe place	

2

Feeding and Nutrition

FEEDING IS PROBABLY THE SINGLE MOST IMPORTANT aspect of caring for a horse. What he eats — the quality and quantity of feed — determines, to a large extent, how well he can perform (or grow) and how healthy he'll be. If he's still young, the feed he eats will make a difference in how well he grows and develops.

A horse inherits his potential for height and size, form and appearance, speed, strength, endurance, and agility. But how fully potential is realized depends greatly on the feed he gets. A horse with good breeding and much potential may develop and perform poorly if fed inadequately. He may have potential for size and strength but end up stunted because of improper or inadequate feed. He may develop joint and bone problems and become lame and permanently unsound if overfeeding while young causes him to grow too fast and carry too much weight on immature bones and joints. Even a sound horse with great ability may perform poorly, lack energy, or suffer problems such as founder or muscles tying up if he is living on insufficient or improper feeds. The conscientious horse person must make sure the diet is balanced and must provide the proper amounts of what the horse needs.

THE WELL-FED HORSE

The horse is unique among domestic animals because he is used for his athletic ability rather than for production of meat, milk, or wool. Whereas a meat animal is fed to gain the maximum body weight in a short time, the horse should be fed for energy and longevity — never to be fat. If he is fat, his health, endurance, longevity, breeding ability, and athletic usefulness are impaired.

Horse Digestion

The horse is a roughage eater like a cow, but unlike the cow, he has a simple stomach and chews food only once. His relatively small stomach holds only 2 to 5 gallons (8–20 L) compared with the cow's rumen capacity of 15 to 30 gallons (57–114 L). The horse eats smaller amounts but more often — continually throughout the day and night if on pasture, stopping only to rest, switch flies, or nap before grazing again. The cow eats more at one time and more quickly, then lies down and rechews it at her leisure.

Ruminants such as cows, sheep, goats, and deer process and digest roughage in the largest stomach, the rumen, where microorganisms help break down cellulose and other indigestible material into usable nutrients. In contrast, horses (and rabbits, kangaroos, and other herbivores that do not have a rumen) break down roughage in the cecum. The horse's hind-gut (cecum and large colon) is the "fermentation vat" that does much the same job as a cow's rumen. Because it is located beyond the stomach and small intestine, the horse cannot process extremely coarse feeds, such as straw, as easily as does the cow.

A cow rechews her food and breaks it down by fermentation in the rumen early in the digestive process, whereas the horse accomplishes the final stages of digestion in the 4-foot-long (1 m) cecum and 12-foot-long (4 m) colon. Together, the cecum and colon have a combined capacity of more than 25 gallons (95 L). They are designed to hold, process, and digest the cellulose and other fibrous parts of the horse's diet that are broken down by bacteria, protozoa, and yeasts.

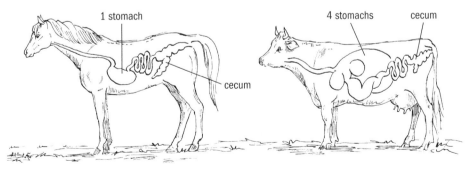

The horse has a small, simple stomach and depends on his well-developed cecum and large colon for fermentation and breakdown of roughages. The cow has four stomachs and breaks down roughages in her large rumen instead of in her less-developed cecum.

CELLULOSE AND BODY HEAT

Cellulose, the fibrous part of roughage, is broken down into glucose and other nutrients, but no animal makes an enzyme that can do this. All grass-eating animals rely on microorganisms to ferment and digest cellulose, changing it to energy-producing fatty acids. The chemical change that takes place during this process also produces heat energy and can help keep a horse warm in cold weather. When temperatures drop, a horse can gain more good from extra hay than from extra grain because digestion of roughage creates more body heat to keep him warm.

Importance of Proper Feed

The horse's fermentation vat works well under normal conditions when he's eating pasture or good hay, but sometimes problems can develop. The cecum can become impacted if a horse is fed finely chopped hay or lawn clippings. In the wild, the horse never ate grain or alfalfa; his digestive system was designed to handle grasses; therefore, these "unnatural" feeds may sometimes create problems. Abrupt changes in feed can upset the population of microorganisms (with some dying off and producing toxins), leading to colic or laminitis because of a change in the rate of feed breakdown and creation of excessive gas or toxins.

The horse depends on microbes and fermentation for digestion of feed and absorption of nutrients and vitamins. The microbes are his allies — essential to his health — but if he is fed improperly, they may become his deadly enemy. Understanding the horse's digestive process enables us to feed him wisely.

The Digestive Tract

The horse's digestive tract is a long tube: esophagus, stomach, small intestine, cecum, large colon, small colon, and rectum. Because the stomach is small, a hay-fed horse should ideally be fed several times a day instead of only once. The daily ration should be divided in two (preferably three) parts and fed at intervals, with the largest portion offered in the evening so the horse can eat it at his leisure during the night. An option for broodmares, as well as inactive horses during winter, is to put out large (half-ton or larger) bales so they can eat as they wish.

The stomach initiates the digestive process, breaking down feed a little and beginning the digestion of proteins. The small intestine continues digestion of proteins and also breaks down starches, sugars, and, to some extent, fats. The pancreas and liver supply enzymes to the intestines so that the digestible parts of the feed can be broken down and absorbed. The horse has no gall bladder (in other animals it releases bile into the intestine for breaking down fats); instead, bile is secreted continuously in the small intestine, but not in large enough amounts to thoroughly digest the fats. Fats are digested when they reach the large intestine.

Many people used to think that watering a horse during or directly after feeding would wash food — especially grain — out of the stomach and into the intestine before it was digested, causing inadequate nutrition. But, under natural conditions, a horse drinks water periodically as he eats. When eating hay, he needs adequate saliva and water to moisten the dry feed. The horse should have free access to water.

The large intestine (especially the cecum) digests roughage more thoroughly. Here, the fibers are broken down by microbes that change the cellulose into energy-producing fatty acids. They also manufacture amino acids — from which the horse gets usable proteins — and B vitamins, which are essential to good health.

Digestion is completed in the large colon, after which the excess water is absorbed from residue in the small colon, forming balls of feces. The horse normally passes from 30 to 50 pounds (14–23 kg) of manure daily.

VOMITING

Unlike a human being, the horse cannot readily vomit because of the way the back of his mouth is constructed (he cannot breathe through his mouth very well for the same reason). If he tries to regurgitate, feed is forced out through his nose; it cannot be ejected easily through his mouth. Material coming back up the throat may go into the windpipe, causing choking or pneumonia if food particles get into the lungs. A horse's stomach may rupture from the pressure of trying to vomit. Because he cannot unload his stomach of bad food, the horse should always be fed the best-quality feeds. Spoiled or moldy feeds, or toxic plants that cause digestive upsets, can be disastrous.

Basic Nutritional Requirements

The horse needs carbohydrates, fats, protein, vitamins, and minerals — all in proper amounts. In many cases, too much can be as harmful as too little. On a natural diet of grasses grown in fertile soil, the horse gets all his nutrient requirements. But, sometimes feeds are grown in geographic areas where soils have mineral deficiencies (check with your county Extension agent or veterinarian for information on your area), and sometimes horses eat unnatural diets (grain, and hay or pellets instead of grass). Many horses are fed supplements as well, to try to provide additional minerals, protein, or calories (in the form of fat) if a certain horse needs more than his pasture, hay, or grain can provide. Because a well-balanced diet is so important, the responsible horse owner must have an understanding of all the horse's basic nutritional requirements.

Carbohydrates

The major source of energy for the horse, carbohydrates, includes sugars, starches, and cellulose. Carbohydrates are derived from cellulose as it is broken down and digested. Oats, barley, corn, and other concentrates (so called because they provide more calories in much smaller volume) may be as much as 60 percent sugar and starch. Cellulose, found in all roughage, is less easily digested and must be eaten in larger amounts to produce the same amount of energy.

Fats

Fats are needed only in small amounts, and natural feeds do not contain much fat. Small amounts of fat, however, do aid the horse in absorption of fat-soluble vitamins and furnish fatty acids necessary to sustain healthy cell membranes. The only fatty acid the horse can't manufacture for himself (linoleic acid) is found in the vegetable fats of roughage and grain.

Some hardworking horses such as endurance animals and racehorses are given fats as a supplement to be metabolized as a source of energy. Adding corn oil or a fat supplement to a hardworking horse's diet can increase the energy content of his ration; fats contain more than twice as much energy as an equal weight of carbohydrate (such as grain) and can reduce the amount of grain needed. This can reduce the risk for colic or founder in a hardworking horse who needs additional calories in the diet. But fat as a supplement should be fed carefully, starting with a few tablespoons and increasing gradually to a cup poured over the grain.

DOES FAT CONTRIBUTE TO A SHINY COAT?

The average horse doesn't need fat added to his diet, although some people like to use vegetable oils and other fats to give a shiny coat. Genetics plays a greater role than diet, however, in determining whether a horse has a shiny coat. This explains why one horse may have a glossier coat than another when both are equally fit and healthy and eating the same feeds.

Research has shown that flaxseed oil (as found in the oil-cake some people use as a protein supplement) may be more healthful than corn oil because it contains the proper essential fatty acids needed by the horse.

Protein

Protein (made up of amino acids) is necessary for growth. Protein helps build and repair bones, blood, skin, hair, hooves, and muscle. A deficiency can impair a young horse's growth, interfere with hoof development, and cause a poor hair coat. Legume hay is a good source of protein, as are soybean, linseed (flaxseed), and cottonseed meals.

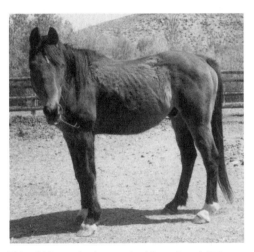

This old horse is slow to shed in spring. A little extra protein in the diet can help a horse shed more easily; the new hair will grow in more quickly.

Horses need extra protein while they are growing or pregnant. For example, a mare needs additional amounts during the last third of pregnancy—when the fetus is making most of its growth—and while she is nursing a foal. A little extra protein in a horse's diet in spring can help him shed his winter coat (as new hair comes in) quickly and easily. Excess protein is metabolized as energy; it also produces heat when digested, so a high-protein feed is a good addition to diet in cold weather but a poor choice during hot weather when a horse is working hard.

Vitamins

Vitamins are organic compounds that are required in small amounts for normal body function. They regulate chemical reactions, keeping the body operating smoothly. There are two kinds of vitamins — fat-soluble (A, D, E, and K) and water-soluble (B vitamins and C). All green forages are rich in most of the vitamins the horse requires. A horse on good pasture should need no additional vitamins for maintenance; pasture provides adequate body requirements in a natural and inexpensive form.

If you have no pasture, limited pasture, or poor pasture (poor soil, poor types of grasses, or mostly weeds), the horse needs good-quality hay. Carefully analyze your situation — pastures, feeds, your horse's condition and use — and then decide whether he needs vitamin supplements. Consult your veterinarian. Vitamins can be costly; make sure that the animal really needs them.

Vitamin A

Vitamin A is important to reproduction, eye health, and hoof growth. A deficiency can cause abortion, birth of weak or poorly developed foals, or vulnerability to digestive upsets, colic, and respiratory ailments (cough, colds, or runny nose).

Vitamin A is made by the horse from carotene in various plants. Green pasture and green, leafy alfalfa hay are good sources of carotene. Grass hay is not as high in this vitamin, especially if cut when mature. Vitamin A is fat soluble and can be stored in the body. Horses on pasture during summer usually store enough to last through winter. If your horse is never on pasture or alfalfa and your grass hay is not very green, or if the horse is under special stress during winter, such as a pregnant mare might be, consider giving vitamin A when supplementing the diet, or make sure your commercial feed mix contains an adequate level of vitamin A.

Vitamin D

Vitamin D plays an important role in the conversion of calcium and phosphorus into healthy bones. A deficiency can cause rickets in growing horses and osteomalacia (soft bones) in mature horses.

Vitamin D is produced in the skin if a horse is out in the sun. Sun-dried hay is also a good source. Under normal conditions, a horse should never be short on this vitamin, but if he is kept stabled all the time, he may need vitamin D supplements.

WHAT'S THE PROPER AMOUNT OF VITAMINS?

All vitamins are required only in very small amounts, and the fat-soluble ones (A, D, E, and K) are stored in body tissues. Feeding excessive amounts of fat-soluble vitamins can be dangerous. For example, too much vitamin D can cause bruises to calcify, and too much vitamin A can cause bone breakdown. Green alfalfa may contain up to 70 times the amount of vitamin A needed by a horse, but this is not a problem unless he is receiving additional vitamin A as well. The maximum daily requirement of vitamin A for a lactating mare is 27,500 International Units (IU). The maximum daily requirement for a mature, nonreproductive horse is 12,500 IU. Always check with your veterinarian before giving vitamin supplements.

Vitamin E

Vitamin E is necessary for the development and function of muscle, as well as for healthy reproduction. It is very important to the immune system and for broodmares, young horses, and horses under heavy use or stress. It is important in neurologic function and can help horses with certain diseases such as equine protozoal myeloencephalitis (EPM), which affects the spinal cord. This vitamin is usually very adequate in natural feeds, but during periods of stress or if natural feeds are not available, horses may need extra.

Vitamin K

Vitamin K is essential to blood clotting and for activating certain proteins in the horse's body. Vitamin K is made in the horse's intestines.

B Vitamins

B vitamins — including thiamine, riboflavin, niacin, pyridoxine, pantothenic acid, biotin, choline, folic acid, and B_{12} — are created by bacteria in the horse's cecum and large colon. The B vitamins are important for energy metabolism and the health of red blood cells.

Vitamin C

Vitamin C is also synthesized by the horse. The horse does not need B-vitamin or vitamin C supplements unless he is ill or anemic and unable to produce them. Vitamin C is an antioxidant that protects fats, proteins, and cell membranes from damage. It also helps in the formation of bones and teeth.

Minerals

Minerals are inorganic elements (not produced by living things) that are required in small amounts for good health and normal body function. Under natural conditions, the horse gets essential minerals by eating plants that store them from soil. The major essential minerals are calcium, phosphorus, magnesium, sodium, chlorine, and potassium. Others needed in minute (trace) amounts are iodine, cobalt, copper, iron, zinc, manganese, and selenium. Supplements are necessary in regions where certain minerals are lacking in the soil. Overdose can be toxic, so check with your veterinarian before giving mineral supplements.

The minerals needed in greatest amounts are calcium and phosphorus. The trace minerals (needed in tiny amounts) that are most commonly supplemented are iodine, selenium, and iron. Salt also has an important role in the horse's diet.

Calcium and Phosphorus

Calcium and phosphorus are important for strong bones and teeth. A lack of either mineral can impair growth or lead to unsoundness. Forages like pasture and hay are usually high in calcium, and grain is high in phosphorus. A mature animal needs a little more calcium than phosphorus (the usual ratio in pasture grass), whereas a growing horse needs the same amount of each. Many breeders give their young horses a little grain because it is high in phosphorus.

KNOW YOUR SOIL

If your horses on pasture eat hay and grasses from your property at any time of year, have your soil tested by your county Extension agent to determine what supplements may be needed in the diet.

Iodized salt should be given in geographic regions that are low in iodine. Lack of iodine in a pregnant mare's diet can result in stillbirths, weak foals, or foals born with contracted tendons. Other trace minerals, such as iron or selenium, are also lacking in some areas. Ask your veterinarian or county Extension agent about possible deficiencies in your region.

Vitamin D is essential for maintaining the proper relationship of these minerals. If there is not enough of it, the calcium cannot be absorbed, throwing off the calcium-phosphorus balance. Too much grain can also upset the balance by providing too much phosphorus and tying up the calcium in a compound that is not absorbed by the body. Simply feeding more calcium does not solve the problem if it can't be properly absorbed. Overfeeding a young horse on grain can result in soft bones, crooked legs, and unsoundness. No horse should be overfed, especially on grain, or oversupplemented with minerals. Salt is the only exception; it can be fed free choice.

Iodine

Iodine helps control body metabolism and also is important for growth. The tiny amount of iodine required by a horse (about 1 mg daily) can be supplied by iodized salt or mineral salt if your region is deficient. But the horse may get an overdose if you're giving trace-mineral salt, plus a grain mix that contains iodine, plus a vitamin-mineral or "shiny hair coat" supplement. Kelp products contain a lot of iodine. Check all your feeds. (See the *Know Your Soil* box for more on iodine.)

Selenium

Selenium is essential for proper nutrition and health. Deficiency can lead to muscle disease and weakness, especially in foals, and can cause some horses' muscles to tie up (the traditional term for severe and painful cramping). A combination of vitamin E and selenium is often prescribed for horses who are prone to this problem. Like vitamin E, selenium is essential to healthy immune function. Horses occasionally get too much selenium (which is toxic) when given too many supplements or when eating certain plants that store it from the soil, including locoweed, almost all vetches, goldenweed, saltbush, snakeweed, woody asters, and some of the other asters.

Iron

This trace mineral is often oversupplemented. Iron is crucial to the formation of red blood cells and hemoglobin, which transports oxygen. If a horse is eating hay, he is probably getting enough iron. If you are also feeding him a commercially prepared grain mix that has an iron supplement, he may get too much. Toxicity from misuse of this supplement is more harmful than iron deficiency. Horse people sometimes give iron supplements thinking it will increase endurance or athletic performance, but this is dangerous. Overdose can be fatal in mature horses, and foals given supplemental iron at birth may die.

Salt

Sodium chloride constitutes ordinary salt — a combination of two important minerals needed by the body. Grass and other roughages do not contain much salt; the horse must get it from some other sources and it should always be available, in blocks or loose. Some feeds have salt added, but it may not be enough if the horse is sweating or lactating. Mare's milk contains 160 to 365 parts per million (ppm) of sodium and 300 to 640 ppm of chloride — the chemical components of salt. You should always provide free access to salt even if a horse is receiving a feed containing salt.

If the horse likes to lick his salt block for something to do and eats more than he needs, the excess is flushed out of his body with urine and does no harm, as long as he always has plenty of water to drink. Salt is the only mineral that can be safely fed free choice. By contrast, many minerals, including trace minerals, are needed in only very small amounts, and overdoses can be quite harmful.

PROVIDING SALT

It is impossible to meet a horse's salt needs by feeding a specific amount per horse or a fixed percentage in a feed ration because horses vary widely in individual salt needs and consumption. The safest way to provide salt is to give free access at all times, along with plenty of water, so the horse can balance his own salt and water intake.

Under normal conditions, mature horses doing minimal work eat about 0.5 lb (0.2 kg) of salt per week, varying with the salt content of prepared feeds being given and how much the horse sweats. Salt can be provided in granule form — in a salt box — or in blocks. Blocks work well if they are not too hard. Rock salt is too hard, causing sore tongues if horses lick them a lot and, consequently, inadequate salt intake because horses don't eat enough.

If there are many horses in a pasture, provide more than one source of salt so timid horses can get their chance at it. If there's heavy rainfall or the pasture is irrigated, put salt in an area where it will stay dry or in a covered salt box. The "roof" over the salt should be high enough for the horse to reach in and eat the salt. Once dissolved by water, the salt is wasted and also kills the grass around it.

This is one reason it's best to offer a horse plain salt rather than mineralized salt. Salt is lost through sweating. A horse who sweats a lot in hot weather or while working hard will need to replenish his salt loss by eating more salt and, if it contains trace minerals, he may eat more of them than is safe.

The sodium and chloride in salt are the main ingredients of the fluid circulating within the horse's body. They help keep the proper water balance in the body and aid in its release through the pores as sweat for cooling during hot weather and periods of exertion. The more a horse exercises, the more salt is lost through sweat and the more he needs to replace it.

Roughages (Forages)

The primary forages fed to horses are pasture and hay. Silage is sometimes used but is more risky. Good corn or grass silage can substitute for one-third to one-half the hay ration, but it should be introduced into the diet gradually and never be the only roughage fed. Because it has high moisture content and weighs more than hay of equivalent nutrient value, it usually takes about 3 pounds (1.5 kg) of silage to equal the food value of 1 pound (0.5 kg) of hay. Silage that is spoiled, moldy, or frozen should never be fed; it can cause digestive upsets or poisoning. Horses are much more susceptible to poisoning from spoiled silage than are cattle or sheep.

Mature straw should not be used as horse feed. If cut as hay (while still green and growing, before grain heads are mature), oat hay is palatable and can be used as part of a ration, but mature straw is too coarse. It has little food value and is inefficiently digested. Grains other than oats or beardless barley have rough beards and seed awns and, if cut for hay, are not suitable for horses because they irritate mouth and throat. Take care in using oat hay; it can sometimes cause nitrate poisoning when grown under certain conditions, such as a period of drought followed by rain. Even though high nitrate levels are more deadly for cattle than horses, any oat hay fed to horses should be checked for nitrates.

Pasture

Pasture is the most ideal feed for horses; it is usually inexpensive and nutritious, and it constitutes the best environment for a horse because he gets sunshine and exercise. The greatest single factor in determining quality of pasture grass is stage of maturity. The older and drier the grass, the less palatable it is and the less food value it has. Mature grasses are coarser and not as easily digested as young, growing plants.

Pasture is the ideal feed for horses.

Some hay grasses (including timothy, smooth brome, and intermediate wheatgrass) are good horse feed when pastured, especially if plants are young. Kentucky bluegrass is good pasture if mixed with other palatable grasses or a suitable legume. Any pasture or dry-land grasses (except downy brome, often called cheatgrass) are preferable to swamp grasses; the latter are usually coarse. Pastures should have adequate drainage to prevent swampy soil. Permanently damp soil grows inferior types of grasses and serves as a breeding ground for horseflies. Fescue pastures should not be used for pregnant mares, especially in late gestation, because fescue grass is often infested with an endophyte fungus that can cause problems in a foaling mare.

Hay

Hay is usually the major part of a horse's diet if he's not on pasture. Hay must be of good quality to supply essential nutrition. It should be cut before fully mature (before bloom stage or heading out of grass seeds), while the grasses are still young, tender, and growing. Good hay is green (not yellow or brown) and leafy, with small, fine stems. Coarse, stemmy hay is hard to digest. Clean, bright, sweet-smelling hay is most nutritious. Dry, dusty hay not only is unpalatable but also can cause respiratory problems. Moldy hay may cause colic.

Legume Hay

Alfalfa, clover, or other legumes should be fed sparingly because horses like them and tend to overeat. Mixed hay, containing both grasses and legumes, is good for horses; the legume adds extra protein. Alfalfa hay should have moderate stems (first cutting). Horses don't need fine-stem, dairy-quality hay; it is not only too rich but does not have sufficient roughage for proper digestion.

It has a higher protein content than is needed, which can lead to problems when horses are working hard in hot weather. At the other extreme, make sure alfalfa is not too mature and stemmy. If it has already bloomed (look for purple flowers), it may be overly mature.

If you are feeding straight alfalfa, don't feed more than 1 pound (0.5 kg) per 100 pounds (45 kg) of body weight; fill in with another roughage, if possible. Too much rich alfalfa can cause digestive upsets, loose bowels, or metabolism problems in some horses when worked hard. It also produces heat during digestion, which is a disadvantage to a working horse. If you have no grass hay, select alfalfa that is not too rich and leafy so that the horse will have adequate roughage.

Feeding Hay

Whether you feed alfalfa, grass hay, or a mix will depend on your horse's needs and the types of hay available in your area. First-cutting alfalfa usually has some grass mixed in and can be a very good hay for horses. Second, third, and later cuttings are usually straight alfalfa and may be too rich for most horses.

Not all hay is the same in weight or nutrients. It may take twice as much poor-quality hay as good hay to keep a horse in condition. Poor hay is never a bargain because it can lead to digestive problems, nutritional deficiencies, or respiratory trouble.

AVOID MOLDS, FUNGI, AND DUST

If they are well cured — dry, not damp and moldy; but not so dry that they are sun-bleached, losing vitamins and food value — both grass and legume hay make good horse feed. Grass hay is less likely to be dusty or moldy than alfalfa and other legumes. The richer hay is a more ideal environment for molds and fungi if it is damp. Never feed moldy hay; it can poison a horse, cause colic, or bring on abortion in pregnant mares. Discard all moldy and dusty hay.

If it's just a little dusty, shake it to remove as much dust as possible, then sprinkle with water to settle the rest. Sprinkled hay should be eaten right away; if it sits wet a long time, it may begin to mold. Extremely dry hay is usually dusty because the leaves have shattered. It doesn't hurt the horse to eat dusty hay; however, it can be harmful if the dust gets into his lungs. Hay particles in the airways and lungs cause irritation and coughing and can lead to lung disease.

HEALTH TIP

Lawn clippings should not be used as horse feed because the animals may overeat and become impacted. Clippings often spoil and ferment if they are left piled for any length of time; if the horse eats them, the consequences can be devastating. If a horse gets access to lawn clippings and develops colic, call your veterinarian.

Weighing Hay

It is easier to estimate the weight of baled hay than loose hay. There can be a lot of difference in size and weight of bales, so weigh a few from each batch you buy to know how much is in a bale. Some are more loosely packed than others and may not weigh much even though they look big. Some hay is light because it is too dry or the bales are too loose, whereas others are heavy because they are moist and moldy. Bales from the top or bottom of the stack may be heavy and spoiled because they have taken on moisture.

Alfalfa hay of the same volume is usually heavier than grass hay. Because it has more nutrients, horses need less total weight per daily feeding than with grass hay, so this means much less volume. Good alfalfa hay is so palatable, nutritious, and easily digested that most idle horses gain too much weight if allowed to eat as much as they want, and some may get indigestion or colic because of it.

Yet if you restrict a horse to the amount of alfalfa needed for maintaining proper weight without providing any other source of roughage, he may start chewing wood. There is so little alfalfa needed that the horse does not feel full and looks for more to nibble on. Alfalfa can be gobbled up in a hurry and he's left with nothing to do. This problem can be reduced if you choose alfalfa with enough stem to provide more roughage.

Grass hay takes longer to eat, occupying more of the horse's time, and provides more fill; it is usually higher in fiber than alfalfa hay. Grass hay more adequately satisfies the horse's need for "chew time," and it provides the roughage needed for the digestive tract to function properly.

Changing Hay

If you change hay, especially from a grass to a legume, do it gradually, mixing the hay for several feedings so the horse's digestive tract can become accustomed to the new feed. A sudden change to alfalfa can alter the microbial environment in the cecum. A gradual change, over several days' feedings, is less apt to cause indigestion or colic.

Grain Concentrates

Concentrates (grain and grain by-products) are low in fiber and high in digestible nutrients. The horse needs digestible nutrients but must also have roughage for proper function of the digestive tract. Roughage provides extra bulk — fiber — as in the stemmy parts of hay or hulls on grain. A bulky feed is one with relatively little weight for its volume. The more weight, the more energy available in the feed, hence the name "concentrates"; grains contain more nutrients per unit of volume than do forages. Grains that are often used for horses are oats, barley, and corn.

Grain provides more energy per pound than hay.

Oats

Oats weigh about 32 to 40 pounds (14.5–18 kg) to the bushel and are approximately 70 percent total digestible nutrients, with 9.4 percent digestible protein. They contain about 11 percent crude fiber. Oats are a standard horse feed, palatable and nutritious. They are the safest grain to feed, even in large quantities, because they are roughly 30 percent hull (outer covering around the kernels), which provides some bulk. Shelled corn, oilseed meals, and wheat (heavy concentrations without much bulk) tend to form a doughy mass in the stomach, which can cause colic, whereas oats form a loose mass that can be easily digested. The bulky oat hulls keep the feed moving through the tract at a normal rate.

ESCHEW WHOLE GRAIN

Crimping or rolling oats increases their food value because they can be more easily chewed and digested. In contrast, whole grain doesn't always get chewed — kernels that get caught in the hollows of the teeth are not ground up but swallowed whole and pass through the digestive tract, wasted. If you examine the manure of a horse who has been fed whole grain, you'll see grain kernels that have passed through intact.

FEED BY WEIGHT, NOT VOLUME

HAY

Not all bales of hay weigh the same. Until you are good at estimating how heavy the "flakes" (bale portions) of a given bale of hay are, weigh them — especially when changing hay — to know how much you are feeding. A 1,000-pound (454 kg) idle horse usually needs 25 pounds (11 kg) of grass hay daily, given in two or three feedings. If the horse's work increases, the amount of hay should be increased. Some horses get by on less hay; others need more.

GRAIN

Not all grains weigh the same, so feeding by quarts or gallons can be confusing. Corn weighs more than oats; a quart of corn may give twice as much energy (or potential body fat) as a quart of oats. Weigh your feed, find out how much your scoop or bucketful holds in terms of weight, and recheck it when changing feeds. Making a change in the ration without adjusting for the weight of the new feed may lead to digestive problems or colic.

A RULE OF THUMB FOR WEIGHING GRAIN

- A quart of oats weighs 1 pound (0.5 kg)
- A quart of barley weighs 1.5 pounds (0.7 kg)
- A quart of corn weighs 1.75 pounds (0.8 kg)

Barley

Barley is 78 percent digestible nutrients — about 10 percent digestible protein and only 5.9 percent crude fiber — and a bushel weighs roughly 48 pounds (22 kg). Barley hulls, like oat hulls, provide bulk to aid digestion, but not as much. Some oats or wheat bran is often added to barley to increase the bulk. Veterinarians recommend that all grains be steam rolled or processed in some manner to kill the organism that causes EPM (see chapter 8). No grains should be ground for horse feed; ground grains are too dusty and fine, tending to dough up in the stomach.

Corn

Corn, with no hulls, is high in energy and low in bulk. Because it's so rich in energy, corn should be fed sparingly or it will make a horse too fat. Being lower in protein, if fed to growing horses or pregnant mares it should be supplemented with legumes or another high-protein feed. Corn weighs about 56 pounds (25.4 kg) to the bushel and is 80 percent digestible nutrients, but only about 6.7 percent digestible protein and 2 percent fiber. It takes 15 percent less corn than oats to keep horses in good condition if the ration is balanced with additional protein.

Wheat

Wheat is usually not fed because of cost, but in wheat-growing areas, it is sometimes economical. It should be rolled or crushed to make it easier to chew and digest, fed only in moderate amounts — less than 20 percent of the grain ration — and mixed with a bulky grain like oats or bran to avoid colic. Wheat is high in energy and protein. It weighs about 60 pounds (27 kg) per bushel and is 80 percent digestible nutrients. It is about 11 percent protein (one of the highest grains in protein content) and only 2.6 percent fiber.

Wheat Bran

Wheat bran, a by-product of milling wheat, is the coarse outer coating of wheat kernels. It can be mixed with concentrates for additional bulk and for a laxative effect. Bran is twice as bulky as oats, poorly digested, and low in energy. Being poorly digested, it increases intestinal fill and amount of manure passed.

Wheat bran can be added to a daily ration (no more than 10 to 15 percent of the ration) or given occasionally as a bran mash (mixed with water). A hot bran mash can be given on a cold winter evening or after a hard day's work. Some people use it as a tonic for a tired or sick horse, to get more moisture into the bowel, or after foaling to keep mares from becoming constipated. Because of its laxative effect, it is often fed to older horses to prevent impaction because of sluggish bowels. Never feed it dry unless it's less than 10 percent of a grain mix and well mixed in. Dry bran will swell in the stomach, pulling fluid from the contents, causing digestive problems and discomfort.

Wheat bran is high in fiber, protein (about 17 percent), and phosphorus, but very low in calcium. It adds bulk to the diet, but its nutrient value is low except for the high protein content.

A drawback to feeding bran (especially on a regular basis) is its calcium-phosphorus ratio — approximately 1 part calcium to 10 parts phosphorus. If

---

RECIPE FOR BRAN MASH

Make a bran mash by adding hot water to about 2 pounds (1 kg) of bran, stirring until it is crumbly but not soupy. Some people like to use boiling water, adding enough to a 2-gallon (8 L) bucket to get the consistency of oatmeal then covering the bucket to let it steam until cooled. Feed the mash to the horse after it has cooled enough to be still warm but not hot.

---

fed frequently, it can cause a serious imbalance in the diet, interfering with proper absorption of calcium. This can cause damage to growing bones in young horses and bone deformity in mature animals. An extreme imbalance in calcium and phosphorus can cause bones to soften and thicken until the skull and jaws are noticeably deformed, a condition called *bighead*.

Sorghum

Sorghum grain, also called "milo," is a good horse feed in areas where it is grown. This is a hard, round grain that should be rolled or crushed, and because it is lacking in bulk, it should be mixed with bran, oats, barley, or some other bulky feed to avoid constipation. Milo weighs roughly 56 pounds (25.5 kg) per bushel and is about 80 percent digestible nutrients and only 2.3 percent crude fiber. It contains approximately 8.5 percent digestible protein.

Molasses

Molasses, a by-product of sugar refining, provides carbohydrates. It is often added to grain to reduce dustiness and to increase the energy content of the ration. Molasses contains about 54 percent digestible nutrients. The proportion of molasses in grain should not be more than 10 percent; too much will give a horse loose bowels. Refined sugars can also be a problem in horses who tie up, making them more likely to suffer this problem because of the way their muscles use this form of energy.

Oilseed Meals

Linseed meal, soybean meal, and cottonseed meal are by-products of extraction of oil from seeds. These meals are high in protein and are often used as supplements, added to a grain ration to increase protein content. Because these are very concentrated feeds, a horse should never be fed more than 1 pound (0.5 kg) per day.

Commercial Grain Mixes

Even though it is generally cheaper to buy grain in bulk (including a ton of oats or barley) or to purchase several types of grain and mix them yourself, many people like the convenience of commercial grain mixes and buy it by the sack. There are many good products available, but be sure to read labels carefully and know what is in the feed and in what quantities. Otherwise, you may overfeed some nutrients, especially if you are also giving the horse a special supplement whose ingredients are duplicated in the grain mix. Today, there are dozens of specialized commercial feeds, formulated and balanced for various classes of horses — mares and foals, weanlings, old horses, hard-working performance horses, and horses with special needs such as insulin resistance or muscles tying up. Those horses, for instance, need a high-fat, high-fiber diet with very low carbohydrate levels.

Pelleted Feeds

Pelleted forages are handy and economical if hay is expensive in your area. They are also helpful if a horse has respiratory problems when eating hay. There are three types of pellets — grains, pelleted hay, and "complete" pellets that contain hay, grain, vitamins, and minerals.

Advantages of Complete Pellets

Pelleted feeds are convenient. Rations can be easily measured; contents are uniform; pellets are easy to store and transport; there is no waste; ingredients are nearly fully utilized; and little passes on as manure, making for easier cleanup in stalls. Pellet-fed horses are trim (they don't get a "hay belly"), and old horses, individuals with bad teeth, or horses with heaves (allergy-caused respiratory disease) often do better on complete pellets than on hay.

PELLET WARNING

If a horse accustomed to a diet of pellets gets accidental access to pasture or hay, he may be in danger of colic or impaction. His stomach and intestines have shrunken and can't manage the extra bulk if he overeats roughages he is not accustomed to eating. His access to forage should be monitored and limited.

Disadvantages of Complete Pellets

The main disadvantage to a diet of complete pellets is lack of bulk. In fact, pellet-fed horses usually chew wood because they crave roughage. Pellets that are complete feed contain hay, but it's finely ground. The horse eats less total bulk and doesn't feel full. He needs some roughage; bulky fiber stimulates the digestive process and keeps the tract moving at a normal rate. Most horses chew wood when first put on pellets, then taper off somewhat after they've become adjusted to the new ration and their stomachs have shrunken a bit.

A horse who doesn't get enough roughage in his diet will chew wood.

Supplements

A feed supplement is anything other than pasture, hay, salt, or water. Technically, grain is a supplement, but the term as generally used connotes any additional preparation of vitamins, minerals, proteins, fats, energy, and so on.

Horses have managed for thousands of years simply by eating grass. But in feeding domestic horses, we've come to rely on unnatural feeds, partly because horses are kept in unnatural conditions — stalls and pens instead of pasture. Some horses are stressed more than their wild ancestors were. The race horse, for example, requires readily digestible high-energy feeds to give "fuel" for maximum performance. Many people try to help their horses perform better or grow faster, so they feed more — and use a wide array of supplements.

Many people overfeed, which can be just as detrimental to health and soundness as underfeeding. Fat horses cannot perform well, foals who grow too fast are likely to develop skeletal problems (especially in feet and legs), and some types of supplements are toxic if overdosed. Part of the overfeeding problem involves misuse of supplements. If a horse doesn't need them, they are an unnecessary cost and a potential danger.

Many commercial feeds and all kinds of supplements are available. Quite a few people add supplements to commercial grain mixes to enhance growth or performance but, as a general rule, a supplement should be added only when

Many horses are overfed. A fat horse like this one has less stamina and more health risks than a normal horse.

the hay (or hay and grain) diet does not supply adequate levels of essential proteins or other basic nutrients.

Use Vitamin and Mineral Supplements Wisely

Few horses need vitamin or mineral supplementation. People who feed or inject vitamin or mineral supplements without knowing whether they are actually needed can cause serious harm or even kill the horse. Injectable supplements are especially dangerous because they are easily misused. Injectables should never be given unless a veterinarian determines that a horse needs them, and feed additives should be given only if an important nutrient is missing in the hay/grain diet.

COORDINATING A BALANCED DIET

The balanced diet for a confined horse with no access to pasture should start with good grass hay. Add grain only if necessary, selecting a mix based on nutrient content of available hay and adjusting the feed amount to meet the horse's needs. You can add alfalfa hay or a protein supplement if a horse, such as a young horse or nursing mare, needs more protein, and grain (or a commercial concentrate feed) if the horse needs more total energy and calories. Beyond that, always be careful when supplementing.

GENERAL RULE FOR FEEDING

The safest feeds are natural roughages — pasture and hay. As a general rule, the young, growing horse or hardworking adult can be fed as much good pasture or grass hay as he will eat.

To determine whether a horse needs a supplement, or what kind, consider the nutritional value of the feed being used as well as the requirements for the individual horse. In the case of a health problem, consult a veterinarian.

Probiotics and Prebiotics

Because horse digestion depends on microbes in the hind-gut to break down and ferment the fibrous portions of roughage, it is essential that the microbe population be healthy and in appropriate numbers and balance. Some people now use commercial feeds or supplements containing "probiotics" (beneficial bacteria and yeasts) and "prebiotics" (indigestible sugars that feed the beneficial bacteria or inhibit the bad ones). Many people feel that these supplements help keep the gastrointestinal tract healthier and more efficient.

Feed Each Horse as an Individual

The amount of feed or supplement a horse needs depends on his age and activities. Old horses, especially older broodmares with the added stress of pregnancy and lactation, may need supplements to keep from becoming thin or deficient because their digestive systems may not be efficient. The amount or type of feed and supplement that keeps one horse fit, healthy, and reproducing or performing well may not be enough or may be too much for a different individual.

Horses have different needs, even when doing the same work. It's not uncommon to find differences of 20 to 30 percent in the amount of energy needed. For example, some endurance horses stay fit on pasture or grass hay alone, with no grain. Others need 6 to 12 pounds (3–5.5 kg) of grain added to the daily ration to maintain the same body condition and fitness under the same work schedule.

Occasionally, a horse needs a special feed to complete his diet — such as an individual being fed bleached or poor-quality hay or dried-out winter pasture, a young horse or broodmare having trouble meeting nutritional needs on feeds available, a horse recovering from severe parasite infestation or debilitating disease, a horse doing strenuous work, or an old horse having trouble maintaining his weight.

An old horse *(right)* may have trouble keeping his weight. This young horse and old horse are on the same pasture, yet one is fat and the other thin.

The Mature Idle Horse

The idle horse needs less feed than one working daily. He needs only enough nutrition for bodily functions and limited activity. A maintenance ration for the idle horse is about 2.5 percent of body weight in good grass hay. A 1,000-pound (454 kg) horse would need about 25 pounds (11.5 kg) of hay daily. This will vary according to the body metabolism of the individual. If hay is of poor quality, feed more pounds of hay or add a small portion of grain. Idle horses often become too fat on good pasture even if they get adequate exercise. An idle horse who is an easy keeper may also grow too fat on a limited amount of good hay and need to have his ration cut.

ENERGY-TO-PROTEIN RATIO

Be careful when adding grain — more corn or oats, for example — to a commercial feed mix, as some horse owners do to try to increase growth or performance. The commercial product is generally properly balanced for minerals and protein. Adding supplements or extra grain can cause an imbalance in the diet, changing mineral ratio (calcium-phosphorus balance) or protein levels. Adding corn to a grain mix to give increased energy may decrease the protein level too much for a young horse. Proper energy-to-protein ratio is important for sound skeletal growth.

NUTRITIONAL REQUIREMENTS DURING PREGNANCY

About 80 percent of the fetus's growth takes place during the last 4 months of gestation. The mare's requirements for the first 7 months of pregnancy are not much more thana a maintenance ration.

The Broodmare

The pregnant mare has greater nutritional requirements than the idle horse. Most owners feed a pregnant mare more but tend to overestimate her needs while pregnant and underestimate them after she has foaled and is lactating. Pregnancy does increase the feed requirements of a mare, but significantly only during the last 4 months of gestation, when the fetus is growing the fastest. Feeding extra as soon as a mare becomes pregnant causes her to become too fat by foaling time.

A 1,000-pound (454 kg) mare in the last quarter of pregnancy — when her feed requirements exceed maintenance requirements — needs about 7.3 pounds (3.5 kg) of total digestible nutrients (TDN; portion of the ration actually utilized) per day. The same mare will need 14.4 pounds (6.5 kg) of TDN daily when she is nursing her foal. That's nearly twice as much as she needed during late gestation. The lactating mare needs almost three times as much nutrition as she would for body maintenance alone.

The Young Horse

A growing horse requires adequate feed for the growth needs of his body. The young foal is growing the fastest. He gains about 45 percent of his mature body weight by the time he is six months old, 58 percent at 1 year of age, and 87 percent by age 2. This varies somewhat with the individual and the breed; some breeds mature more slowly than others.

A foal will grow well on good green pasture if his mother milks well. Some mares need grain in order to milk well, and if a mare is not on green pasture, she will need alfalfa hay or some other high-protein feed to gain the protein necessary for lactation. If a mare does not milk well, foals should be started early on grain and a little alfalfa hay.

For proper bone growth and muscle development, the young horse needs adequate protein, vitamins, and minerals. Good pasture (or alfalfa hay) and a grain ration are usually adequate. Young horses should never be overfed. Their immature bones, tendons, ligaments, and joints are more easily injured when they carry excess weight. Also, too much grain or too many vitamins and minerals can result in skeletal problems that lead to crippling and unsoundness, including decreased absorption of calcium, causing soft bones (rickets). Fast-growing foals fed too much rich feed are especially prone to bone and joint problems. Overfeeding can cause permanent damage.

The Stallion

A stallion should be well fed for fitness and fertility, but never overfed. Excessive fat is detrimental to fertility, stamina, and sex drive, but so is lack of adequate nutrition. He needs a balanced diet, with the right amount of protein, vitamins, and minerals — essential for production and quality of sperm. A program of good feed and exercise will keep him in fine condition for breeding.

The Working Horse

A horse being ridden or driven needs more feed than the idle horse. As his work increases, he may need grain if he is unable to maintain proper body condition and energy output on pasture or hay alone. Feed requirements vary for individuals, but in general, a horse doing light work (2 to 3 hours daily) requires about 50 percent more energy than one doing no work. A horse doing moderate amounts of work (4 to 5 hours daily) uses about 70 percent more energy than the idle horse. A horse doing heavy work (6 to 8 hours daily) requires even more energy.

A horse doing strenuous work, such as sorting or working cattle all day, needs more feed than an idle horse or one doing light work.

The energy required depends on the type of work being done. For example, a horse being ridden for 3 hours at a walk along a trail won't need as much energy as a horse being ridden for 3 hours to move and sort cattle — galloping, stopping, and turning at high speed, continually exerting strenuously.

A horse with efficient metabolism may stay in good condition on pasture or hay, without grain, even when working hard; but many hardworking horses need a little grain to boost energy and keep from losing weight. Find the amount of feed that seems right for each horse. Some are gluttons, some more picky eaters. Some need more roughage, others more concentrate.

The Old Horse

The old horse eats more slowly and has less-efficient digestion than a younger one; he needs a good diet with ample vitamins and minerals. He may become thin on the same feed that used to keep him fat a few years earlier. If he's not

FEEDING TIPS

Conscientious feeding can make a big difference in the health and well-being of your horse. Take care to:

- **Maintain feeding regularity.** Abrupt changes in feed can cause digestive problems. Small amounts given regularly are far better than a large amount all at once. This is especially true of grains because too much can cause a horse to founder. The horse's small stomach can handle only a certain amount at one time.
- **Never feed large amounts of grain all at once.** The horse's stomach is small and, if overloaded, sends part of the grain on through undigested. Starches and sugars finish digestion in the small intestine and if too much gets into the hind-gut before being digested, the fermentation of this material changes the microbial environment and can lead to colic or founder.
- **Never overfeed.** A horse who is too fat cannot perform well and will tire easily when you ride him. Fat horses founder more readily than horses in fit condition. Feeding too much grain when a horse is not working can lead to serious problems, such as muscles tying up (PSSM — *polysaccharide storage myopathy* — which used to be

on pasture, feed him by himself so he doesn't have to compete with more aggressive eaters.

Feed him frequently in small portions. Several small meals during the day are better than one or two large ones. Keep his diet somewhat laxative so if his digestion is a little sluggish he won't become constipated or impacted. Green pasture is ideal, provides adequate nutrition, is easy to eat, and is somewhat laxative — and he can nibble on it as often as he likes.

If he can't be on pasture, give him alfalfa along with his grass hay, and add an occasional bran mash if he tends to become constipated. If alfalfa makes him cough, use pellets instead — they are less dusty than hay. Prepared feeds and supplements designed especially for older horses can be beneficial to individuals who lose weight on traditional feeds. Old horses do not absorb vitamins and minerals (especially zinc) as well as younger ones and are more likely to need supplements.

---

called azoturia in horses with this genetic tendency), when you do ride him. To prevent this, eliminate the grain ration on days he is not ridden or exercised.

- **Gradually change from one feed to another.** Change the diet over a period of several days or a week.
- **Keep feed boxes and tubs clean.** This will keep moldy or spoiled feed from accumulating.
- **Limit pasture for a horse who's been fed only hay.** If you are putting a horse on pasture after he's been on a diet of hay, let him graze only 30 minutes the first day, 1 hour the second, 2 hours the third (while still giving him hay), until, after 4 or 5 days, you can leave him out at pasture.
- **Make sure the horse has plenty of fresh water, salt, and exercise.** Exercise is just as important as feed when it comes to developing good feet, legs, and lungs in the growing horse and in keeping the mature horse in good condition. "Feed the horse to stand the riding, then ride the horse to stand the feed" is an old horsemen's adage that is very appropriate — muscle and fitness are never built by feed alone.

Make sure his feed is always fine and palatable, never coarse, dry, or dusty. Because coarse feeds are hard to chew, he may not clean up his hay. If he has trouble eating hay, check its quality and also have a veterinarian check his teeth; he may need dental care to relieve a painful eating problem. Grass is usually easier to eat than hay, and most older horses maintain good body condition on pasture. But if the horse is retired and on pasture, watch him closely to make sure he doesn't become overly fat and founder.

The Insulin-Resistant Horse

Some horses have metabolic problems such as insulin resistance (similar to type II diabetes in humans) or equine Cushing's disease (which is different from Cushing's in humans and dogs). Insulin resistance can occur in horses of any age, whereas Cushing's is most common in individuals more than 15 years of age. Both of these abnormalities hinder the horse's ability to handle sugars, and he may be prone to laminitis. These horses also develop fat deposits in abnormal places such as the rump, sheath, and neck (cresty neck). They do much better on diets high in fat (to supply the needed calories) and very low in sugar and soluble starch. There are now a number of commercial high-fat/high-fiber, low-sugar/low-starch feeds designed for horses with these problems. See chapter 9 for more information about Cushing's disease.

3

Seasonal Care

CERTAIN ASPECTS OF HORSE CARE vary by season and by region. The responsible horse person prepares for weather conditions and problems related to climate at all times of year.

Winter Care

Winter care may be minimal in mild climates or more complex in regions with subzero temperatures. Care depends on factors such as whether the horse is outdoors or in a barn and the amount of cold weather, snow, rain, and mud the horse must endure. Even if a horse is on vacation until spring, this doesn't mean he needs less care. In fact, some aspects of care become more important in cold weather to ensure the horse stays healthy and comfortable.

In winter, he needs additional feed, access to unfrozen water, and some kind of shelter or windbreak — and blanketing if he doesn't have a winter coat of hair. He should be regularly dewormed so he can utilize feed efficiently and won't harbor internal parasites that rob him of nutrients. Shoes should be taken off if he's not being ridden, and hooves trimmed periodically so they won't grow too long. Horses on winter pasture need extra feed if snow gets deep and covers the grass. Old and young horses need special care: the horse-keeper should make sure they get adequate nutrition.

Cold weather in itself is not a problem if a horse has a chance to prepare for it gradually by growing winter hair. Wind and wet weather create more hardship than does cold by itself. And wind makes blankets or shelter necessary.

Windchill Index

Knowing the windchill can be useful, whether you are planning to ride (and need to know how warmly to dress) or need to know how "cold" it actually is for an outdoor horse — whether he might need shelter or a blanket.

WIND TEMPERATURE SPEED (FAHRENHEIT)

Wind Speed (mph)	Temperature (Fahrenheit)																																			
	-20	-18	-16	-14	-12	-10	-8	-6	-4	-2	0	2	4	6	8	10	12	14	16	18	20	22	24	26	28	30	32	34	36	38	40	42	44	46	48	50
0	-20	-18	-16	-14	-12	-10	-8	-6	-4	-2	0	2	4	6	8	10	12	14	16	18	20	22	24	26	28	30	32	34	36	38	40	42	44	46	48	50
2	-23	-21	-19	-17	-15	-13	-11	-9	-7	-5	-3	-1	1	3	5	7	9	11	13	15	17	19	21	23	25	27	29	31	33	35	37	39	41	43	45	47
4	-25	-23	-21	-19	-17	-15	-13	-11	-9	-7	-5	-3	-1	1	3	5	7	9	11	13	15	17	19	21	23	25	27	29	31	33	35	37	39	41	43	45
6	-28	-26	-24	-22	-20	-18	-16	-14	-12	-10	-8	-6	-4	-2	0	2	4	6	8	10	12	14	16	18	20	22	24	26	28	30	32	34	36	38	40	42
8	-30	-28	-26	-24	-22	-20	-18	-16	-14	-12	-10	-8	-6	-4	-2	0	2	4	6	8	10	12	14	16	18	20	22	24	26	28	30	32	34	36	38	40
10	-31	-29	-27	-25	-23	-21	-19	-17	-15	-13	-11	-9	-7	-5	-3	-1	1	3	5	7	9	11	13	15	17	19	21	23	25	27	29	31	33	35	37	39
12	-33	-31	-29	-27	-25	-23	-21	-19	-17	-15	-13	-11	-9	-7	-5	-3	-1	1	3	5	7	9	11	13	15	17	19	21	23	25	27	29	31	33	35	37
14	-35	-33	-31	-29	-27	-25	-23	-21	-19	-17	-15	-13	-11	-9	-7	-5	-3	-1	1	3	5	7	9	11	13	15	17	19	21	23	25	27	29	31	33	35
16	-37	-35	-33	-31	-29	-27	-25	-23	-21	-19	-17	-15	-13	-11	-9	-7	-5	-3	-1	1	3	5	7	9	11	13	15	17	19	21	23	25	27	29	31	33
18	-38	-36	-34	-32	-30	-28	-26	-24	-22	-20	-18	-16	-14	-12	-10	-8	-6	-4	-2	0	2	4	6	8	10	12	14	16	18	20	22	24	26	28	30	32
20	-41	-39	-37	-35	-33	-31	-29	-27	-25	-23	-21	-19	-17	-15	-13	-11	-9	-7	-5	-3	-1	1	3	5	7	9	11	13	15	17	19	21	23	25	27	29
22	-43	-41	-39	-37	-35	-33	-31	-29	-27	-25	-23	-21	-19	-17	-15	-13	-11	-9	-7	-5	-3	-1	1	3	5	7	9	11	13	15	17	19	21	23	25	27
24	-46	-44	-42	-40	-38	-36	-34	-32	-30	-28	-26	-24	-22	-20	-18	-16	-14	-12	-10	-8	-6	-4	-2	0	2	4	6	8	10	12	14	16	18	20	22	24
26	-49	-47	-45	-43	-41	-39	-37	-35	-33	-31	-29	-27	-25	-23	-21	-19	-17	-15	-13	-11	-9	-7	-5	-3	-1	1	3	5	7	9	11	13	15	17	19	21
28	-52	-50	-48	-46	-44	-42	-40	-38	-36	-34	-32	-30	-28	-26	-24	-22	-20	-18	-16	-14	-12	-10	-8	-6	-4	-2	0	2	4	6	8	10	12	14	16	18

Note: Thermometers with both Fahrenheit and Celsius scales are available. To convert Fahrenheit to Celsius, subtract 32 from the Fahrenheit number. Divide the answer by 9. Multiply that answer by 5.

How Horses Keep Warm

Horses have a normal body temperature of about 100°F (38°C). They maintain this temperature in cold weather using several mechanisms that include shivering, change in hormone levels to increase body metabolism, increased digestion of fiber, growing long thick hair that can "stand up" on the skin and provide insulation, increased feed consumption, and increased activity. On a frosty morning, cold horses often run and buck to warm up.

A well-fed horse can manage temperatures down to −30° or −40°F (−34° or −40°C) if there's no wind and he's not wet. Wind ruffles the hair and eliminates its insulating quality. A wet horse loses body heat up to 20 times faster than a dry horse because moisture flattens the hair and reduces air spaces between the individual hairs, greatly diminishing the insulating effect.

The horse's best defenses against cold are long hair and a layer of fat beneath the skin to protect against loss of body heat. The insulating quality of winter hair is lost if the horse is wet or covered with mud; therefore, he needs a dry, sheltered area during cold or wet weather, as well as regular grooming (especially if he's muddy) to keep his coat from matting.

Shelter

Your horse will prefer the warmth of winter sun to a shady shed. Most horses, if given a choice, stay outdoors even in the coldest weather but use a windbreak or shed to get away from cold wind, blizzards, or rain. They don't like to stand in mud; if horses are in muddy paddocks, they need a high, dry spot where they can stand or lie without being in wet, cold mud. Horses that are constantly standing in mud are also vulnerable to hoof problems such as thrush (see chapter 4).

Many horses in cold climates spend winter outside, using canyons and creek bottoms, timber, or brushy areas as windbreaks. If a horse is accustomed to being outdoors, don't put him in a barn just because of foul weather. A horse in good condition, given a chance to toughen naturally as winter approaches, is best left outdoors. In fact, horses kept outdoors with access to a shed or windbreak have fewer respiratory problems than do those kept in a poorly ventilated barn that is full of dust from hay and bedding.

Body Condition

Make sure horses go into winter in good flesh. A thin horse won't winter as well as a fat one. Horses should never be too fat but should have a layer of insulation under the skin. Horses in cold climates need to go into winter with extra flesh if wintering outdoors for energy reserves as well as insulation.

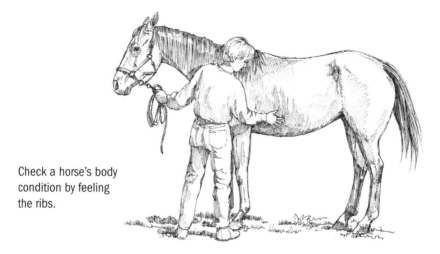

Check a horse's body
condition by feeling
the ribs.

Some horses gain weight in winter when not working; others can't handle the cold and lose weight unless feed is increased. It's sometimes hard to tell how fat a horse is and how much feed he needs because long winter hair can hide body contours. Check your horse's body condition periodically by feel — running your fingers over various parts of his body to determine how much flesh is under all that hair. Feel along neck, withers, hips, backbone, and over the ribs.

If you can't feel each rib individually he has plenty of fat, maybe too much.

If you can feel a layer of soft tissue between ribs and skin but can still feel each rib, he has about the right amount of fat.

If you can feel the ribs quite prominently, like running your fingers over a washboard — and there is no soft layer of tissue over them — the horse is too thin.

Give Roughage to Create Heat Energy

If the weather gets cold or is continuously wet, increase feed for all horses kept outdoors; they need more total calories to keep warm. If a horse doesn't get enough food to generate heat, he burns fat and muscle tissue to keep warm and loses weight. As temperatures drop, you should feed more hay, not more grain. You can even increase the amount of hay to all he will eat. More body heat is created by digestion of roughage than by digestion of grain. Remember that hay is digested in the large intestine and cecum by means of fermentation — which produces heat in the process — whereas grains are digested in the small intestine. A pound of hay generates more heat than a pound of grain.

HEALTH TIPS FOR OLDER HORSES

Old horses should not start winter thin. They may have poor teeth, making it hard to eat well and keep weight on. They generally have less resistance to cold and may have arthritic joints that become painful and stiff in cold weather. These horses should not winter outside. If an old horse is fat and an easy keeper, however, or if he has respiratory problems when indoors, he's better off outside with a windbreak shed and a blanket. Many older horses have breathing problems in an enclosed barn because of years of stress on air passages and heightened sensitivity to dust and molds. If your horse has this condition, provide dust-free bedding in his outdoor shed or where he likes to sleep. Dry, clean bedding will keep him warmer and also make it easier on his old bones when he gets up and down.

Digestion of protein also produces heat, so alfalfa hay can help a horse keep warm in winter. Although corn is higher in energy content than hay, it produces less heat when digested. Corn is more useful for weight gain than for heat production. If you are feeding grain in winter, it is often better to use oats than corn. Because of their higher fiber content, oats produce more heat than a similar weight of corn. Two pounds (1 kg) of oats and plenty of hay usually produce adequate body heat for keeping warm.

Give a Larger Feeding at Night

Horses should be fed at least twice a day. During cold weather, the larger feeding should be in the evening because winter nights are not only colder but longer. In northern regions, there may be 16 hours of darkness and only 8 hours of daylight, and if the horse gets his evening feeding before dark, it's a long time until breakfast! The evening feeding should include as much hay as the horse will eat.

Don't shortchange your horse on roughage at night or he may start nibbling fence posts, bedding, or anything else within reach — even his pasturemate's tail or mane — because he is cold and craves more roughage. Having access to plenty of hay at night will

Feeding on Cold Nights

A rule of thumb is to increase the hay ration by 10 percent for every 20 degrees below freezing.

keep him busy and warm. If horses are bedded on straw, they'll eat the bedding, which can lead to impaction or other digestive problems. It's better to give the horse extra hay so he won't resort to eating the bedding. Alfalfa can be fed during cold weather; horses metabolize the extra protein into heat energy. Take care to avoid dusty or moldy alfalfa, which can lead to coughing, heaves (similar to asthma in humans), or colic.

If horses are fed in groups, make sure the feeding area is roomy enough for all to eat without bossy ones chasing timid ones away from the hay. If a horse is a slow eater, keep him separate from the herd so he gets his share. Some horses like to eat their meal periodically throughout the day (or night), whereas others gobble theirs in an hour or two. If penned together, fast eaters get more than their share and slow eaters are shortchanged. Older horses with arthritis may not move well in cold weather, hanging back from the feeding area to avoid being chased by younger, bossier herdmates.

Water

Horses generally drink less water in cold or wet weather. They don't like ice-cold water, but they still need an adequate and constant supply or they may become impacted. If the water tub freezes and a horse doesn't get enough fluid, he won't eat enough and may lose weight or be unable to maintain body heat.

Check manure to see if he's drinking enough. If fecal balls are hard, dry, and small instead of soft and moist, he's not getting enough water. Other signs of dehydration include loss of skin elasticity, failure to clean up hay (it's too difficult to eat much dry feed without some water), or "drawing up" of the flanks and abdomen.

If the weather is cold, check the watering place twice daily and break ice if necessary. If the horse waters at a stream, pond, or ditch, he may be reluctant to step on ice to reach a water hole. You may have to spread dirt or gravel over the ice to give him safer footing.

DON'T DEPEND ON SNOW

Horses will eat snow, but if being fed hay, pellets, or grain, they may need more fluid in the digestive tract than snow alone can provide, especially if the snow becomes crusty and hard to eat. A horse may become impacted if he doesn't get enough water to maintain his body functions.

BENEFICIAL WINTER HAIR COAT

Long winter hair traps a layer of body heat between skin and the cold air. When weather is cold, tiny muscles in the skin make the hair stand up fluffy, increasing the insulating effect. A winter hair coat is better insulation than most blankets. Adding a heavy blanket or piling on several light ones can actually make the horse colder, flattening down the hair and severely reducing its insulating effect.

If he is watering in a tub or tank, make sure the water doesn't freeze too quickly. If possible, give warm water or use a tank heater. This way, the horse will drink more and the water won't freeze as fast. Mature horses need 6 to 10 gallons (23–38 L) a day in winter; if a continuous supply is not available, give fresh water (45 to 65°F [7.2–18.3°C]) twice a day, with the largest portion in the morning. Younger horses will drink less, with the actual amount depending on their age and size. Many horses don't drink much at night when temperatures are lower, drinking more during the warmer daylight hours. If the water tub is empty or frozen throughout much of the day, the horses will be seriously shortchanged.

Foot Care and Safe Footing

Shoes should be removed unless pasture is rocky or snow is deep and the horse must paw for grass. In that case, the front shoes can be left on but should be reset periodically so hooves don't grow too long. Usually, a horse is better off barefoot during winter because snow won't ball up under his feet so badly. He'll have better traction on ice and frozen ground — and less strain on feet and legs than when walking on four slippery balls of ice.

Pay attention to ice and slippery footing on paths, in gateways, near watering areas, and so on. Serious injury to feet and legs can occur if a horse slips and falls or spraddles (sprawls out) — possibly stretching a tendon or tearing a muscle. You can prevent such accidents by putting gravel, sand, or rock salt on icy places to improve traction.

Blankets

A horse grows thick hair as protection against cold and wind, rain and snow. If he's blanketed to prevent extra hair growth, or clipped for ease in cooling out (so he won't have a long hair coat drenched with sweat after a ride or workout), he'll need protection from winter weather. Don't clip a horse who has to spend time outside unless you are prepared to blanket him.

Blanketing may be necessary for a clipped horse or one who is moved to a colder climate without having a chance to grow a heavy coat. Even if a horse has a good hair coat, once he becomes so wet and cold that he has to shiver to maintain body temperature, he will burn more calories and need extra feed to keep warm — otherwise he'll lose weight. Under these conditions, he's better off with a blanket.

What Kind of Blanket?

There are many kinds of blankets; some are warmer and more weather resistant than traditional wool blankets. Some have an inner and outer shell surrounding an insulating filler material. The outer shell (usually made from a durable synthetic fiber) is waterproof and windproof. The filler provides warmth without bulk or heaviness. The inner shell is smooth and nonabrasive so it won't chafe the skin. These blankets are more durable than cotton or wool and light enough to not press down the horse's hair.

Outer shells are usually made of nylon (such as cordura) or canvas. Nylon is strong and fairly indestructible; it won't snag, tear, rot, or mildew. Canvas is durable and economical, giving partial protection from wind and rain because of its dense, tight weave, but it isn't truly waterproof. It eventually soaks through in heavy rain. A net cooler (usually cotton) under the blanket can wick away moisture and provide an air space for insulation.

All-cotton canvas is susceptible to mildew and color fading. Some canvas outer shells are treated with a special coating to increase water resistance. Advantages of canvas outer shells are ease of cleaning (they can be brushed off, hosed down, and hung on a fence to dry), durability and toughness, water resistance (fibers swell when wet, making less spaces for moisture to soak through), and ability to breathe. They hold in body heat but wick away moisture such as sweat. Canvas makes a good top layer over other blankets or sheets, and the weight helps hold everything in place. Disadvantages of canvas are its weight, slowness in drying, and tendency to tear fairly readily and to soak up urine and moisture from manure. Cordura nylon is actually more durable and foolproof and is the fabric of choice for most of the newer blankets.

Some blankets have a waterproof finish that keeps moisture out and holds heat in. The disadvantages of these coatings are that they may crack in temperatures below −20°F (−28.9°C) and are not good during changeable weather because the seal does not allow heat to escape; a horse may overheat under the blanket.

See the next two pages for tips on using blankets and proper blanket care.

Spring Care

Spring brings longer days, warmth, melting snow, and rain instead of blizzards. Frozen ground or deep snow gives way to mud. If spring is wet, horse pens (and places in a pasture where horses congregate) may become a deep sea of mud. If muddy conditions persist for long periods of time, they can be a health hazard for your horse.

Mud Problems and Health Concerns

Mud can make it hard for a horse to keep warm. If he stands in it, he will chill more easily in a storm than if he were standing on drier ground. A hair coat matted with mud loses its insulating quality. Mud also creates ideal conditions for problems such as rainrot and other skin diseases, thrush, and scratches.

A muddy paddock provides no place for a horse to lie down for a nap or even for a good roll. Being reluctant to lie down in deep mud, he may spend most of his time standing or try to lie down in a dry but dangerous place, such as on a built-up manure pile next to the fence. If he lies too close to the fence and tries to roll, he may become cast — on his back, against the fence, and unable to get up.

In wet, muddy weather, pay special attention to your horse's feet and skin. Be aware that moisture makes an ideal environment for bacteria and fungi and that long hair can hide a lot of skin problems. During a wet spring, you may encounter rainrot, scratches, lice infestations, or foot problems such as thrush or abscesses. (See chapters 12, 10, and 4, respectively, for more information.) If a horse has shoes on, he may lose a shoe in the mud. While struggling in deep footing, he may step on one foot with another and pull the shoe right off.

BEWARE OF MUD

Mud makes bad footing, which is hazardous when horses run and play. They may pull muscles or strain joints and tendons, especially if they are out of shape from winter vacation.

TIPS FOR USING BLANKETS

If you are going to blanket your horse, here are some rules of thumb to ensure that blanketing will be beneficial, rather than a hindrance, to your horse's health and safety.

- **Keep track of temperatures. Don't load a horse with blankets he doesn't need or leave him bare when he needs protection.** If he's shivering, he needs more covering. If he's sweating under the blanket, he's too warm. You don't want him to be sweating and wet under the blanket and then chilling at night when the temperature drops.

- **Don't blanket a horse for turnout in the cold early morning and leave him blanketed throughout the day, especially if it gets warm in the afternoon.** Feel his chest or just behind the elbow at his girth; these are the first places he'll start sweating. It's a good idea to have a heavy blanket for cold nights and a lighter one for daytime.

- **Put the blanket on correctly.** Start with it well forward and then pull it back into place so it won't ruffle the hair.

- **No matter what type of blanket you use, make sure it fits the horse.** A blanket that is too tight may rub and chafe. A high-withered horse kept blanketed all the time may develop skin irritation and infection where the blanket rubs the withers. Blanket rubs are a common wintertime problem and can cause sores. Some people sew a band of cotton (such as a leg wrap) around the shoulder edge of the blanket to minimize rubbing. A blanket that is too large or loose may slip and end up under the horse's belly. If his legs become entangled in the straps, he may seriously injure himself while struggling. The blanket should fit properly, with straps well adjusted so there is no slipping.

- **Make sure the blanket is designed to fit well and stay in place while still allowing freedom of movement.** This is especially important if your horse is quite active and playful when outdoors with a blanket on. There should be a strap that goes around each hind leg (below the stifle and above the hock) to keep the back end of the blanket in place. These leg straps are easier to keep clean if they are not made of leather. Some blankets have elastic leg straps or straps with elastic portions that hold well yet "give" when the horse moves. Others have replaceable straps or swivel snaps that don't get twisted up. Some people cross the leg straps under the horse when putting on the blanket (running the straps between the hind legs, then crossing them before attaching them at the sides) to keep straps from chafing the legs and also to ensure the blankets stay in place even if one strap comes loose.

  Most blankets have a *surcingle* strap under the belly that helps hold the blanket in place; some have double straps that cross under the belly. Elastic inserts on these straps are nice; the extra give can keep a horse from tearing out of his blanket. Some styles have an option for closed or open front (with fasteners), cut back at the withers or covering withers and shoulders (to provide extra warmth and to keep the blanket from shifting). Other styles have shoulder gussets for more freedom of movement, reducing the rubbing or pulling across the chest and adding "give" at the shoulders.

- **Don't use the same blanket on different horses.** This will prevent the spread of skin problems.

- **Keep blankets clean.** Wash them at least twice during winter if used often. If a horse constantly gets his blanket dirty, choose a type that can be washed frequently. Remember that buckles on the blanket straps are hard on a washer and drier. You may want to wash them in the special heavy-duty machines at a laundromat that caters to horse blankets, saddle pads, and the like. Most horse blankets should be washed in cold water with soap and a disinfectant, then very thoroughly rinsed. You don't want soap left in the blanket or it may irritate the horse's skin. Dry cleaning is not recommended. It won't remove odors, solvents tend to make the waterproof coatings disintegrate, and heat makes the bindings shrink. Choose a good blanket that fits your horse well. Keep it clean, and it will last a long time.

Daily grooming — keeping him free of mud and preventing the hair coat from becoming matted — and feet cleaning can give you a good handle on what's happening. In this way you can minimize or prevent certain wet-weather problems. If you catch a skin problem or foot problem early, it will respond much more quickly to treatment.

Mud Management

Mud that freezes on cold nights can make treacherous footing, as the humps, bumps, and ruts in pens and around water tanks or travel areas and gateways become solid. Horses may strain their legs or cut themselves on frozen lumpy ground and ice pockets created by deep hoofprints. The frozen ridges and holes make it difficult to walk, let alone run and frolic, and injuries can easily occur.

The problem of frozen humps on cold mornings can be alleviated by hauling in gravel or coarse sand to build up main traffic ways around the barnyard and to create smoother, drier areas in paddocks where horses can safely walk to water or lie down. Keeping a stockpile of sand or crushed rock for winter and springtime emergencies will help you reduce hazards of ice or mud.

Pens should be situated on high ground, if possible, rather than in low areas that collect rainwater and melting snow. You can resolve most drainage problems in a boggy pen or paddock by hauling in sand or gravel to create high spots. Where there are problems with spring run-off, dig a ditch or make a swale (with tractor and blade) to channel excess water away from or out of pens, paddocks, and riding areas so moisture does not pool.

HORMONES AND SHEDDING

Hormones affect hair coat and shedding. A stallion usually has a sleeker, shinier coat than a gelding, sheds faster in spring, and retains a short sleek coat longer in fall. Thyroid imbalance or Cushing's disease can affect shedding; some horses with these problems don't shed winter hair until late spring and others don't shed at all — they have to be clipped to prevent discomfort in hot weather.

Older horses who don't shed may have Cushing's disease, which mainly affects horses in their late twenties or thirties; a malfunction in the hypothalamus in the brain affects the pituitary gland and interferes with the proper balance of hormones, making the adrenal glands produce too much cortisol. This, in turn, affects shedding. Mineral deficiencies, especially selenium, can also cause slow shedding.

Shedding

Hormonal, environmental, and nutritional factors trigger hair growth and shedding. The lengthening days of spring stimulate shedding, whereas the decreasing hours of daylight in fall stimulate growth of winter hair.

A healthy horse, well fed and eating a balanced diet, sheds more rapidly in spring than does an undernourished horse. Because hair is made up of protein, the horse must have adequate protein for a healthy coat. A horse infested with worms — especially a young horse — often has a dull hair coat and is slow to shed because these parasites rob him of essential nutrients. Stress, illness, starvation, or other abnormal conditions also make a horse slow to shed.

The key to promoting healthy hair and speeding up shedding is a combination of balanced diet, regular grooming, and massage. Blanketing also stimulates shedding; the sweating process initiates getting rid of the extra hair.

Horses often rub on fences or trees or roll a lot when shedding to work the old hair loose. They also enjoy being groomed and scratched. If the horse exercises and sweats, he's even more itchy and eager to rub off the extra hair. If a horse seems too itchy — rubbing himself to the point of damaging the skin — or has a lot of dandruff or greasy skin, this could be a sign of lice or skin disease. Take a closer look. See chapter 12 for a more in-depth skin condition discussion.

Grooming regularly each day helps a horse shed faster.

If spring is rainy, winter hair may harbor fungus or bacteria. The combination of moisture, dirt, and dead hair can lead to skin problems that require medical attention. At this time of year, a daily grooming will speed up shedding and minimize the chances for skin disease. Use a rubber curry comb or shedding blade to remove the dead hair, and do a lot of brisk brushing and massage, which increase circulation and promote skin health by stimulating the oil glands. "Elbow grease" is the best key to a healthy hair coat.

Summer Care

Summer brings heat and humidity, flies and other parasites, and the potential for problems caused by dry climate and bright sunlight.

Dry Climate

In dry conditions, the hoof dries out and gets harder, but if feet are too dry, they become brittle and likely to split and crack. Your veterinarian or farrier can advise you on types of hoof dressing that are best for dry, brittle feet. If hooves are brittle, trim and smooth them often (if the horse is barefoot) to prevent splitting.

Bright Sunlight

Horses with white markings or light-colored skin are sometimes bothered by bright sunlight. Dark skin is the best protection against the sun's rays. Pigment is nature's sunscreen. Horses who have light-colored or pink skin are prone to sunburn and peeling, which makes them miserable.

Sunburn often occurs on white markings (especially on the nose) or where hair is thin and gives little protection. If a horse sunburns, use zinc oxide to protect the skin from ultraviolet rays, or use human sunscreen lotion as long as it does not contain para-aminobenzoic acid (PABA), which can sensitize a horse's skin to ultraviolet light and create more problems. Another way to protect pink skin is to paint those areas with an organic dye such as methylene blue or gentian violet. These dark colors block the sun's rays.

Photosensitization is much more serious than sunburn. It occurs when a horse eats certain plants (including Saint John's wort, perennial ryegrass, dried buckwheat, and alsike clover) containing a photodynamic agent that causes a toxic reaction in the unpigmented skin; it is absorbed from the gut into the bloodstream and travels to the skin, where it is activated by sunlight and kills the skin cells (see chapter 12). A horse with this problem should be turned out only at night.

Eye irritation from sunlight often occurs in horses with white markings around the eyes. Bright summer days (and bright winter days when sun reflects off the snow) can make the eyes water. Dark skin absorbs the sun's rays, but light skin reflects it into the eyes. Sunlight can irritate and damage the eye and surrounding tissues. In fact, cancer of the eyelid is common in horses with unpigmented eyelids. If a horse has a white face and is bothered by bright sunlight, apply something dark around the eye such as mascara, the dark purple coloring used in pinkeye medications for cattle (obtain this from your veterinarian), or black theatrical greasepaint used by actors to reduce the glare of stage lights and by athletes to reduce sun glare. The greasepaint is actually a cream that can be easily applied around a horse's eyes, especially if it is warmed for easier spreading. Anything used near the horse's eyes should be nontoxic and nonirritating.

Insects

Biting insects such as flies and mosquitoes are a nuisance to horses in summer; many species cause annoyance and injury. (These are discussed in detail in chapter 10.) Shelters or sheds can help protect your animals from horseflies and deer flies; these flies don't like to go into an enclosed building or shady, dim place. If flies are bad, you may want to stable your horses during the day. It also helps if pastures are well drained and not swampy; mosquitoes and horseflies require wet areas to complete their life cycles.

A fly mask can protect the horse's eyes when flies are at their worst.

GROOMING TIP

When grooming your horse in summer, take into consideration the fly problem before you cut off his mane or make his tail so short that it doesn't reach his hocks. Mane and tail are the most effective weapons against flies.

Insecticides for controlling flies on the horse can be applied as dusts or sprays or wiped on. (These are discussed in chapter 10.) Fly masks can give relief from those that chew on the horse's eyelids.

Health Concerns

If humidity is high, horses have difficulty cooling themselves by sweating. Sweat does not evaporate when air is already full of moisture, so the horse stays hotter. Temperatures above 80°F (26.7°C), especially if humidity gets above 50 percent), increase the chances for trouble, particularly if a horse is being worked or ridden.

The horse gets rid of extra body heat created by exertion through air exchange in the lungs, radiation effect from the skin, and evaporation of sweat. The hotter the air, the less efficient are his cooling efforts by means of air exchange or radiation effect. The horse can efficiently cool himself if the environment around him is cool, but as air temperatures rise and approach that of body temperature, he must rely on the evaporation of sweat to dissipate heat. If the air is humid, the sweat does not evaporate very well.

Hyperthermia

Extremely hot weather can cause problems even in a horse who is not exerting. When he has abnormally high body temperature, he suffers from hyperthermia (as opposed to hypothermia, which occurs when body temperature is too low). It's normal for a horse's temperature to rise slightly during a stretch of hot weather and humidity or when he is exerting on a hot day. But if he must continually endure heat and humidity in a shadeless paddock, remain

BE CAREFUL WHEN TRAILERING

Heat stroke can occur in horses who are confined in a hot, stuffy stall or trailer with no ventilation. In hot weather, transport horses at night when temperatures are lower, and never let a trailer sit in hot sun with horses inside. The trailer should spend as little time as possible standing still when horses are in it; the movement of travel creates air flow that prevents heat buildup and helps with evaporation of sweat to cool the horses. Make sure the trailer has adequate ventilation for good air flow. Don't leave horses tied to a trailer for long periods in the hot sun (as at a horse show or trail ride). If they must be tied to the trailer, use the shady side and provide some sort of awning.

enclosed in a hot trailer, or work hard in the heat, he may develop hyperthermia and heat exhaustion. If the problem is not treated immediately to bring down his temperature, he may go into shock. This type of shock is called heat stroke, and it can be fatal.

Heat Stroke

Heat stroke is more the result of the horse's inability to cool himself than a direct effect of hot sun. Dehydration from sweating can interfere with the body's ability to cool itself. Anything that inhibits the horse's ability to cool himself (such as a long hair coat, too much fat, a heavy blanket) can lead to heat stroke.

Signs of heat stroke may begin as restlessness and anxiety, progressing to erratic or irrational behavior, depression or excitability, disorientation, rapid breathing, weakness, and dry skin when the horse stops sweating. Body temperature may rise as high as 106 to 110°F (41.1°–43.3°C). The affected horse won't eat, and, in spite of the heat, he stops sweating. If you suspect heat stroke, call your veterinarian immediately and start trying to reverse the condition at once while waiting for him or her to arrive. Without prompt treatment, the horse may die. (For instructions on how to take a horse's temperature, see chapter 6.)

HEAT STROKE REFERENCE GUIDE

Symptoms	Treatment	Prevention
▪ Restlessness	Call veterinarian immediately and then do the following while you wait for him or her to arrive:	▪ Provide adequate water
▪ Anxiety		▪ Provide shade
▪ Erratic or irrational behavior	▪ Move horse to shady area	▪ Feed appropriately (avoid large amounts of protein)
▪ Depression or excitability	▪ Create breeze with fans	
▪ Disorientation	▪ Spray body with cold water or apply ice packs, especially to insides of legs and thighs, lower side of neck, head, and areas with large veins	▪ Monitor feed and water intake, as well as behavior
▪ Rapid breathing		
▪ Weakness		
▪ Dry skin (inability to sweat)	▪ Back off on rapid cooling after body temperature comes down to 105°F (40.6°C) to avoid causing a chill	
▪ Body temperature as high as 106 to 110°F (41.1°– 43.3°C)		
▪ Lack of interest in eating	▪ Have veterinarian administer intravenous fluids and medications to control shock	

Treatment for mild cases of heat stroke involves reducing the high body temperature as quickly as possible by putting the horse in a shady area where there is a breeze (even if you have to use an electric or manual fan) and spraying him with cold water or using ice packs. Apply cold water or ice packs to the head, neck, and areas with large veins, especially the inside of legs and thighs, the lower side of the neck, and the head. When the horse's temperature drops below 105°F (40.6°C), back off a little on the rapid cooling; too drastic a reduction in temperature at this point can chill the horse and make the condition worse.

In serious heat stroke the horse needs intravenous fluids and medications to control shock. If he recovers from the shock, he will probably survive, but his ability to endure hot weather or hard work may be permanently impaired. It is better to prevent this problem in the first place.

Insufficient Water Intake

Lack of water during hot weather can cause heat stroke. Problems can arise if an automatic waterer quits working; you might not know it until horses have been without water for a while. It's hard to monitor water intake or tell if the waterer is working unless you are watching when the horse is trying to drink. It's better to use a large tub so you know whether a horse is drinking and how much; you can monitor his intake because you are personally involved in the daily filling of the tub.

When necessary, cool an overheated horse with water.

┌───┐

KEEP 'EM COOL

Fat and heavily muscled horses don't dissipate heat well and should be watched for signs of heat stress. Limit a horse's exercise when temperatures and humidity are dangerously high.

OVERHEATING RULE OF THUMB

When temperature and humidity are added together and the total exceeds 130 (as when temperature is 80°F [26.7°C] and humidity over 50 percent), there is greater risk of horses overheating. This is also true whenever the temperature exceeds 100°F (37.8°C), even if humidity is low.

└───┘

If temperature is lower than about 85°F (29.4°C), a 1,000-pound (454 kg) horse will drink about 6 to 12 gallons (23–45.5 L) of water a day. At higher temperatures, he will sweat a little to cool himself (or sweat a lot if exercising); water consumption may increase by as much as three times. A lactating mare in hot weather may drink even more.

Inappropriate Feed

The type of feed a horse eats can make him susceptible to heat stress because horses generate more body heat when digesting large amounts of protein (grass hay is better than alfalfa when a horse is working hard in hot weather). During hot, humid weather, a horse may become impacted if he eats dry hay without access to abundant water. Any horse recovering from heat-related illness should be on a diet of green grass or other feeds with high water content.

Inability to Sweat

Sometimes horses in hot climates (especially tropical countries or Gulf Coast states) suffer from *anhidrosis*, which means "without sweat." This problem usually develops during the hottest part of the year and may come on quickly or gradually over several weeks. If a horse has to sweat continuously to keep cool during a hot, humid summer (as when confined in a hot, stuffy stall), his sweat glands work overtime and may shut down. The horse is then dry-skinned and his temperature climbs. He may have a few patches of sweat under the mane, behind the ears, under the elbows, and at the flanks, but most of his body stays dry. Prolonged inability to sweat can be fatal.

Horses appreciate shade on a hot summer day.

If discovered early, the condition can be corrected by cooling the horse — bathing with cool water to bring his temperature back down to normal (99 to 100.5°F [37.2°–38°C]), and keeping his temperature low enough that he won't need to sweat (about 100°F [37.8°C]). He should be kept out of the sun but not in a hot, humid stall. Use a portable fan to help stabilize his body temperature until his sweating ability recovers.

Adequate Shade

On the hottest days of summer, your horse will greatly appreciate shade. A few trees in a pasture can provide him a cool place to fight flies on a hot day. If there is no natural shade, consider building an inexpensive sun roof or run-in shed (see chapter 1).

Good Care Is Essential

Your horse's health is very important, and the seasonal care you give him will make all the difference in keeping him healthy and in good condition. This includes having a yearly vaccination program for infectious diseases (see chapter 7); regular deworming, including treatment for bots after the first hard frosts in fall (see chapter 10); fly control (see chapter 10); and foot care (see chapter 4). The horse also needs regular exercise and a safe environment. In some respects, summer can be the easiest time for horses, especially if they can enjoy green grass and nice weather, but in other ways it brings challenges and problems; good care and management are essential.

4

Foot Care

THE OLD SAYING, "NO FOOT, NO HORSE," couldn't be more true. There is a clear relationship between the working ability of the horse and the soundness of his hooves.

If you handle a horse's feet often, he grows accustomed to it as part of his daily routine and learns to mind his manners for trimming and shoeing. Also, you become more apt to notice problems. As you clean his feet daily, you can remove any rocks stuck in the feet and prevent buildup of packed mud or manure that can lead to thrush. You are also likely to notice any heat in the leg or hoof (or swelling of the lower leg) that could be a sign of infection or injury.

Because the horse is an athlete, his feet and legs are crucial parts of his structure. Care of the feet is essential; neglect can lead to unsoundness and pain.

HOW TO PICK UP A HORSE'S FOOT

To pick up a front foot, face the rear, run your hand down the leg, and gently squeeze just above the fetlock joint — in the indentation along the back tendon. This will encourage the horse to pick up his foot. Hold it between your legs for cleaning. To pick up a hind foot, stand close to the horse and run your hand down his leg in the same way, squeezing gently at the fetlock if he does not pick it up readily. Rest the leg across your thigh for cleaning the foot. If the horse does not want to pick up his foot, lean against him a little to encourage him to shift his weight to the other three legs. (See chapter 5 for more on safe hoof handling.)

What Makes a Sound Hoof?

Horses are individuals; the feet of each one differ slightly from those of others in shape, hardness, and rate of hoof growth. The hoof is a specialized horny shell that covers sensitive living tissues — bones, blood vessels, and nerves. The outer shell is a unique covering that grows continuously to compensate for wear and tear.

The sole should be somewhat concave to allow for expansion when weight is placed on it. The hoof wall is designed to carry most of the weight and the bars serve as a brace to prevent overexpansion or contraction of the foot. The V-shaped "frog" serves as a cushion in the middle of the foot, helps absorb concussion, and regulates hoof moisture. If any of these outer tissues are injured or abused

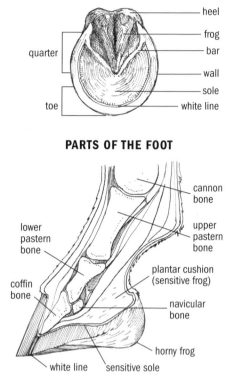

PARTS OF THE FOOT

Labels: heel, frog, bar, wall, sole, white line, quarter, toe

Labels: cannon bone, upper pastern bone, lower pastern bone, coffin bone, plantar cushion (sensitive frog), navicular bone, horny frog, white line, sensitive sole

---

HOOF GROWTH

The hoof wall grows down from the fleshy coronary band that goes around the hair line at the top of the hoof. Any injury to the coronary band affects growth of the hoof. The average rate of hoof growth in a normal horse is about ¼ to ⅜ inch (0.6–1.0 cm) per month. The entire wall is usually replaced by new horn every 8 to 12 months.

A horse doing normal activity usually wears his hooves at the same rate they grow. If the horse is on pasture or in a paddock without much riding, his feet don't wear fast enough to keep up with growth and need periodic trimming. If hooves are allowed to grow too long, they may split or splay out and hinder his movement. Toes that are too long put strain on the leg tendons. If a horse is shod, his shoes should be reset or replaced every 3 to 10 weeks depending on the rate of hoof growth and shoe wear (some endurance horses wear out their shoes within 3 to 4 weeks).

by excessive trimming, the normal functions and soundness of the hoof are impaired.

Foot Conformation

How the foot is built plays a role in its ability to hold up. Front hooves should be larger, rounder, and stronger than hinds because front legs support about two-thirds of the horse's weight. Hooves should be wide at the heels, not narrow or contracted. The sole should be slightly concave in the front feet and even more so in the hinds. A horse with flat feet is likely to suffer stone bruising or develop navicular disease. (See chapter 9 for more information.)

Ideally, the hoof wall should be thick, pliable, and resistant to drying out, and should grow at a normal rate. The sole should be thick so it can't be easily bruised, and the bars should be strong and well developed. The frog should be large and healthy and centered in the foot. An off-center frog is an indication of crooked feet and/or legs; the hooves probably do not wear evenly.

Abnormalities and Malformations

Few horses have perfect feet and legs. When breeding horses, you should select individuals with the best possible conformation because this characteristic is passed on to offspring. When acquiring or keeping a horse with less than ideal feet, you may decide to overlook the foot faults. In spite of a foot problem, plenty of folks have purchased or held on to a child's dependable old horse or pony, or an animal with malformed feet because of an earlier injury, or a horse purchased for his wonderful disposition or exceptional ability. These horses may need special foot care or more than routine foot trimming to stay comfortable and reasonably sound.

Contracted Feet

A contracted foot is narrower than normal, especially at the heels and quarters; the frog becomes small and atrophied, shrinking to such an extent that it no longer has contact with the ground. A horse with contracted heels or feet is likely to go lame because of the inability to properly absorb and dissipate concussion — making him susceptible to problems such as navicular disease or concussion-related breakdowns in other structures of the foot and leg.

Contracted heels are often caused from improper shoeing (the shoe too narrow at the heels with no room for hoof expansion when weight is placed on the foot), injury, a diseased frog, or unnecessary shoeing — leaving shoes on too long or keeping the horse shod year round. If shoes are left on too long

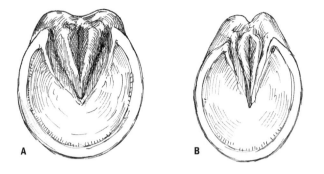

A. Normal foot
B. Contracted foot (heels too close together; frog small and atrophied)

without trimming the foot and resetting the shoes, the hoof wall grows long and heels become underrun, inhibiting expansion at the heel and quarters.

Lameness from any cause may make a horse put less weight on a foot and, over time, the lack of frog pressure results in contraction. Once a foot is badly contracted, it may take more than a year to become normal again, even with special shoeing.

Contracted feet are more common in front feet than hinds, especially if the condition is from improper shoeing. Sometimes only one foot is contracted because of an injury. It is easy to tell the difference between the normal foot and the contracted one: the normal foot has healthy frog and heels whereas the heels of the contracted foot are too close together.

Foot contraction is often accompanied by a dished or concave sole. The foot no longer flattens when weight is placed on it because the heels cannot expand; the sole becomes more concave, arched upward. If contraction becomes severe, the hoof wall may start to press against the coffin bone inside the foot, making the horse lame and unsound — a condition called "hoof-bound."

Foot expansion helps counteract concussion; the normal foot has a concave sole that flattens when weight is placed on it, the heels springing wider apart. The coronet narrows and drops backward as the overall height of the foot decreases.

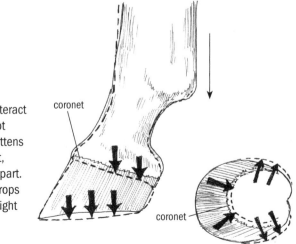

coronet

coronet

Treatment. Corrective trimming and shoeing can help reverse contraction (with the help of a good farrier), but the primary cause — such as lameness, improper shoeing, or dry feet — should be identified and corrected. If feet are hard and dry, efforts should be made to restore proper hoof moisture by using a good hoof dressing daily. A dry foot lacks elasticity for proper foot expansion.

Flat Feet

Flat feet lack natural concavity of sole — an inherited condition. There is not much you can do to correct this, but you can keep the horse from stone bruising by using Easyboots or some other brand of protective hoof boots over the shoes when riding among rocks or by having your farrier attach hoof pads under the shoes.

> ## APPLYING IODINE
>
> An easy way to apply iodine to the sole is with a small syringe. Then you don't get it all over your hands or spill it on the horse. Draw up a little into the syringe and squirt it gently, a little at a time, covering just the sole area.

Some flat-footed horses get by without hoof pads if the soles are kept toughened so they don't bruise easily. This can be done by applying a little iodine over the sole of the foot, taking care not to spill any of it on the horse's skin (because it burns) or let it run over the hoof wall (because it dries out the tissues). The iodine can be applied on days when the horse will be traveling on rocks or gravel.

Club Foot

A club foot is one that has a steep pastern, with hoof and pastern angle of more than 60°. The hoof angle is often steeper than the pastern angle, and the heels grow faster than the toe. If just one foot is this way, it may be because of an old injury (lack of use can cause contracting and shortening of the tendons, making the foot more upright). Sometimes the problem is genetic; in certain family lines, several individuals may have one abnormal foot (always on the same side), possibly because of conformation of that leg and the way the foot

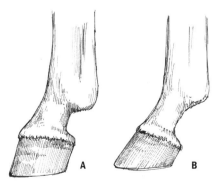

Club foot (**A**) compared with normal foot (**B**)

and pastern grow. Other horses may inherit the club-foot trait in both fronts. This disability makes a horse less agile, with a rough and stumbling gait.

Treatment. Frequent and proper foot trimming can help a horse with a club foot, especially if corrective trimming is begun while he is still young and growing, before the condition gets worse. Letting the hoof grow too long between trimmings can aggravate the problem, causing the horse to have a very long heel and upright foot and pastern.

Basic Foot Care

You may wish to leave all foot trimming to your farrier; however, if your horses have normal feet and legs and do not need corrective trimming or special work, you may wish to learn how to trim. Ask the farrier to give you lessons on trimming so that you can take care of the routine matters. Even if you never trim a foot yourself, be sure to handle and clean all feet regularly. Then, you'll know if there are any problems and be able to appropriately schedule trimmings with the farrier.

When cleaning a foot, a hoof pick or the blunt edge of a hoof knife is the best instrument for getting out all dirt or embedded rocks. *Never use a very sharp or pointed object for cleaning a foot.* You could injure yourself or the horse if he happened to jerk his foot at the wrong time.

Pulling a Shoe

Once in a while, it becomes necessary for a horse owner to pull off a shoe. Even if a farrier does all your trimming and shoeing, you may, at some time, face a situation when you can't wait for him or her. Even if the shoe was well clinched to start with, it may become loose — get hooked on a fence, caught in a deep bog, stepped on by a hind foot and pulled loose, or loosened when traveling through rocks.

In these instances, it's best to pull off the shoe. If it is not immediately removed, it may injure the

When cleaning a foot, use a hoof pick.

TOOLS FOR REMOVING SHOES

It is fairly easy to remove a shoe without breaking the hoof wall if you have a few appropriate tools or use adequate substitutes. Shoeing hammer, clinch cutter, nippers, and rasp make the job easy, but you can use a flat-edged screwdriver in place of a clinch cutter and a regular carpenter's hammer, if necessary. Pulling nippers or hoof cutters are best for pulling a shoe, but if you don't have them (and the shoe is fairly loose) you can use a pair of vice grips or regular pliers to hold on to the shoe and gain leverage for pulling it.

horse if hanging loose on one side, catch on something and get the horse into trouble, or cause a corn or bruise because it has slipped out of position and is putting pressure on the sole or bars. If the shoe is accidentally pulled off, it may break the hoof wall, taking out a chunk and making it harder for the farrier to reshoe the foot.

Unclinching the Nails

The shoe is easiest to remove (and pieces are less likely to break out of the hoof wall as you pull it off) if you first unclinch the nails that are still holding. You can use any kind of hammer to drive the clinch cutter or screwdriver under each nail end to pry up the clinched end and straighten it out. This is easiest while the horse's foot is on the ground. Be careful not to cut into the hoof wall if you are using a screwdriver. Once each nail is unclinched, cut off the straightened nail end with nippers if you have any. If you don't have a clinch cutter or screwdriver, rasp off the clinched nail ends with a rasp or file.

Removing the Shoe

If you don't have a rasp or file, you can still pull the shoe with nippers or vice grips, especially if the shoe is already partly loose. Hold the hoof in regular shoeing position — between your legs for a front foot, across your thigh for a hind. Place the nippers or vice grips between shoe and hoof at the heel (starting on the looser side to make it easier), and use a downward force, pushing the handles slightly toward middle of the foot, to pry and loosen the shoe. Work alternately along each branch of the shoe, starting at the heels and moving toward the toe as the shoe comes loose.

If you are unable to undo the clinches, you can still remove the shoe in this manner, although it takes a bit more strength and leverage because you must pull the clinches loose and on through the hoof wall. The nails will straighten out as you pull the shoe; they will come out with it.

If some of the nails are still tight, the hoof wall may break unless you take each nail out as you go. This may happen when you pry on the shoe because the clinches are still secured in the wall. To get hold of a nail head, you may have to gently pound the shoe back down against the hoof so the loosened nail head protrudes up enough to grasp with hammer claws, nippers, or pliers. Pull it out, then loosen the shoe enough to take out the next nail, alternating down each side of the shoe.

Safety Tip

If a nail breaks off in the hoof wall, grasp it with pliers or nippers and pull it out. Don't leave nails or pieces of nails in the wall sticking out; the horse may hit and cut himself with them.

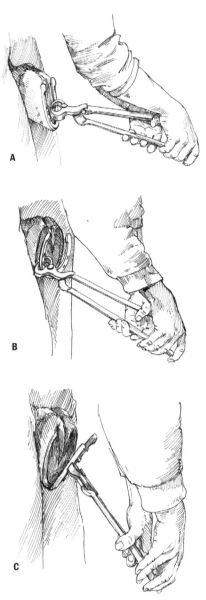

A. Pull a nail out to help loosen the shoe. **B.** Work down each branch of the shoe with alternate pulls and take each nail out as you come to it. **C.** Pull the shoe.

Going Barefoot

For many horses, going barefoot can be healthier than having shoes on all the time. Whether a horse is a good candidate for leaving shoes off depends on the horse's lifestyle, use, hoof conformation, and so on. For feet to be healthy and strong while barefoot, the horse must be in a natural environment—reasonably dry ground rather than continually wet areas—with room to exercise. A horse kept in a stall can't keep his feet as healthy and strong as the horse roaming a 300-acre rocky pasture. The latter will be wearing his feet normally, about the same speed they grow, and getting regular exercise that aids the blood supply to the hoof.

The more nearly a horse can approach totally natural conditions, the healthier his feet will be. Barefoot life won't work for a horse who is confined or kept in a wet pasture where his feet stay soft. A horse with soft, bare feet will quickly go tender-footed or lame from bruising if you try to ride him on a gravel road. However, if you allow his feet to toughen gradually—put him in a drier, larger area and limit your rides to very short distances at first—and he's not ridden excessively on rocky ground, he may get by just fine without shoes.

Bare feet must also be properly balanced so the hoof walls on each side bear equal weight and stress and the toes are not too long. Otherwise, the hoof walls will split and crack from abnormal pressures. Many horses must be kept shod to keep the foot from cracking and chipping, and to keep the wall from wearing away too fast if the horse is ridden regularly on rocky ground.

Thrush

Thrush is an infection of the frog, caused by bacteria (and sometimes fungi) commonly found in barnyards or pastures. Because these organisms thrive in wet, decaying material such as manure or mud, thrush is common in horses that live in muddy pens, wet pasture, or dirty stalls. If a horse's feet are frequently packed with dirt, mud, or manure, the lack of air next to the frog and the constant moisture in the hoof make ideal conditions for the organisms to flourish, rotting the frog tissue.

The telltale foul odor of thrush is unmistakable when you clean a horse's foot. There are also black secretions along the edges of the frog, and it may be soft and eaten away at the edges. The early stages of thrush are indicated by just a little dark coloration and grime around the frog, or dark spots along the white line of the sole, plus the bad odor. You can quickly clear up the problem at this point by keeping the foot cleaner and applying iodine, bleach, or a commercial preparation for thrush daily to the affected area to kill the bacteria. Just be careful not to let any of these medications run down the hoof wall to the coronary band or they may burn the skin.

WINTER REQUIRES EXTRA DILIGENCE

Winter can be a hard time to prevent thrush if pens and pastures are wet and muddy. It may take diligence — daily foot cleaning and care — to prevent thrush or keep it from getting worse. A bad case that invades the inner tissues of the foot must be treated as a puncture wound would be; seek advice from your veterinarian.

If a horse is kept in a muddy pen, boggy pasture, or dirty stall, and his feet are rarely cleaned — never having a chance to dry out — thrush can progress to the point of lameness as the infection penetrates and spreads to sensitive parts of the foot. If thrush is long-standing and deep, the horse will flinch when his feet are cleaned or trimmed because the frog is undermined with infection.

Prevention and Treatment

Prevention is the best "treatment": keep the horse in a clean environment, clean his feet often, and ride or exercise him regularly. If he can get out of the wet paddock and travel on dry ground, his feet will have a chance to dry out and air will get to the bottom of them, inhibiting thrush. No hoof dressing or medication can keep thrush from recurring if the horse is constantly kept in dirty surroundings and his feet are packed with mud and manure most of the time.

If you detect the beginnings of thrush when you clean a foot, don't ignore it. Clean the foot thoroughly, then swab the affected areas (usually the edges of the frog and any black spots in the sole) with iodine-soaked cotton or squirt on a little iodine with a small syringe. Daily treatment with strong (7 percent tincture) iodine for 3 or 4 days will clear up an early case. Clean feet often and keep track of their condition — then, thrush won't get a head start.

White Line Disease

This is the common term for a progressive infection and subsequent separation of the hoof wall resulting in the wall coming loose from the foot. The problem usually starts at the bottom of the foot at the white line between sole and wall and travels upward, creating a hollow area between wall and foot. Earlier terms for this condition were "seedy toe" and "hollow hoof."

The cause of this problem is an opportunistic fungus that "eats" hoof horn, entering the foot through damaged tissue. If the foot is out of balance or too long, with a flare on one side or a dished toe, the extra stress on the hoof wall may create a separation at the white line. Each time the horse places weight on the foot, it stretches the white line area, enabling the fungal spores to enter. They may also enter through any break in the hoof wall, or through an old abscess.

The fungi become established in the area between the outer wall and the inner, sensitive tissues and gradually eat it away. Tapping the outside of the hoof produces a hollow sound. The hoof horn residue inside this area is like chalky, dried, and crumbled cheese. If a lot of the inner wall is damaged, the foot loses some of the attachment that binds the coffin bone to the hoof wall, and that bone may drop, just as it does in severe cases of laminitis (founder).

Prevention and Treatment

The only successful way to treat white line disease is to trim out all the diseased tissue to get rid of most of the fungus. Topical medication can be applied to get rid of the rest. You will probably need help from your farrier, especially if much of the hoof wall must be cut away. The area must be opened up to the air because these fungi thrive in an airless environment. If the hoof wall attachments have been so seriously damaged that the coffin bone moves, the horse will need special shoeing for support. If caught early, white line disease is fairly easy to treat and clear up; however, it may take 6 months to a year for the horse to grow a new hoof wall in advanced cases in which a lot of the wall must be cut away. Those horses may need the foot to stay in a special boot or wall cast to protect it until they can grow more wall.

White line disease can be prevented by keeping the feet well trimmed (not too long) and in proper balance so there are never any extra stresses on the foot to stretch or damage the white line. Because some horses tend to get repeated fungal infections, once the problem is cleared up, your farrier may recommend soaking the feet every couple months or so in a product containing chlorine dioxide. Studies have shown this treatment to be most effective against redevelopment of this fungal infection. Since the fungi can be transmitted from one hoof to another, farriers should disinfect their trimming and shoeing tools after working on a horse with white line disease.

Foot Injuries

The horse's foot is strong and durable, but it can be injured if he steps on a sharp rock or nail. The tough outer covering of hoof wall and sole protect the delicate inner tissues from most types of trauma, but sometimes an exceptionally deep penetration, severe blow, or constant pressure causes bruising of the underlying tissues.

Corns and Sole Bruises

Corns are bruised areas on the sole, usually involving the tissues at the angle where the hoof wall meets the bars. A corn appears as a slightly reddened area and is especially noticeable when feet are trimmed and the sole pared down. The red area may be hot and tender in the early stages of the bruise. Corns are most common in front feet because they bear more weight and are more subject to bruising than are hind feet. Horses with flat feet are more likely to get a corn or sole bruise from stepping on sharp rocks or gravel.

Corns are most often caused by improper shoeing (ill-fitting shoes that are too narrow) or by leaving shoes on too long. The hoof wall begins to grow down around the outside of the shoe at the heel area, and the shoe puts pressure on the sole at the angle between hoof wall and bars, bruising the sole. Trimming a horse's heels too low may also cause corns as a result of increased pressure at the angle of the wall and bar. Corns in this area are rare in a horse that goes barefoot, but he may suffer sole bruising (a similar sore spot) from stepping on rocks. Horses with thin soles or chronic laminitis (founder) with dropped soles are very susceptible to sole bruising. Bruises in the sole generally occur in the toe or quarter area.

Locating the Problem

Corns or sole bruises can make the horse lame. If the problem is a corn, the horse will favor the heel and put more weight on the toe. If the problem is a

---

DEFINITIONS

- A **dry corn** is bruised tissue in which there is red staining caused from bleeding inside the sole.
- A **moist corn** is caused by severe injury to the sole, resulting in seeping fluid beneath the injured horny tissue.
- A **suppurating corn** is one that has become infected and abscessed.

HOOF TESTERS

A hoof tester looks like large hoof nippers, with curved parts large enough to fit around the entire hoof. The hoof tester puts pressure (by pinching) on the sole to check for soreness. When you put pressure on the sore spot, the horse will flinch. Your veterinarian or farrier can show you how to use a hoof tester.

Use a hoof tester to pinpoint the area of soreness or to locate a corn or sole bruise.

bruise in the toe area, he will try to land on his heel to keep the weight off his toe. Your veterinarian or farrier may use a hoof tester to locate the corn or bruise. The horse will flinch when the sore area is pressed. If the sole at that spot is pared down with a hoof knife, a reddish (or bluish if an abscess is developing) discoloration may be seen in the area where the horse shows pain.

Treatment

When improper shoeing causes corns, removal of shoes may be all that is necessary for healing. The horse should not be ridden or reshod until the corn and lameness have disappeared. To make sure shoes do not cause corns, check that the heels of each shoe extend well back and cover the hoof wall completely at the heel and quarters to allow room for hoof expansion. The hoof wall in this area must rest on the shoe. If the hoof extends beyond the shoe when the foot expands, the shoe is not large enough.

If a horse develops an abscess from a corn or sole bruise, your veterinarian will pare down the sole at that area until the infection in the sensitive tissues begins to drain. Then soak the foot daily, packing and protecting it between soakings, as you would for a puncture wound.

Puncture Wounds in the Foot

Puncture wounds in the body are always serious because they can lead to tetanus if a horse has not been vaccinated, but a puncture in the foot can be very bad because it may damage the inner tissues and cause any one of several serious problems—inflammation or fracture of the coffin bone, decay of the bone, or decay of the digital cushion. A puncture in the middle third of the frog may puncture the navicular bursa.

A puncture in the bottom of the foot is not always easy to locate if the foreign object (nail, stick, or rock) is no longer embedded. A puncture in the frog can be difficult to find if the spongy tissue has closed up again after the object makes its hole.

Treatment

If you suspect a puncture (or the object is still embedded), consult your veterinarian. You may not realize there is a problem until an abscess develops within the foot and the horse goes lame—from infection putting pressure on surrounding tissues. It's best to start treatment immediately after the puncture occurs.

Hoof Abscess

An abscess may be caused by a puncture wound, infected stone bruise, misplaced horseshoe nail that "quicked" the horse (entered the sensitive inner tissues), or a case of deep, neglected thrush that penetrates into sensitive inner tissues. Sometimes, a hoof crack becomes deep enough to allow entry of bacteria that start an infection.

Treatment

The area should be opened to drain, soaked, cleaned with disinfectant, and the hole plugged with a wick that will allow drainage. A puncture wound in the sole should be opened so there is at least a ¼-inch (0.6 cm) hole into the infected tissue, with the walls of the drainage hole widening toward the ground surface of the sole so it won't become obstructed. When the wound is in the frog, the frog should be trimmed away at the site of the puncture until adequate drainage for the abscess is established.

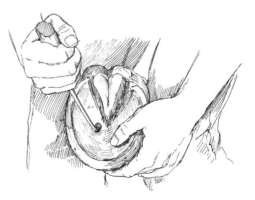

Open up a sole abscess or puncture wound with a hoof knife for infection drainage.

Part of the treatment for a puncture wound or foot abscess consists of soaking the foot. After the veterinarian has dealt with the original condition — opening the area for drainage and flushing it out — and given the horse a tetanus shot and possibly antibiotics, he may tell you to soak the foot daily for several days to draw out infection and promote rapid healing. Soaking the foot in a warm water-and-salt solution not only pulls out any remaining infection but also makes the sore foot feel better.

Soaking a Foot. If a horse has never had a foot soaked, get him used to the idea carefully and gradually so it will be a good experience — so he will cooperate rather than resist. Put his foot gently into an empty rubber tub. Once he stands there with his foot in the tub, wash the foot thoroughly with water — scrubbing with a rag, if necessary, to remove mud or dirt — and replace it in a clean, well-rinsed tub. Carefully add a little warm water, slightly warmer than the horse's body temperature (100°F [37.8°C]). After he accepts it and is relaxed, you can add several tablespoons of Epsom salts (magnesium sulfate) to the water. Don't add the salt to the water until you're sure he's comfortable about the soaking in case he moves around and spills the first batch.

After he's standing with his foot in the warm salt solution, periodically add hotter water until the water in the tub is quite warm but not burning hot. Keep the water as warm as the horse will tolerate. It's easiest with two people — one to keep the horse from spilling the tub water if he picks up his foot (hold the foot and guide it gently back down into the water so he won't put it on the ground and get it dirty), and one to manage the tea kettle and refills. Usually 20 to 30 minutes of soaking per day is adequate. Most infections clear up after 3 or 4 days of this.

Protecting a Drainage Hole. If there is a drainage hole in the sole or on the bottom of the foot, protect it between soakings by wrapping and bandaging the foot to keep out mud and dirt. An easy way to bandage is to wind Vetrap

---

TUB OR BUCKET?

A rubber tub or bucket is best — if you don't have a special soaking boot — because it won't injure or startle the horse as much as a rigid (noisy) metal one. If the injury is low (at the coronary band or heel) or on the bottom of the foot, a rubber tub works fine. If the area to be soaked is higher (at the fetlock joint), the tub may not be deep enough to cover the area in warm water. In this case, you'll need a rubber bucket.

(stretchy material that sticks to itself) around the foot to cover the bottom, then use duct tape from mid-hoof on up to the pastern to hold everything in place, and put the bandaged foot into an Easyboot or some other type of protective boot so the bandage stays clean and dry. If the horse can be kept in a clean stall, a treated abscess will often be healed enough to turn the horse out again in 3 to 4 days. If you don't have a stall and the horse's pen or pasture is wet or muddy, use a waterproof boot to keep the foot and bandage clean between soakings.

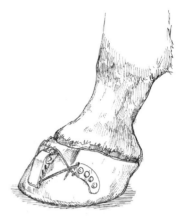

An Easyboot and other similar products can help keep the foot clean and dry.

Coronary Band Injuries

Sometimes a horse suffers injury to the coronary band at the top of the hoof by running into something, stepping on himself with a shod hoof, or having his foot stepped on by another horse. A deep wound at the coronary band that results in scar tissue may disrupt the hoof-growing cells and make a weak spot in the hoof as it grows — an area that may easily crack or disrupt hoof growth completely.

Treatment

A serious or deep injury at the coronary band requires immediate veterinary attention so it can heal properly and prevent "proud flesh" (excessive granulation tissue; see chapter 16).

Hoof Cracks

Occasionally, a crack develops in the hoof wall from splitting and chipping, concussion, or injury. Barefoot horses often develop chips and cracks, but these are usually not serious unless neglected. In a long, untrimmed foot, cracks can become worse quickly, traveling up into sensitive tissues. These can be hard to correct. A deep crack may make the horse lame.

Toe cracks, quarter cracks, and heel cracks start in the hoof wall at ground surface, usually from a chip or split in an overly long hoof, and travel upward. Sometimes, a crack starts in the heel or quarter area and travels horizontally around the foot because of weakness in the wall at that area (from a blow or injury, as when a horse strikes the hoof against a rock). Sometimes, a crack starts at the coronary band. This type of crack travels downward because of weakness in the hoof wall at that area.

A barefoot horse with a serious crack must be trimmed often to relieve pressure on the crack and keep it from splitting further. It is difficult to grow out a bad crack on a barefoot horse because it's impossible to take all pressure off the crack and keep it from spreading. The horse may have to be shod.

Treatment

When trimming a foot with a crack, the farrier will cut away the hoof wall at the ground surface of the crack so it does not bear weight (which would expand the crack). He may also try to stop the progress of the crack by burning a small notch across its highest point with a crescent shaped iron. This helps stop the splitting because, if weight is placed on the crack, it sends the force along the notched groove instead of up the hoof wall.

Corrective Shoes. A crack that originates in the coronary band caused by an old injury that affects horn growth, may be a more persistent problem. In this case, corrective shoeing may be necessary for the rest of the horse's life. Regular use of a hoof dressing, lanolin ointment (such as Bag Balm or Corona), or olive oil may keep the coronary band and hoof wall more pliable in that area and less apt to crack.

A shoe can help take the stress off the hoof wall and keep a crack from splitting further, enabling it to grow out. A clip on each side of the crack helps keep the hoof from expanding there when weight is placed on the foot.

Fast-Drying Glue or Plastic. Some cracks can be repaired with a strong, fast-drying glue or plastic that fills the hole and holds the cracked area together so it can't keep expanding. This protects sensitive tissues from contamination and infection if the crack is large or deep. There are several types of acrylic

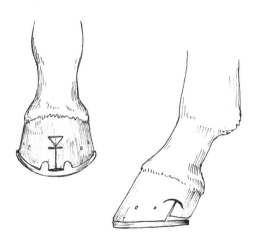

Halting the progress of a crack by shoeing with clips (to keep the hoof from expanding at that point and spreading open the crack) and by burning a notch at the top of the crack with a bar and triangle or crescent design sends the force of impact along the notched grooves instead of up the hoof wall.

> ### BRITTLE FEET
>
> Brittle feet — caused by dry conditions, poor nutrition, or genetics — make hooves very susceptible to cracks. Brittleness is more common in white hooves than dark ones, but actual hoof strength depends on the individual horse. Some have tough, resilient feet regardless of hoof color. If a horse has brittle feet that crack easily, a good hoof dressing can help, as can certain feed supplements (see chapter 2). Consult your veterinarian.
>
> A serious hoof crack should receive immediate attention to halt its progress and protect the underlying tissues from infection. If the crack reaches sensitive tissues, the horse may need a tetanus shot, antibiotics, and foot soakings. If a horse ever develops lameness in conjunction with a hoof crack, consult your veterinarian.

bonding agents that your farrier or vet may use for repairing cracks. The foot will still need to be trimmed often or the shoe reset — until the crack grows out — so the glue or bonding agent may have to be reapplied several times.

Diagnosing Lameness

Pain is nature's way of keeping an animal off an injured leg or foot. If a horse limps, try to determine the cause. A sore front leg (or foot) will cause him to bob his head more than normal as he hurries to get off the painful limb and takes more weight on the good leg. A sore hind leg or foot will cause a hitch in his stride as he puts less weight on the bad one and hits the ground harder with the good one.

To tell which leg is lame, watch his head and hips. The horse's head goes higher when he steps on the bad front leg and drops lower as he takes more weight on the good front leg. His hip stays higher on the lame side if he's got a bad hind leg and drops lower on the good side that is taking more weight.

Check his feet and legs to see what the problem might be. It may be as simple as a rock caught in his foot, or it may be a stone bruise or leg wound. If there is infection in a leg, there will be heat and swelling. A pulled tendon or injured joint will also cause heat and swelling. If you cannot determine the exact problem or how to treat it, call your veterinarian.

5

Horse Handling

THE HORSE IS AN INTELLIGENT, sensitive animal with great athletic ability. Part of our fascination with him comes from the realization that he's a wild animal that has been domesticated. When a good horseman or horsewoman works closely with a well-trained horse, there's practically no limit to what they can accomplish as a team. To develop that unity, or to handle any horse safely, we must understand how horses think and react. The horse owner needs to be mentally in tune with the horse.

Horse Sense

Horses have better memories than humans, but they don't think the way we do, and often people make the mistake of assuming they do. The horse is a prey animal. Most of his natural reactions are geared toward flight unless he's at ease with the situation at hand. When he feels alarmed, he becomes nervous and ready to flee. He has a strong herd instinct; he's happiest and feels most secure with other horses.

Horses are most at ease when their human handlers act like the dominant herd member (boss horse) without being too aggressive or submissive. As a herd animal, the horse is accustomed to being bossy or bossed.

Horse and human can develop a strong loyalty bond — the horse transferring his needs for safety and comfort to the person he trusts.

LEVELS OF COMMUNICATION

You are always sending signals to the horse — through your emotions as well as actions and body language — whether you realize it or not. Every time you handle him you are training (or untraining) him, making things better or worse. Everything you do with the horse should be with thought as to how he will respond. With every action you take, ask yourself whether you are aiding or defeating your goal of training him to be well mannered and easy to handle.

It comes naturally to him to submit to a more dominant individual, so this is the role the horse owner must take when handling the horse. The extremes in horse handling methods are force and bribery. Force may make an aggressive horse resentful, always fighting back, or a timid horse more afraid. Bribery will quickly spoil a horse; he soon learns that he (not you) is in control. The most effective handling methods fall in between, with the horseman taking an assertive leadership role in tune with that particular horse's ability and nature.

Understanding the Horse's Temperament

The horse in nature is a member of a group. Most horses become frantic, lonely, or neurotic with no herdmates. The lone horse is in a situation contrary to his natural instincts; his attitude is healthier if he has some company — even a goat or a cow or some other animal if there are no other horses nearby; however, some adjust more readily to an isolated existence than others.

There is a direct relationship between temperament (a horse's emotional characteristics — whether he is bold, calm, easygoing, timid, or insecure) and ease of handling. A horse who is readily upset or distracted can be more challenging to handle than a calm one.

When handling horses, it's important to work with their temperament, making sure bold ones don't try to dominate you, lazy ones don't cheat and try to avoid doing what they are supposed to do, and timid ones don't become more afraid and skittish. Use good judgment and keep the horse's individual personality in mind when catching, leading, handling feet, and so on, and your horse handling will go a lot more smoothly.

Cultivating a Good Relationship

Cultivate the horse's respect and cooperation so he knows what to expect. If you use consistency and quiet confidence in all your actions, transmitting a message of benevolent dominance, the horse feels more secure and trusting than he would if handled by a nervous, inconsistent, timid, or angry person.

Working with a horse is much easier if you are tuned in to what he is doing and thinking. He should also be comfortable with signals he picks up from you. Remember that you control and communicate not only by means of physical cues and actions but also with your mind. This type of control over a horse comes with familiarity (knowing the horse well) and your understanding of one another, gained through working together — the horse knowing what you want and being conditioned (through proper handling and training) to obey you. He senses your relaxed but dominant attitude rather than fear, anger, or nervousness. He respects you as the dominant one (boss horse); it's part of the relationship you develop with him as you work together and you handle him daily. He can read what you are thinking by your body language and the way you act. If you give mixed signals — act aggressively even though you are nervous or afraid of the horse, for example — this will make a horse very nervous and untrusting.

Relax

Horses can read a person well. They sense changes in physical or emotional tension that tell them a person is frightened, nervous, or angry. Indeed, they can sense our mood and feelings no matter how much we try to disguise them. Animals have the ability to communicate with each other through a combination of body language and empathy, and they are able to apply this to human beings to the extent that they can read us like a book.

You must truly be at ease in your relationship. Even if you go through the motions of being the dominant member of the team in your words and actions, if in your mind you are afraid of the horse, he can sense it. Your fear will overshadow your actions and will make him more nervous or more aggressive.

The best way to get a nervous horse to relax and cooperate is to be relaxed yourself, both mentally and physically. If you are tense and nervous, the horse becomes agitated also, and because he thinks that whatever is alarming you is also potentially dangerous to him, he may become very upset.

A combination of body language and confident attitude gives you dominance over the horse and puts you in the position of boss and teacher, enabling the horse also to be at ease in the relationship — not feeling the need to aggressively disobey you through lack of respect or being unduly afraid of you. If you have confident dominance over him, the horse will respect and trust you, accepting you as boss. You will rarely have to resort to a loud voice, punishment, or any physical contest of strength while handling him.

Use Appropriate Tone of Voice

The sound and tone of your voice are important. An approving voice can be a powerful reward or positive reinforcement for good behavior. A disapproving tone is often punishment enough for bad behavior; the horse realizes you are displeased with him. Horses don't have to fear us to submit to our wishes; punishment and reward by use of body language, voice, and attitude are enough when there is good communication in your working relationship.

Punishing Bad Behavior

If a physical reprimand is needed, it should be instant and appropriate. The horse may need just a jerk on the halter rope or a tap with a whip (to remind an overeager individual not to drag you along too fast, or to punish an act of bad manners). For example, if the horse is trying to charge forward too fast, a tap on the chest with the butt of a whip can remind him to slow down. Or, a rap on the leg when he shoves into you and steps on your foot can remind him to stay back and not invade your personal space. One short application is enough. Most horses understand one swat — after all, in the herd situation one bite or a single kick is adequate to keep a member in line. They don't understand beatings or continual punishment.

Any punishment should be of short duration, immediate, and forceful enough to fit the circumstance. The psychological aspect (realizing he has stepped out of bounds and offended the dominant member of the team) is more effective than physical discomfort or pain, although the punishment should be sufficient to effect a change in the particular horse's behavior. For some, vocal disapproval is enough. For others, especially bold or highly aggressive individuals that test your authority, a sharp spanking (on the body, never the head) may be necessary. If the horse is testing your dominance, as he might do in a herd situation trying to work up to being boss horse himself, you must remind him that you are still on top of the pecking order.

AVOID EXCESSIVE PUNISHMENT

Excessive punishment makes a horse lose respect for the handler, replacing trust with fear or strong resentment. Continually pecking at a horse can thoroughly confuse him. He will quit trying to do the right thing or become afraid of everything you do.

> ### SAFETY TIP
>
> Never take any horse for granted. Even the most dependable one might move suddenly if startled and hurt you unintentionally if you are in the way. Have safety-conscious working habits even around horses you know and trust; it makes for fewer stepped-on toes, bumped heads, or more serious mishaps. Always be aware of which way the horse might move, and be prepared to move with him.

Use Caution

The horse is a large, strong animal. If he becomes upset or frightened, or moves suddenly, the person handling him may be injured if he or she is in the wrong place at the wrong time. Most accidents can be prevented if you make a practice of handling yourself and the horse in such a way that there's less chance of unexpected trouble. If safety is foremost in your mind and you are thinking ahead, prepared for anything the horse might do, you are less likely to be hurt.

An important factor in minimizing accidents is having a good working manner and quiet confidence — being firm but gentle. Then, the horse is less apt to try you out (if he's bold) or be afraid and flighty (if he's timid). You want him to relax and trust you. A nervous horse who is uneasy about the way you handle him is more likely to become unmanageable than the one who feels secure with you.

Handling horses safely is a matter of awareness (being tuned in to what the horse is doing and thinking, being prepared for what he might do next) and body position. You dominate a horse through confident attitude and body position. This type of mental control comes with understanding one another. The horse must know what you want and be conditioned to obey; he knows *whoa* means stop and stand still, that he must accept restraint by the halter, and that he must behave when you pick up his foot.

The horse is stronger than you are, but through training and your confident attitude, he accepts your dominance. If, however, he becomes momentarily frightened or upset, or worried about the pain you or the veterinarian may cause when he must be vaccinated or treated, he may forget his manners. At those times, you must be able to calm and restrain him and keep from being hurt by him, even though you are not as strong as he is. You can use body position and leverage (and physical contact with the horse) to your best advantage to keep him under control and avoid being kicked, bumped, or injured.

APPROACHING A HORSE SAFELY

When you catch and halter a horse — whether in paddock, pasture, or stall — speak to him as you approach; make sure he knows you are there. Even a gentle horse may kick if you startle him while he's napping. Talk to him before you touch him. Don't approach from directly behind him (in his blind spot) but at an angle, so he can see you better. Speak softly and move slowly.

Always keep in mind that the horse's instinct is to protect himself from danger — to run away, strike, or kick to defend himself. The veterinarian's needle, medication spray, or any procedure the horse isn't accustomed to may provoke a defensive response that is dangerous to people handling him. Be prepared for evasive or defensive actions and handle the horse in a way that will minimize or prevent these actions, or be in a position where his actions won't make physical contact with you or inflict harm.

Holding a Horse

At times, you must hold a horse for the farrier or the veterinarian for treating wounds, vaccinating, and so on. The person holding the horse can make a difference in whether the procedure is successful, difficult, or dangerous. Keeping the horse from moving at the wrong time and keeping him calm or preventing him from kicking can be crucial.

Use a properly fitted halter. With some horses, a bridle may be necessary for more control. Some need a chain over the nose to keep them restrained (or even a twitch or lip chain — see subsequent discussion), but if a chain is used, it should not cause pain or annoyance. Otherwise, the horse may become more unmanageable, fighting the restraint as well as the procedure being done. If you use a chain, never allow your fingers to get accidentally wrapped in it.

Always choose a safe place to hold the horse. Sometimes an open area is best, with no obstacles to bump if he moves around. If he must be still, hold him next to a stall wall or solid fence, never next to a wire fence. If he usually tries to rush backward when confronted with something he doesn't like, plan in advance to back him against a fence or wall so he can't use this tactic. For most procedures, if there is a solid fence or wall on one side and the handler and person working on him are both on the other side, the horse will stand still.

When holding a horse for the farrier, keep the horse calm and still so he won't move at the wrong time.

The person holding the horse for medical treatment should stand on the same side as the person treating the horse. Stand at the horse's shoulder, with one hand on the halter and the other against the neck or withers, ready to move with the horse if he moves. With his front leg turned up, he is unable to kick on that side.

Don't Stand in Front

When holding a horse, don't stand in front of him. If you're at his shoulder you'll be out of the way if he lunges forward or strikes out with a front foot. Stand facing his shoulder with one hand on the halter and the other on his neck or withers, with feet positioned so you can move when he moves and you won't be stepped on or jumped into. Body contact is crucial. If you are braced against him, you can move with him rather than being bumped. Holding the halter, facing his shoulder, you can read his expression and intentions.

With your other hand rub his neck or withers to help calm and distract him. Often, a soothing voice, soft whistling, and rhythmic rubbing will help keep his mind off the procedure, and he'll tolerate the situation instead of fighting it.

Prevent Kicking

Maintain proper contact with the horse's head to gain leverage over his entire body and his movements. This is important for the person at the halter, who should be tuned in to the horse's state of mind and possible reactions to prevent movement at a crucial time or keep the horse from kicking. Raising the head or pulling it toward you at the right moment can make it awkward for the horse to kick the person who is working on him. If the halter is too loose, hold it snugly under the horse's chin (keeping the loose part together) so your hand has contact with his jaw while holding the halter. The more contact you have with him, the better control you have and the less chance there is for him to bump you.

Lead Safely

When leading, walk beside the horse's left shoulder, holding the halter rope or lead shank a few inches from the halter or grasping the halter itself if more precise control is needed. This will afford the best control over his movements.

Always have a rope or lead shank attached to the halter; never try to lead a horse with just a halter. The rope should be at least 6 feet (2 m) long, preferably 8 feet (2.5 m). If you use a lead rope, the horse is less able to pull away, hurt your arm, or dislocate your shoulder should he jump, rear, or turn quickly. With a rope attached, you can play him like a fisher plays a fish; he cannot jerk you as hard as if you were hanging on to the halter alone. Never wrap the end of the rope around your

Proper position when leading a horse: walk beside his left shoulder, and hold the lead rope close to the halter with your right hand, with the extra length of rope looped (not coiled) in your left hand.

hand. Keep extra length in neat loops, not a coil that might get caught around your hand or arm if he bolts.

When leading a bridled horse with loop reins, take the reins down over the head. If the reins are still up on the neck and he balks, backs up, or bolts, you have no control and they only make the problem worse because one side will be pulling at his mouth.

Don't Walk in Front

Don't walk in front of a horse you are leading. If he spooks and leaps forward, he'll bump into you or step on your heels. You have no control of his movements if you are in front of him. When you walk beside his shoulder you move with him and have control over his head and all his actions. You can keep him going at the speed you wish and halt him when necessary. If he tries to go too fast or bolt, you can use body leverage to halt or slow him by leaning into his shoulder and pulling his head around so he has to circle. When you have this kind of contact and body leverage, you can be as strong as he is, making him pivot around you so he can't bolt forward. Having your arm or body against his shoulder gives you more control than just holding him by the halter.

Tying a Horse

When tying, always keep safety in mind; an improperly tied horse can be potentially dangerous.

Use strong halters and ropes, and keep them in good repair. Nylon web halters should be at least three-ply with

Use body leverage to your advantage when leading an overeager horse. Keep your hand close to the halter and your arm braced against his neck when necessary. With your body against his, you are strong enough to spin him around and make him circle if he tries to bolt.

sturdy hardware. Remember that a halter rope is only as strong as its snap. It's often best to have a rope permanently braided onto the halter ring or tied with a secure knot rather than depending on a snap to hold if a horse sets back. If you don't leave halters on horses (and you shouldn't — it's a dangerous practice because a horse may catch the halter on something in his pen or pasture),

USE A SAFE KNOT

Always use a knot that will securely hold yet be easy to untie in a hurry. The "manger tie" is a good quick-release knot that can be undone with just a pull on the loose end. If a horse nibbles at the rope or tries to untie himself, you can put the loose end through the loop so he can't undo the knot.

you don't need a snap-on lead rope and can keep your ropes securely and permanently fastened to the halter.

Always tie to something solid. Careless tying, use of a flimsy rope or halter, and tying with bridle reins or to an insecure object invite disaster. If a horse sets back (as even a well-trained one will do if suddenly startled) and breaks the rope or bridle reins, he could fall over backward, pull the top pole or board off a fence, or bolt down the road dragging whatever he was tied to. If this happens, he may seriously injure himself, people, or other horses. It may also be such a traumatic experience that he can never be dependably tied up again.

Never tie to a wire fence. The horse can get a foot caught if he paws. Tie to the post itself or to a pole or board only if it is nailed to the other side of the post and cannot be pulled off. When tying, don't leave so much length of rope that the horse might get a foot over it. Tie short, but with enough rope to give some freedom of head and neck movement, and level with his head or higher. Then if he pulls back it won't damage his neck muscles as readily as it would if he were tied too low.

When tying to a tree, make sure the rope won't slip down the

Properly tied, the horse has some freedom of head and neck but never can get his foot over the rope. The rope should be tied level with his head or higher.

When tying a horse to a tree, tie high, make sure the rope cannot slip down the trunk, and be sure there are no branches to create a hazard for his eyes.

trunk. Choose a tree with no branch ends that might injure his head or eyes. When tying to a horse trailer, be sure the trailer is attached to the pulling vehicle or securely blocked so the wheels can't move. When tying near other horses, keep the horses far enough apart so they can't kick at each other. Never tie with bridle reins; this is a sure way to end up with broken reins or an injured mouth.

Handling Feet

When handling a foot, the more body contact you have with the horse, the better. Not only are you braced against him for holding the leg, but you can also sense his mood (tense or relaxed) and anticipate any movement he might make. (See chapter 4 for more information.)

Picking Up a Foot

To pick up a front foot, stand at the shoulder, facing to the rear, and slide your hand down the leg. Press your fingers just above the horse's fetlock joint. If he doesn't immediately pick up the foot, press a little harder in the area just in front of the back tendon — the indentation between tendon and cannon

SAFETY TIPS FOR HORSE WORK

Use basic rules of safety when working around a horse for any purpose — proper positioning and body contact will work to your advantage. You are less apt to be bumped, kicked, or stomped if you're close to the horse and braced against him in such a way that you keep him from moving improperly. If he does move, you'll be moving with him.

When working around front feet, take care never to be right in front of a leg as you bend over to examine it or apply medication or hoof dressing while the foot is on the ground. The horse can unintentionally bump you just by picking up his foot; his knee can hit you in the face. Always watch out for the knee.

When working around hindquarters, be aware that maintaining close contact is always safer than being a short distance away where you'd get the full force of a kick. Close contact can prevent a kick if you are braced against the leg. It also allows you to move with the horse and not be hurt.

bone. If necessary, lean into his shoulder at the same time to encourage him to shift weight off that leg and pick up the foot. Pinching just in front of the back tendon generally makes even the stubborn horse pick up his foot and is easier than trying to wrestle with him.

To pick up a hind foot, stand next to his hindquarters facing to the rear and run one hand down the leg while you lean against the horse with your other hand to encourage him to shift his weight and pick up that foot. If he does not want to pick it up you can pinch the back tendon gently.

Run a hand down the back of the front leg and prepare to pick up the foot. Press your shoulder against the horse's to encourage him to shift his weight off that foot. Squeeze gently at the fetlock joint if he doesn't pick the foot up for you.

To pick up a hind foot, lean against the horse to help him shift his weight. Run a hand down the leg, pinching the back tendon gently if necessary, then pick up the foot.

Once you've picked up the hind foot, prepare to rest it against your thigh for examination or cleaning.

If the horse tries to jerk the foot away while you are cleaning or trimming it, you can thwart this by bending the fetlock joint, firmly pulling the toe up. He cannot kick at you or take the foot away as readily with the joint flexed. This added leverage works in your favor.

Holding a Foot

When holding a front foot to clean, trim, or doctor, be sure to keep your upper arm or shoulder in contact with the horse's side. This affords more leverage if you have to lean against him and hold tightly to the foot to keep him from pulling it away. When holding a hind foot, try to keep almost the entire hind leg in contact with your own leg. If the horse tries to jerk his hind foot away, firmly bend the foot by pulling the toe up toward you. This makes it harder for him to pull it away and kick you.

Holding Up a Front Foot

Sometimes a horse fidgets or kicks so much that certain tasks become awkward or impossible to complete. In many instances, holding up a front leg can minimize these actions and immobilize him. The safest way to do this is with another person holding the horse. On the side from which you don't want the

THE IMPORTANCE OF TOUCH

Safety around horses is a matter of knowing the individual horse well and being alert to his potential reactions. It means intuitively keeping yourself in the position relative to him that puts you at least risk. This is usually a position very close to the horse, touching him and transmitting your confidence to him through that touch.

horse's hind foot to kick, pick up the front leg. Hold up the leg with one hand, with maximum contact and leverage, and brace yourself against the horse's shoulder with the other. Be prepared to move with him in case he tries to lunge and rear. Don't be in a position where you would get bumped or hit by the leg, and be ready for any movement the horse might make.

Establishing Body Contact

Body contact with a horse not only gives better leverage for holding or working on him but also enables you to tell whether he is relaxed or tensing up for sudden action. In addition, close contact and confidence on your part — being relaxed and having a

When holding a front foot, keep your upper arm and shoulder in contact with the horse.

matter-of-fact attitude — help the horse to be more relaxed and secure. Because he knows where you are and what you are doing, he is less apt to be jumpy or try to kick. This is especially important when giving injections or treating a wound. Sudden contact from a needle or medication spray, for example, may startle or alarm him. If you are touching him in that area and reassuring him, and he's aware of and tolerating your presence, you can usually accomplish the procedure without much reaction on his part.

Using Restraints Properly

There are several ways to restrain a horse if he does not want to submit to a particular procedure. Extreme measures involve the use of ropes (scotch hobbles or sidelining — tying the hind leg to the front leg) to keep the horse from kicking, or tranquilizers. The average horse person should not try these; improper use of ropes and hobbles can put you and the horse at risk for injury, and improper use of tranquilizers can be hazardous and should be given only by a veterinarian. It's better to rely on other methods, if possible. Should these fail, seek the help of a professional or a veterinarian.

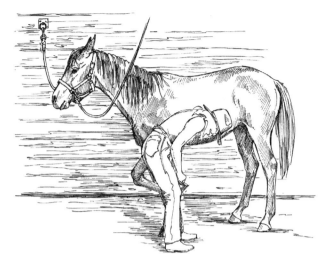

Cross-ties often help to keep a horse from moving around while you are working with him.

Cross-Ties

Cross-ties are two ropes with snaps that secure the horse in the middle of an open space; the ropes are snapped to each side of the halter. This approach is often used in a box stall or barn aisle. If the horse is accustomed to being cross-tied, this is often adequate restraint. He is held from both directions but with open space on each side for you to work in — he can't crash you or himself into a fence or wall. Yet, the restraint of a rope on both sides of his head limits his movement.

Chain over the Nose or through the Mouth

For some horses, a chain over the nose gives better restraint than a halter. Whereas they root and pull and misbehave with just a halter, the chain over the nose gives more pressure when you need to remind horses to stand still. In some cases, the chain works better through the mouth (like a bit) rather than over the nose. The halter rope or

A chain over the nose gives better restraint than just a halter if you need to remind a horse to stand still or to gain more control when leading him.

lead strap has a short length of chain at the snap end to put over the nose or through the mouth, if needed. When using a chain, be careful not to get your fingers wrapped in it.

Lip Chain

Some people prefer to put the chain under the upper lip and over the gum. This works well for some horses but not others. The lip chain is effective in making a fractious horse behave and stand because it utilizes pressure points that stimulate release of endorphins. If properly used, without jerking it too hard, it does not hurt the horse. However, it can be a severe restraint that injures the tender gum and upper lip if the horse rears and plunges. A piece of hay twine or small rope or heavy cord works as well as a chain and is less likely to be abrasive to the tender gums.

Twitch

The traditional twitch is made from a length of wooden handle 15 to 30 inches (38–76 cm) long, with a loop of rope or chain attached to one end. Newer metal versions clamp onto the upper lip and the halter. This enables the horse handler to have both hands free to work on the horse after adjusting the twitch if there is no one to help. The disadvantage of this "one person" design is that if the clamp twitch loosens and the lip comes free, the horse is no longer under control and is left with the tool dangling from his halter. If it's too tight and he breaks away, he may start running blindly to try to escape the clamp stuck on his upper lip.

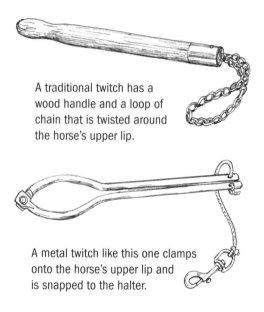

A traditional twitch has a wood handle and a loop of chain that is twisted around the horse's upper lip.

A metal twitch like this one clamps onto the horse's upper lip and is snapped to the halter.

Not All Horses Can Be Twitched

Some horses will not tolerate a twitch. Others, such as foals or young horses, should never be twitched to begin with. A youngster who has not had a lot of handling may react with fright and resent any further handling of the muzzle.

The best candidates for twitching are emotionally stable adult horses who are accustomed to human contact; these animals are basically trusting and tractable and don't mind having a person handle the upper lip.

Why a Twitch Works

Twitching increases the horse's ability to handle pain by stimulating receptors in the skin that activate a pain-decreasing, morphine-like substance that causes heart rate to decrease and sensations of pain to be reduced. These substances, called *endorphins*, also make the horse calmer. After application of the twitch, he becomes quieter and appears somewhat sedated, with drooping eyelids. He stands as if in a trance, with greater tolerance for pain and discomfort. A lip chain works in the same way. A twitch relaxes a horse quickly and easily and only temporarily, but you can usually keep the horse twitched for up to 10 or 15 minutes with no after-effects. As soon as you release the twitch or lip chain, he's back to normal. There's no waiting for the effects of a sedative or tranquilizer to wear off.

If done correctly, twitching is not resented by most horses and won't make them headshy. The twitch should never be painfully tight, just enough to put a little pressure on the upper lip. When the procedure is finished you should release the twitch slowly and gently massage the upper lip. Then, there are no bad memories for next time.

To put a twitch on a horse, hold on to the handle with one hand and place your other hand through the loop and with that hand take hold of the horse's upper lip then gently slide the chain or thong over your hand and around the lip. Next twist the handle until the loop tightens around the lip. It must be tight enough to put pressure on the tissues, but not so tight that it causes pain. You may have to tighten it more during the procedure being done with

A twitch increases the horse's ability to handle pain. It should be tight enough to put a little pressure on the upper lip.

the horse if the horse is not "sedated" enough, or release a little pressure if it's too tight. The first time you try to twitch a horse, it's best to do it under the direction of an experienced person.

Hand Twitching. A horse who tolerates a twitch can often be "hand twitched" by gently grasping his nose and squeezing. This can make him stand still and calm for a few minutes, rather than fighting the procedure being done to him.

The Stableizer

A more recent tool for restraining nervous, fractious horses, the Stableizer is a modern adaptation of the Native American war bridle but is more effective, humane, and versatile, often working better than a twitch, lip chain, or tranquilizer. One person can put the device on a horse, and it stays in place. The Stableizer relaxes and sedates the horse, yet the horse can still move and be led while wearing it, which is not the case with a twitch.

Tightening the device activates pressure points behind the ears, releasing endorphins that tend to block out pain and make the horse feel calm and relaxed. The pressure point beneath the upper lip blocks release of adrenaline, relaxing the horse even more. The portion of the device that goes under the lip is covered with plastic tubing, so it never cuts into the gum and is more humane than a lip chain.

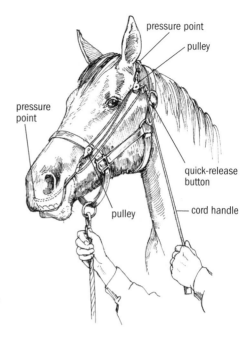

When tightened by pulling the cord handle, the Stableizer stimulates pressure points behind the ears and beneath the lip, releasing painkilling endorphins and blocking adrenaline, causing the horse to relax and feel good. The Stableizer can be set at any pressure with the quick-release button.

When Physical Restraints Won't Work

In extreme cases, when a horse is already in pain from serious injury or colic, or in an unstable emotional state, a chemical restraint such as a sedative or tranquilizer given by your veterinarian may be necessary to safely handle the horse for examination or treatment. Twitches, lip chains, Stableizers, or hobbles are not adequate in these instances.

Dealing with Bad Manners

Some horses develop bad habits if they are allowed to get away with misbehaving, and they may become dangerous to work with. A young horse just learning how to interact with people should never be allowed to continue bad behavior or he'll grow up spoiled and unruly, and an older spoiled horse should be reprimanded and retrained or he will think he can always do as he pleases. Some horses have bad habits because of improper handling in their formative years, or a bossy attitude that was not curtailed by a previous owner. As the old saying goes, "There are no problem horses; just problem people." A horse's habits and manners are a product of his reaction to people and the way he's been handled.

Pushy horses readily take advantage of a timid or softhearted person who allows them to do as they please instead of handling them with consistent firmness. Bold horses spoil easily if they feel they can have their own way in the relationship. Regardless of the horse's temperament, aggressive behaviors such as biting, kicking, rearing, walking on you instead of respecting your space, or rushing backward are not only annoying but dangerous.

The Biter

If you have a horse who bites, try to figure out why he bites. His biting may be a defensive action if he thinks you are a threat or if you cause him pain. He may also bite in playfulness, since this is how he interacts with other herd members. Alternatively, the biting may be an aggressive action if the horse is trying to establish dominance, not being content to accept you as boss. Within the herd, the horse uses his teeth to reprimand other horses, establish dominance, and attack intruders. Some nervous horses bite when frustrated or bored; confinement may lead to irritability, and they take out their frustrations on the person who comes to handle them.

Prevention

Halt nipping so it won't become a serious problem. With a young playful horse, the solution is to teach him to respect you; make it clear that you are dominant and must be obeyed. The biter needs firm discipline every time he tries, just as he would get in the herd from a more dominant horse.

Be constantly alert to his intended actions so you can prevent them or reprimand him instantly. All disciplinary action, especially for biting, should be given as the act is happening or about to happen, not afterward. Otherwise, it may become a game for him — trying to sneak in a nip, then jerking his head away before you can bump him. If you bump him on the nose after he nips, he just jerks his head up and may become headshy or play games to see who is quicker.

Punishment

If you can punish him before he nips, he'll lose interest in this game. Position yourself so you can always prevent a nip by meeting him halfway. When grooming, bending over to clean a foot, saddling, and so on, have some body part or tool ready to bump him in the nose whenever he turns toward you to nip. Meet him with your hand or elbow or grooming tool. When you bump him in the nose, do it in a nonchalant manner so he thinks he's bumping himself.

Don't jab at him, reprimand him with your voice, make a fuss, or get angry. Stay casual so he thinks he's doing it to himself. If you remain alert and catch him every time he tries to bite, this "self-punishment" will discourage him and he'll quit; he'll soon tire of initiating his own discomfort.

If a horse is a persistent biter, it may take more than bumping him in the nose with your elbow to discourage him. Use a nail or hoof pick — something he'll pay attention to. You already know when he's most liable to nip at you, so for his lesson, you can set him up, putting yourself in whatever position inspires him to take a bite. Then, you can catch him at it. Don't jab; let him hit himself on the nose with your nail or hoof pick. Just be ready for him. After a few instances of self-punishment, he'll stop.

The Kicker

Many kickers, especially those who kick when startled or upset, can be trained to stop this bad habit. A horse who kicks viciously with intent to hurt needs handling by a professional trainer (unless you decide to sell), but most horses who kick on occasion are just nervous or sensitive individuals who have not been fully acclimated to handling. They can usually learn to tolerate the situations that previously caused them to kick.

A kick toward a human being is generally an expression of nervous fear or self-defense: the horse is protecting himself from a perceived threat. Horses in groups also use kicking to keep subordinates in line or to express annoyance. Often when a dominant horse merely threatens to kick, the subordinates get out of his way.

Working Close to the Horse

When handling any horse, kicker or not, approach his head and work back to his hindquarters if you need to handle that end, touching his body all the way. Always work close to the horse, especially around his hindquarters, while touching him. A kick is less apt to hurt you if you're close; he can only push you or bump you with his leg. If you are a little distance away, you're more apt to get the full force of the kick, which can be powerful enough to cause serious injury.

Watch his ears and tail, not his feet, for signs of mood (see box on next page, Warning Signals). Stand on the opposite side from the leg you're working on, making the inner part of the leg easier to reach when applying medication, fly spray, or anything else that might prompt a kick. When handling a young horse, get him accustomed to having hind legs handled early on so he won't be as defensive about his hindquarters. This makes future procedures easier.

Prevention

A horse who tends to kick must be trained to stand still and trust his handler, learning that kicking doesn't work and that what he earlier perceived as

<hr>

WARNING SIGNALS

Most horses give warning signals before actually kicking: laying ears back, rolling eyes, presenting the rump, clamping or angrily swishing the tail, lifting a hind leg threateningly, or backing toward the intruder. When handling horses, be alert to these signs. A startled horse may kick explosively without warning, but many times a kick can be avoided if you are alert to his mood and can see a kick in the making — and prevent it or move out of the way.

<hr>

threats are not so terrible. He must become less touchy about having his hind legs brushed or bumped so that he's less likely to kick when startled.

Stand at the horse's shoulder and position him against a solid fence or wall so he cannot move away from you. Then, use a folded cloth feed sack/gunny sack or old saddle blanket (something soft, with a familiar smell he won't be afraid of) to rub along his back and over his hindquarters. Talk to him, praise him when he stays calm, and scold him if he tries to move or kick — but don't yell at him.

One way to teach the horse to let you touch his hind legs — while staying out of kicking range — is to use a small "pole" 4 to 5 feet (1–1.5 m) long, like a broom handle. Old-time harness horse people used poling for getting "green" horses accustomed to the feel of shafts and harness before hitching them to a wagon the first time. Let the horse smell the broom handle. Then rub and touch him with it, over the sides and rump and along the flanks. Don't raise it high or make any threatening movements. When he is relaxed about it, slide it down around his hind legs, then between his hind legs. It may take several sessions to convince him that having his hindquarters and legs touched is no big deal.

Have a helper hold a temperamental horse rather than tying him. Being tied by the head makes him feel more trapped. If a horse is tied, the post or fence may be a danger for you if he pulls back and then lunges forward, smashing you against it. You do not want him setting back when tied; you don't want him to become a halter-puller.

The Halter-Puller

A horse who sets back when tied is a danger to himself and to people around him. If he rushes backward while you're still tying him, you could lose a finger or hand. The best cure is to tie the horse often as part of his daily routine, leaving him tied for an hour or two to something safe and solid, with a rope and halter he cannot break, until he accepts the restraint.

Preventing Injury to the Horse

To avoid injury to neck muscles, and to teach the horse he cannot pull free, it helps to tie him to an inner tube securely fastened to a post; this has some give when he pulls back. You can also use a body rope (around his girth) so that his whole body feels much of the strain, not just his head and neck. Put the rope around his girth and tie it under his belly, then run the free end between his front legs and up through the halter ring. Tie with that free end of the rope to a solid fence post that will hold him even if he sets back with all his strength. Always tie level with his head or a little higher; tying too low puts too much strain on his head and neck when he pulls back.

The Hard-to-Catch Horse

Solving this problem requires time and patience. Many horses never improve because their frustrated owners continue to chase and corner them, instead of beginning again with this aspect of handling.

One thing that helps is a reward system. Catch the horse and then give a treat as you groom and saddle him: a carrot or a few bites of grain, or some nibbles of green grass if the horse is on a diet of hay. Green grass will then be considered a treat. Don't take the grain with you to pen or pasture. Some horses play games and try to get a bite or two, yet still keep their freedom. The horse has to learn to go by the rules and be willingly caught — then he'll get his treat.

The Timid Young Horse

The young, inexperienced horse may be hard to catch if he is timid or afraid, but this can be resolved with more handling and patience. Once you have gained his trust, he'll allow himself to be caught. Weaning time is a good opportunity to gain his trust because he is missing his mother and must rely on you for food and companionship. Even the most independent and stubborn youngster will eventually come around to seeing things your way if you catch and handle him every day, making his experiences pleasant.

TYING THE YOUNG HORSE

A young horse being trained to tie will generally pull back only a few times; once he's discovered he cannot pull free, he will respect the restraint. Some headstrong individuals or confirmed halter-pullers need daily tying with the body rope until they learn the rope is stronger and resign themselves to standing patiently.

The Old Spoiled Horse

The confirmed spoiled horse can be a challenge because, to persuade him that being caught results in a reward, you first have to catch him. If you must run him around the pasture each time, you've defeated your purpose. You have to outsmart him, and he must decide he wants to be caught.

Using a Small Pen

It usually helps to keep the horse in a small pasture or pen for a while until he gets used to being regularly and easily caught. Horses who run from habit rather than fear sometimes don't bother to run if there's no room. Putting a horse by himself can also help. Horses in a group may be hard to catch if one or two are elusive or habitual runners. Keep the problem horse by himself and catch him several times a day until he has been rehabilitated.

Catch him to take him to water, feed him, or give him a bite of grain or another treat, so he realizes that he is totally dependent on you and that catching is associated with good things. Catch him occasionally just to brush him and turn him loose again. When he finds that catching is the most pleasant event in his day, he'll look forward to it instead of avoiding it. Feel your way with each problem horse, and find a way to work around his attitude and overcome it. The key to retraining the hard-to-catch horse is to spend enough time convincing him that being caught is just part of his daily routine, and a pleasant part, at that.

Reforming the Mannerless Horse

Some horses are frustrating to handle. They fidget when you try to groom them, step on your toes, bump into you, root and tug at the halter, drag you along when you lead them, or have no respect for your personal space. Often, these problems develop when you treat the horse as a pet or lack consistency in handling. The horse thinks of himself as the boss.

Good manners are crucial to control of the horse and to your safety. A horse who is disrespectful or inattentive can be very dangerous — whether he stomps on your toe or takes a parting shot at you with a hind foot as you turn him loose in his paddock.

Establish Ground Rules

Establish consistent rules for the horse so he can understand, in no uncertain terms, the allowable limits of behavior. He must learn that human beings have special status. You not only give the orders but must be respected at all times. He must understand that you are not to be touched without permission: never

---

MANNERS WHILE LEADING

Some horses root and pull, kick at the person leading them, or wander about, paying little attention to the handler. Some become rooters because of the improper way they've been led. Most horses resent being pulled. They don't like being dragged along, and they don't like being held back if they want to go faster than you do. If you must continually restrain the overeager horse by pulling back on his halter, he will tug and root even more. It takes two to have a tug of war and, if you pull on the horse, you invite retaliatory behavior. Soon this becomes a habit.

When you apply pressure on the halter to slow him down, it should be intermittent, not continuous. A led horse should have a loose rope so he can accompany you in a relaxed manner. If you have to slow him down, give well-timed short tugs rather than a continuous hard pull.

---

bumped — even by accident — or bitten, kicked, or tail-swatted. If you spoil a horse by allowing him to nuzzle, rub, play, or push, he will consider you an equal, a buddy to roughhouse with; this can lead to dangerous consequences because human beings are much more fragile than his herdmates.

Enforce the Pecking Order

Social ranking is an important aspect of herd life: who eats first, who leads, and who follows. Pecking order is a major element of horses' interactions with each other and with people. Much of their behavior is based on dominance and submission, and the horse must figure out how human beings fit into that picture — who is higher on the pecking order and who must submit to whom. If a horse is successful at dominating people, he becomes aggressive and does what he pleases unless he comes up against a person who can set him straight.

Maintain a Respectful Distance

One of the basic rules of behavior the horse must learn is to keep a respectful distance. Whenever he presses his limits or violates rules, don't ignore it, even if it seems like an accident. Through your own body language, let him know you have a personal space that he must not violate.

Even the gentlest horse can be dangerous if he doesn't respect your personal space. When handling any horse, make sure he knows where his place is. Allow him to be comfortable in it, but always make him uncomfortable if he pushes into your space. Let him run into your hand, the rope or lead strap, a stick or a whip; create an imaginary space around you (about 10 to 12 inches [25–30 cm] or so) that he is not allowed to penetrate.

Discourage Rubbing

Don't let a horse rub on you, look for treats, or nibble your hand or clothing. Be especially firm and consistent with young horses still learning how to relate to people and experimenting with dominance. After all, little nippers grow up to be big biters. If you rub on a horse (some people like to rub the forehead), he will rub you back, just like he does with a herdmate, and become pushy. By doing this, you set yourself up to be challenged as to who will be dominant in the relationship. Stroking is fine; rubbing is risky unless the horse is older and well trained.

The Overly Aggressive Horse

For the horse who wants to drag you around much faster than you want to go, or one who ignores your signals to halt, you may have to use a nose chain when leading. It is placed over the horse's nose and works in much the same way a choke chain works on a dog. It doesn't inhibit movement unless it is engaged by the horse (by going too fast) or the handler (by asking the horse to stop, with a short tug). For proper use, the chain must be attached to the halter correctly. Wrapping it around the nose band decreases its effectiveness.

When holding the lead line, grasp the end of the chain with your right hand where it meets the line, having the horse on a relaxed lead as you walk beside or just ahead of his shoulder at arm's length. There's no set rule for position because the size and neck length of each horse vary. Be somewhere behind his head and in front of his shoulder. If you're ahead of his cheek, you'll be dragging him; if you are behind the shoulder, he'll be dragging you.

If he tries to bolt, you'll be able to hold the line adequately if you have made a knot just in back of the place where the chain fastens into the nylon or lead rope. This makes the lead line jerk against the chain over the horse's nose with the same effect of a choke chain on a dog. If the chain is properly adjusted on the halter, it presses into the bridge of his nose when he lugs into the halter or bolts forward, and it releases and loosens automatically when he stops. You can also engage the chain with one firm downward jerk if the horse misbehaves while being led. After the jerk, you should immediately give slack to release chain pressure. This jerk-and-release pattern is an effective form of discipline — much better than a futile tug-of-war.

The Body Basher

Some horses fidget around, stepping on you, bumping you, or even trying to smash you against the stall wall or fence when you try to groom them, saddle them, or do anything else while they are tied. Distinguish between behaviors triggered by discomfort or fear of abuse (from a prior bad experience), and deliberate bad manners. A horse who is jittery about some aspect of grooming or saddling needs patient, gentle handling to get him over his fears, but a horse who is trying to bully you needs firm and immediate discipline.

Usually when a horse is trying to get the best of you, he doesn't show any indication of fear or alarm but is quite calm and purposeful, or even angrily aggressive if he's used to bullying people and getting his own way, and is waiting to see what you do about it. A deliberate show of bad manners should be punished with an immediate and firm verbal reprimand or an appropriate smack on the offending part of his body, delivered without anger. Use something other than your hand, such as a brush, hoof pick, or whatever you are holding; using your hand will probably hurt you more than the horse. If a horse steps on your foot, try stepping on his with your boot heel. After making your point go on about your business so the horse knows he hasn't gotten the best of you, he must behave as you continue to groom or saddle him, and he cannot get away with invading your personal space.

Correction

Horses who habitually bump into a person must learn to stay back or move back on command. Using the blunt end of a whip, you can tap the horse on the chest to make him back up or move over, rewarding him with praise when he responds properly. Anytime a horse steps uninvited into your space, react immediately to get him back where he belongs. It's not enough to command him to halt; he must back up or move over. This makes him realize that he must submit to you and that you are in control, not him.

For the aggressive horse who tries to mash you into the wall, be prepared with a short, stout stick. Hold it so he runs into the stick instead of you. After a few instances of self-punishment, most horses learn it's not pleasant to move into your space.

The Turnout Terrorist

Some horses have bad manners when you take them out of a stall, put them at pasture, or turn them loose after working with them. People may unwittingly foster dangerous habits by letting the horse have his way. Often the person is unaware of his own negative influence, which may be mere inconsistency, timidity, lack of discipline, or inappropriate discipline.

If a horse has a habit of taking off the instant he is unhaltered at pasture, perhaps bucking and kicking, the handler may try to quickly unsnap the lead shank while hurrying through the gate, hoping the horse will continue on through and go a few steps more before starting obnoxious and dangerous behavior. This timid, undisciplined turnout approach leads to loss of control of the horse and even more trouble.

Prevention and Correction

To help prevent shenanigans when handling high-spirited or aggressive horses, think ahead and stay in control. Always use a halter and lead shank when taking a horse to and from pasture or paddock, no matter how gentle or well trained the horse or how short the distance. Lead only one horse at a time, pay attention, and convey a calm and confident attitude.

To thwart or prevent the "charge" at turnout time, make the horse stop and relax before you let him go. Loop the lead rope around his neck before unsnapping him or taking off the halter so he can't go rushing off the instant he thinks he's free. Keep him guessing. Make him realize you are still in control — that you are the one who will leave him, not the other way around. When he is calm and relaxed, you can quietly walk away. With persistence, you can correct the horse who tries to immediately gallop off.

BUILDING TRUST

Every time you interact with a horse, you are making him either easier or more difficult to handle, depending on how you do it. If you use firm tact, consistency, and appropriate discipline, working on manners as part of the horse's daily routine, you can usually nip bad actions in the bud before they become bad habits. Your horse will be well mannered and pleasant to work with, not a danger to you or others. He will respect you and your personal space and accept you as the dominant force in his life, someone he can feel at ease with and trust.

PART II

Health Care

6

Reading the Signs of Health and Sickness

YOUR HORSE'S HEALTH IS YOUR BIGGEST CONCERN. All aspects of care and management are aimed toward keeping him healthy and sound. Give him the best of feed and water to meet his individual needs, vaccinate and deworm to keep him free of infectious diseases and internal parasites, and protect him from pests such as flies and mosquitoes, which not only annoy him but also spread disease. Monitor him closely to make sure he is healthy, comfortable, and happy.

A conscientious owner is in tune with each horse, knowing how he is feeling by reading the signs of health or sickness and recognizing what is normal or abnormal for that individual. If you know what to look for, you can interpret the clues that the horse gives you and detect early signs of illness.

Vital Signs

Always check vital signs if you suspect a problem. A horse's temperature, pulse, and respiration (TPR) can give a good indication of whether he is healthy or sick. Your first impression of him as you approach tells a lot — whether he is perky or dull — but a quick check of his vital signs can confirm or lay to rest your suspicions. Because fever and pain are evidenced by elevated temperature, breathing, and heart rate, a quick check of these can tell you if he might be suffering from illness, colic, or some other problem.

Normal Rate

Check each horse when at rest and not exerting. Expect the rates to vary somewhat by horse. Numbers given for "normal" rates are averages, such as a pulse of 36 or a temperature of 100.5°F (38°C). If you take each horse's TPR rates a few times when he is at rest and healthy, it is good practice in case you have to check him someday in an emergency situation. Being already familiar with the procedure, the horse will be less likely to resist or try to kick you.

> **Normal Vital Signs for a Horse at Rest**
>
> **Temperature:** 99–100.5°F (37.2°–38°C)
> **Pulse rate:** 36–40 beats per minute
> **Respiration rate:** 8–10 breaths per minute

Temperature

Normal average body temperature for a horse ranges from 99 to 100.5°F (37.2°–38°C). It is generally lowest in the cool of early morning and a little higher in the evening after he's been active, especially if the day has been warm. A horse might have a normal, healthy temperature of 98°F (36.7°C) on a cold December morning and 101.8°F (38.8°C) on a hot August afternoon.

For taking temperature, any rectal thermometer will do, but an animal rectal thermometer is best; it's sturdy and has a ring in the end for a string. Never put a thermometer into a horse's rectum without a string attached or it may get lost in the rectum. If you don't have an animal thermometer, securely attach a string to the end of a human rectal thermometer with masking tape or duct tape. You can also use one of the newer electric thermometers, some of which have soft, flexible tips and are less likely to break.

Have the horse tied or have someone hold him. Shake down the thermometer to below 96°F (35.6°C) so you can get an accurate reading. Have the end well lubricated with petroleum jelly or your own saliva (a little spit works

KNOW YOUR HORSE'S VITAL SIGNS

Check every horse now and then to determine the normal rates for each individual. If you know what a certain horse's normal TPR numbers are at rest, you have a better idea whether they're abnormal when you suspect he's sick. In consulting with your veterinarian about a problem, you can mention the horse's vital signs. This information, coupled with your description of other symptoms, helps the veterinarian determine the seriousness of the problem.

When taking the rectal temperature of a horse who's not used to the procedure, you can prevent trouble if you stay relaxed and work close to the horse. Stand directly beside his hindquarter with your body against his hip and stifle, left arm over his rump. With the left hand, gently rub his tail area until he relaxes. Most horses like their tails rubbed and will be cooperative if you don't rush them.

After the horse relaxes, he'll raise his tail a little instead of clamping it down, and you can slide a lubricated thermometer into the rectum. If your body is pressed against his hindquarter and stifle, it's hard for him to kick forward, and you are safely out of the way if he kicks backward. Most won't kick, however, if you get them relaxed before inserting the thermometer.

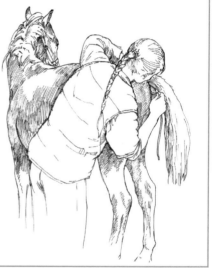

When taking a horse's temperature, stand close to the horse with your body against his stifle and hip so he can't kick you.

fine) so it will slip in easily and not cause discomfort. A dry thermometer is not as easy to insert and the horse may object. When you are ready to insert it, stand close to the horse and gently rub alongside the top of his tail. Don't abruptly pick up his tail because he may clamp it down and become unhappy about what you are doing. If you gently rub the sides of his tail, he'll relax and raise it a little on his own. Most horses enjoy the rubbing because this is a hard place to reach an itch.

Once the tail is raised, it's easy to slip the thermometer into the rectum, aiming it slightly upward and rotating it a little as you go. The twisting motion helps the thermometer go in more easily. If the horse tries to clamp his tail down, gently hold the tail to one side. If you insert the thermometer properly, without poking the sides of the rectum or causing any discomfort, he won't fret about it. Even a nervous individual will become accustomed to it after a few practice sessions and submit to it willingly because he likes to have his tail rubbed. Insert the thermometer all the way in, leaving only the very end (with string attached) visible.

For accurate reading, the thermometer must rest against the rectal wall. If it's stuck in a fecal ball, the reading will be low, which can be deceptive when you're trying to determine if the horse has a fever. If the thermometer doesn't go in easily or gets stuck in manure (which will make it hard to push in), take it out and try again. Leave it in for a full minute or more. The most accurate reading will be obtained if you leave it in for 3 minutes.

Aim the lubricated thermometer slightly upward, and rotate it as you gently push it into the rectum. This technique allows it to slide in more easily.

Pulse

The pulse, or heart rate, of a healthy horse is, on average, 36 to 40 beats per minute. Every horse's normal rate is different. One may have a normal resting pulse as low as 28, and another's normal may be 44. This is why it's a good idea to check a horse's pulse now and then to know what his "normal" is. A foal's rate is a little higher than an adult's. Also, athletic, fit horses tend to have lower resting pulses than horses who have never been in top physical shape.

Take your horse's pulse when he is quiet and resting to find out his normal rate. It's easiest with a stethoscope (placed at the heart girth behind the left elbow), but it's not difficult to take with your hand if you know where to feel. A good place is along the lower jaw where the big artery runs across under

---

SHORTCUT TO TAKING THE PULSE

An easy way to determine pulse is to count the pulsing of an artery for 15 seconds if you have a watch with a second hand, then multiply by four. This is a quick and handy approach, especially if the horse doesn't want to stand still for a whole minute. If you wish, you can do it a couple of times (15 seconds each) to double-check your figure.

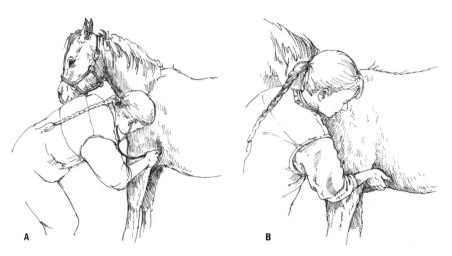

Heart rate can be checked with a stethoscope positioned at the heart girth behind the elbow on the horse's left side (**A**) and with your hand at the girth area (**B**).

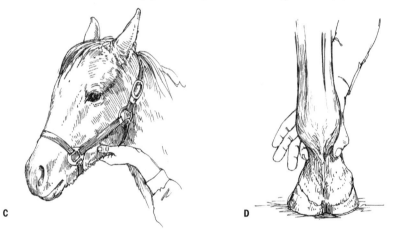

Heart rate, or pulse, can also be checked at the big artery under the lower jaw (**C**) and at the fetlock joint by pressing a finger against the artery that runs just under the joint at either side of the foot (**D**).

the bone. Move your fingers along the bottom of the jawbone until you find the artery. It feels like a small, firm cord. Press lightly on it with your finger to feel the pulsing blood.

You can also check pulse directly over the heart, just inside the left elbow, by feeling with the flat of your hand against the rib cage. The actual heartbeat is two beats in one — very easy to distinguish if you are listening with a stethoscope. Count each *lub-dub* as one beat. Be aware that pulse can become elevated with exertion, excitement, nervousness, pain, fever, or other serious problems.

Respiration

Respiration is easy to determine: just watch the movements of the horse's nostrils or flanks. Count as he inhales or exhales, but not both. Just as in taking the pulse, it is often easiest to count for 15 seconds and then multiply by 4. Normal respiration rate in the average horse at rest is about 8 to 10 breaths per minute, if weather is cool and he's not excited.

Body Language

The person who works closely with horses usually develops an intuitive feel for their well-being. An important part of this skill is being able to sense when a horse is sick or uncomfortable by reading subtle signs from body language.

Posture

The observant horse person can tell at a glance whether all is well by noticing the way a horse stands or moves. General bearing; leg position; and head, neck, and tail carriage (for example, droopy instead of jaunty) are good indicators of something amiss. If he's standing with head down and looking dull rather than bright and alert, you should be immediately suspicious. Remember that when a horse is not feeling well or is in pain, he becomes less perky than usual, less aware of what's going on around him. He is tuned inward instead, focused on his own misery. It would be wise to take his temperature; dullness may indicate a fever. In some instances, such as when a horse has a nervous system disorder or disease, he may be restless and showing abnormal behavior.

Mild Abdominal Pain

Every horse person should know about signs of colic, or obvious abdominal pain (the term "colic" simply means "abdominal pain"): pawing, rolling, sweating, elevated heart rate. Mild abdominal pain may be harder to detect unless you know your horse well and can tell when he isn't quite himself. A

IF YOU AREN'T SURE, CHECK TPR

A horse won't necessarily have rapid respiration with a fever. He may just seem a bit dull. If he isn't quite himself, take his temperature. While you're at it, check his pulse. An elevated pulse can be a sign of pain. If you decide the horse's condition warrants a call to the veterinarian, mention these vital signs so he or she can decide whether a thorough examination and diagnosis are necessary.

horse with mild abdominal pain may be just a little dull, slightly restless, or off feed. He may get up and down more than usual or spend too much time lying down. He may lie with his head tucked around toward his belly or flank.

He may also stand in a corner away from other horses or display slightly abnormal posture. He may stretch or stand with his head up high and front legs forward, with hind feet farther back than normal — trying to ease discomfort or gas pains in his abdomen. If he acts strange or looks dull, check vital signs and abdominal sounds. Use a stethoscope if you have one, or your ear pressed to his side, to listen for gut sounds (such as rumbles and gurgles). Constant rumbling may mean an overactive gastrointestinal tract. Absence of gut sounds (ominous silence, except for the gentle "whish" of lung sounds as the horse breathes) could mean a blockage or shut-down gut. Call your veterinarian and keep the horse under observation. If he has mild colic, it could worsen, depending on what is causing it.

A horse with mild abdominal pain may lie with his head tucked around, or he may keep looking at or nosing his belly.

A horse with front legs forward and hind legs back (standing in a stretched position) may have gas pains.

Back Humped Up

A horse standing with his back humped up and stomach muscles tense is showing signs of severe body pain — injured back or ribs, chest pain, peritonitis (inflammation and infection in the abdominal cavity), or serious digestive-tract pain. He may be reluctant to move because any movement hurts. He needs immediate medical attention.

Position of Legs

How the horse is standing — position of legs and overall body stance — can give clues to other problems as well.

Pointing

Standing with one front leg in front of the other (pointing) usually means he's trying to relieve pain in that more foreword leg by not bearing much weight on it. He may have bruised the heel or sustained a more serious injury — to the deep flexor tendon, flexor muscles, or related ligaments. Pointing can also be indicative of navicular disease (see chapter 9). Or the horse

Standing with one front foot forward (pointing) may mean that the horse's leg is sore and he doesn't want to put weight on it.

may be trying to ease the discomfort of a strained biceps muscle, bruised shoulder, or injured elbow. Check his leg for heat and swelling, and check his heel for soreness (see chapter 4). If this does not reveal the cause, have your veterinarian do a more complete examination.

Standing on Three Legs

You should suspect more serious injury when the horse is standing on only three legs, not putting any weight at all on one leg and hobbling if he has to move. If this continues for more than a few minutes, get professional help to diagnose and treat the problem. The horse may have a foot abscess, strained or sprained fetlock joint, nerve or muscle injury, or even a fracture. Close examination of foot and leg gives clues. Check for heat and swelling; exaggerated, unusually strong pulse (felt in the lower leg when feeling fetlock joint area and pastern for heat); punctures or embedded foreign objects in the hoof; or hoof tenderness (check with a hoof tester; see chapter 4).

If a horse paws with his front foot, it may be an indication of colic or of muscles tying up.

If, however, the pain is temporary and the horse soon walks normally again after a few minutes, he probably just banged the leg during playful antics, was kicked by a pasturemate, or landed the wrong way while running and bucking but did not suffer serious damage. In this case, there is no need to call the veterinarian, but it's a good idea to keep an eye on the horse for a day or two to make sure he's all right.

Hind Legs

Hind legs are not injured as frequently as fronts because they carry less weight, but accidents sometimes happen. Standing on three legs, resting a hind foot, is usually nothing to worry about but might indicate pain and trouble. If the horse's stance seems abnormal, check the foot and leg more closely. Make him walk a bit to see whether he favors the leg while moving.

If he's standing with one hip dropped and hind leg dangling, or if fetlock joint and pastern are knuckled under (the front of the pastern resting on the ground with the bottom of the hoof pointed up behind him), suspect serious injury. A few horses rest a hind foot this way, but if yours generally does not, you'd better check it out. Odd hind-leg stance could mean a broken leg or a ruptured ligament or muscle high in the leg. If he's unwilling to move or his leg will bear no weight, do not try to make him move because this could make the problem worse. Seek prompt veterinary help.

Rigid Hind Legs

Another instance in which a horse may be reluctant to move is when major muscles, especially in the hindquarters, tie up (see chapter 9). Because of painful muscle cramping, he stands with hind legs rigid, often farther back

IS IT COLIC OR MUSCLES TYING UP?

The sweating, pawing, pain, and anxiety from muscles tying up might be mistaken for colic, but you can tell the difference by noting the horse's stance, actions, and reluctance to budge from the spot. The horse with tied-up muscles does not want to move, and his gait is stiff if he does, especially in the hind legs. Moving will make the problem worse; don't try to make him walk. Your veterinarian can give medication to relieve the condition.

than their normal position, and paws the ground with his front feet. This reflects his frustration at not being able to move without discomfort. He will probably break out in a sweat, especially around the flanks.

A horse's muscles generally tie up during exercise, early in the workout when a horse is being ridden, or right after a ride. The problem can occur in some horses any time they overexert. You might find the horse standing this way after he's been running or bucking in mud or snow, or racing around with pasturemates.

Most cases of severe muscle cramps involve horses who have an inherited genetic defect in their muscle metabolism, making them vulnerable to the condition, especially if they are on a diet containing grain or other feeds containing high levels of starch and soluble sugars. An abnormal amount of sugar is stored in the muscles because they are programmed to continuously make glycogen. Yet the muscles cannot access it properly for energy when the horse is working and they start cramping up when the horse exerts strenuously, creating this painful condition (see chapter 9).

Legs Bunched Together

If a horse is standing with front legs too far back and hind legs more forward than usual (front and hind end bunched together), it is generally a sign that his body hurts. He may be trying to relieve pain in his back or discomfort in his chest. A dull horse standing "camped under" like this could be seriously ill.

If he seems normal in other respects — alert, eating, drinking, with usual attitude about everything — he may have a simple strain or bruise. If he is dull, sweating, trembling, unwilling to move, wobbly, staggering, or in any way acting strangely, call your veterinarian. The horse may have a seriously injured

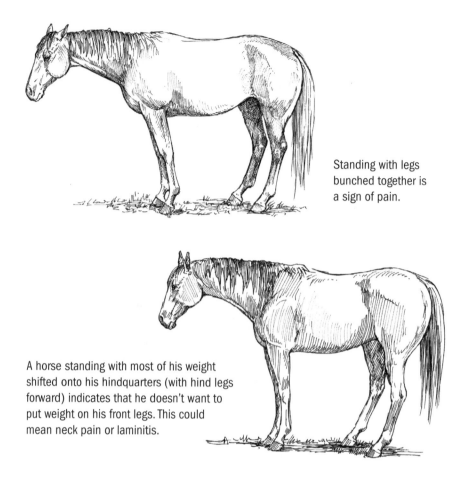

Standing with legs bunched together is a sign of pain.

A horse standing with most of his weight shifted onto his hindquarters (with hind legs forward) indicates that he doesn't want to put weight on his front legs. This could mean neck pain or laminitis.

back, severe intestinal problem, or respiratory condition such as pleuritis — inflammation of the chest lining and lung covering accompanied by fever, fluids in the chest cavity, and painful breathing.

Weight Shifted Back onto Hind Legs

A horse standing with weight shifted onto his hind legs is showing that bearing weight on his front legs is painful. This could indicate founder (laminitis; see chapter 9) or neck pain. Founder should be your first suspicion, so check his feet for heat or a pounding pulse. See whether the horse will move. If he tries to carry most of his weight on his hind feet and is very reluctant to move or turn to the side, you are probably dealing with founder (this causes terrible pain in the front feet — sometimes in all four feet). Call your veterinarian.

Be Observant

Know your horse well; you can't tell what's wrong if you don't know what is right and normal for that animal. Look at each horse every day, as many times as possible. Be alert to any changes in normal behavior. "Acting funny" is often the first hint that something is seriously wrong. If the horse is showing unusual alarm, alertness, agitation, or personality change, or if he is dull and preoccupied, droopy, unresponsive, disoriented, or abnormally aggressive, take heed. Changes in behavior and posture may signal an insignificant or temporary problem, but they might also be the first signs of serious injury or illness such as rabies, encephalomyelitis (sleeping sickness), or tetanus (all discussed in chapter 8).

If a horse is sweating, trembling, straining to urinate or defecate, drooling, grunting, wheezing, panting, or showing other obvious and alarming changes, call your veterinarian. Even subtle changes, while they may be nothing serious, should be carefully checked out because they could be early warning signs of a major problem.

It's also a good idea to groom a horse often and run your hands over his body. This may alert you to any unusual lumps or swellings or any heat in a leg. There's no substitute for constant observation and hands-on checking, to be able to tell whether everything is well and good or something is wrong.

OBVIOUS SIGNS OF TROUBLE

Something's probably wrong when you see any of the following signs:

- Rolling or pawing
- Sweating when at rest (unless weather is very hot)
- Having rapid respiration when at rest
- Not eating at mealtime
- Not drinking the usual amount of water
- Spending too much time lying down
- Lying in an abnormal position
- Displaying unusual posture or stretching (legs bunched under the body or stretched out)
- Trying to carry most of the weight on hind legs
- Refusing to put weight on a foot
- Being reluctant to move when asked
- Deviating from usual personality or behavior

7

Disease Prevention

MOST EQUINE DISEASES CAN BE PREVENTED through good care and regular vaccination. Many of the most deadly infectious diseases — tetanus, influenza, strangles, West Nile virus, and sleeping sickness — can be prevented with annual vaccinations. You should definitely protect your animals in this way.

Not every horse needs to be vaccinated for every preventable disease because some are a problem in certain regions but not others. For example, Venezuelan equine encephalomyelitis (VEE) occasionally threatens horses in the Southwest when this disease heads north from Central America, but it is not a concern in the rest of the United States. Likewise, Potomac horse fever is a threat in some regions, primarily where there are rivers and streams, but not in others. The same is true of rabies and equine viral arteritis (EVA: of special concern if you are breeding horses). Ask your local veterinarian which diseases to vaccinate against in your horse's annual schedule of shots.

Some diseases are spread only by horse-to-horse contact. If any of your horses go to trail rides, shows, or to another horse farm to be bred, or if other horses come to your place for any reason, or your neighbors have horses, or you ride with other people, you should vaccinate all your horses against the common infectious diseases — especially influenza, rhinopneumonitis, and strangles. If you have a backyard horse who never comes in contact with other horses, or your ranch horses never leave your property and no other horses come in, you may not need to vaccinate for all diseases. If there's any chance of contact, though, play it safe.

What Is Disease?

Disease is defined as an unhealthy condition of the body. There are several types of disease — primarily infectious and noninfectious. An infectious dis-

DEFINITIONS

- **Antigens** are foreign substances, such as toxins, that stimulate the body to produce antibodies. A vaccine contains a modified antigen that will trigger an immune response in the body without causing the disease.
- **Immunity** is the body's ability to fight off disease-causing organisms. A healthy horse with a strong, healthy immune system is less likely to become sick than one with a weak immune system and poor immunity.
- **Pathogens** are microbes that cause disease. Illness from an infectious disease occurs when the body is overwhelmed by an invading pathogen.

ease is caused by a pathogen, a tiny organism that invades the body and stays there. Most of these invaders are microscopic. Bacteria (as in the case of strangles), viruses (as in the case of influenza), protozoa (one-celled organisms), fungi, and parasitic worms can cause infectious disease. (Infectious diseases of horses are discussed in chapter 8.)

Noninfectious diseases are caused by conditions within the body itself, such as cancer, respiratory disease caused from an allergy, hormonal imbalances, genetic disorders, and metabolic or stress-related problems that throw the horse's body chemistry out of balance (see chapter 9).

How the Immune System Works

The immune system is complex; it involves generalized mechanisms for fighting disease as well as the ability to develop immunity to specific pathogens. When a virus or bacterium enters the body, it causes damage by multiplying and creating toxins, which are poisons produced by living organisms. The presence of the pathogen (which the body recognizes as a foreign protein, or an *antigen*) or its toxins stimulates the body to create an antibody, which reacts with the invader and neutralizes it. An animal with a healthy immune system can mobilize a strong defense very quickly, creating antibodies to attack the invader.

Antibodies and Vaccines

Antibodies are blood proteins that are produced in response to a foreign substance (antigen) in an effort to neutralize its harmful effect. Produced by one type of white blood cell, antibodies are carried throughout the body in the bloodstream. If an animal already has antibodies against a specific disease organism, when that organism invades the body again, the white blood cells and their antibodies converge on the site to kill the invader. Ability to recognize and respond to foreign material is the body's immune response; this is the basis for vaccination.

Vaccination stimulates the production of antibodies by serving as the antigen, but in altered form or small amount so as not to create the actual disease. When a vaccine (a foreign substance) is injected into an animal, the body reacts by producing immune substances — antibodies — to fight the foreign invader. The end result is protection for the body (immunity) against subsequent exposure to that disease. If the horse eventually comes into contact with the actual infection, the antibody is already in the bloodstream, able to inactivate the pathogen.

Vaccines are often created from killed viruses, modified live viruses, or other inactivated disease organisms that still retain their characteristic so the body's immune system can recognize the invader and create antibodies to fight the real thing. For example, antibodies produced in the horse's bloodstream by vaccination against influenza lie in wait, searching for the antigen (the flu virus) that originally stimulated their production. If the horse is exposed to influenza — through contact with a sick horse or its cough droplets — the white blood cells "programmed" by the vaccine to make a particular antibody will quickly respond and produce enough to inactivate the invaders. As a result, the horse will not get sick.

Developing Immunity

The vaccination process exposes the injected horse to specific antigens. To develop immunity, his body must respond by producing antibodies and other immune substances. This takes about 7 to 10 days in a healthy horse if the immune system is intact and fully functional. If the immune system doesn't work properly, the horse does not develop strong immunity. Most

---

HOW VACCINATION WORKS

Vaccines are small doses of disease — not enough to cause the disease, but enough to stimulate the production of antibodies so the body can recognize and fight the real disease if exposed to it.

horses respond to vaccination successfully, but some have only a partial or poor immune response and may not be protected against the disease despite having been vaccinated.

Many factors influence the development of immune response — or lack of it. A stressed, undernourished, or heavily parasitized animal may not make a good response to vaccination. Stress factors typically include pregnancy, illness, fatigue, and extremes of weather.

Vaccination Failure

Sometimes vaccination fails because the horse is already incubating the disease at the time of vaccination (he has been exposed and is coming down with the disease). In foals, vaccination often fails because of interference from maternal antibodies in colostrum (the mare's first milk). If foals are vaccinated, they usually need a booster shot a few weeks or months later, or several boosters, to gain full immunity. There are different times to start vaccinating for different diseases, and starting too early (especially for influenza) may hinder a foal's future immune response. Ask your veterinarian about this.

Another cause of vaccination failure is the mishandling of vaccines. The most common example is improper storage. Most vaccines must be refrigerated; if left too long unrefrigerated, they may not be effective. In addition, sunlight can be detrimental to any live- or modified live-virus vaccines. Using a vaccine at improper dosage or beyond the expiration date can also contribute to inadequate immune response.

Stress

Stress can be defined as anything that poses a threat to the physical or mental well-being of the animal. There are many kinds of stress — some temporary, others more prolonged and harmful. Among the factors that contribute to stress are unaccustomed noise, upsetting changes in management that lead to emotional insecurity, crowding, temperature extremes, lack of water, toxic reactions, allergic reactions, pain, fear, anxiety, strenuous exercise, and exhaustion.

The way the body deals with stress can compromise the efficiency of the immune system to ward off invading pathogens. If the stress is severe or prolonged, it can lead to illness. It can hinder the development of immunity when a horse is vaccinated, and it can cause a previously vaccinated horse to lose his immunity.

The respiratory system is especially vulnerable to the effects of lowered resistance because of stress. This is partly because harmful organisms are a natural part of the horse's environment, almost always present in the air he breathes. Pathogens are constantly being drawn in, so the immune system must be continuously active to protect the respiratory system and keep the pathogens under control. This is one reason horses tend to more readily develop flu, rhinopneumonitis, "colds," and other respiratory problems when hauled to shows, competitions, races, and the like. The stress of transport, unfamiliar surroundings, and strange horses weakens the respiratory system's resistance; moreover, the horse is more likely to come into contact with infectious agents in places where horses are brought together from many areas. Stress and contact with other horses give infections a prime opportunity. Stress at weaning and during long periods of extreme weather also gives rise to respiratory problems and illness.

Disease Prevention

You can't just give horses their annual vaccinations and expect them to remain disease-free. Vaccinations should always be given when the horse is healthy and not stressed, well ahead of stressful situations like weaning, training, conditioning, or transporting so the horse can build peak immunity. Some types of vaccinations should be given more than once a year if the horse's immunities could be compromised. Flu and rhinopneumonitis are two diseases that may require more than one annual booster, as do equine encephalomyelitis and West Nile virus if mosquito season lasts a long time in your area. Discuss a vaccination schedule with your veterinarian.

There are vaccines for most of the common diseases that affect horses. Develop your horse's vaccination schedule according to the diseases prevalent in your area and how you use your horse. Some vaccinations should be given even if your horse never comes in contact with other horses. For example, tetanus, West Nile virus, and equine encephalomyelitis are a risk in almost all locales; tetanus is caused by a bacterium that lives in the soil and encephalomyelitis and West Nile virus are spread from birds to horses through mosquitoes. These diseases are deadly, and vaccine is inexpensive compared with the costs of treatment or death of your horse. Don't risk losing your horse.

Tetanus Shots

It's a good idea to give a pregnant mare tetanus toxoid so there will be antibodies in her colostrum (first milk) to give the foal temporary immunity. The foal himself should have a series of tetanus shots (how early these should begin will depend on whether the mare was vaccinated), and once he becomes a yearling each horse should have an annual booster, for the rest of his life. Even then, if a horse suffers a serious wound or deep puncture and has not had an annual booster within 90 days preceding the injury, he should have another booster to make sure protection is adequate.

West Nile Virus

Every horse should be vaccinated annually for West Nile virus before the mosquito season begins because this deadly disease is spread to horses from birds through mosquitoes. In northern regions with short summers, one vaccination is adequate. In southern regions with a longer mosquito season, many veterinarians recommend vaccinating twice a year. Consult your vet for proper timing.

SAMPLE VACCINATION SCHEDULE FOR AN ADULT HORSE NEEDING ONLY BOOSTER SHOTS

Season	Vaccine Administered
Spring (before mosquito season)	▪ EEE, WEE, WNV, influenza, and tetanus ▪ *Optional* (if a horse will be in contact with strange horses): Rhinopneumonitis (modified live-virus vaccine), strangles ▪ Recommended in some locales: Potomac horse fever, rabies
Fall (if mosquito season is long)	EEE, WEE, WNV

EEE = Eastern equine encephalomyelitis; WEE = Western equine encephalomyelitis; WNV = West Nile virus
- *Note:* Young horses being vaccinated for the first time will need a series of injections, several weeks apart, for all vaccines.
- Pregnant mares may need several vaccinations during gestation to protect the growing fetus from certain diseases and to maximize antibodies in the colostrum.
- *Important:* Discuss each horse's vaccination needs with your veterinarian. What you ultimately give will depend greatly on the individual animal, his use, and your location.

Equine Encephalomyelitis (Sleeping Sickness)

In most areas of North America, bird (and sometimes reptile) populations serve as reservoirs of infection for equine encephalomyelitis. After biting an infected bird or reptile, mosquitoes carry the disease to horses and, sometimes, to human beings. Migratory birds and mosquitoes can transport the disease a long way, especially in wet years with storms that carry the insects long distances. Vaccinate in spring before mosquito season begins. Immunity from vaccine lasts only about 4 months, so revaccinate later in summer if mosquito season lasts a long time.

Other Inoculations

Your vet may recommend vaccination against equine influenza, rhinopneumonitis, strangles, Potomac horse fever, equine viral arteritis, botulism, rabies, or leptospirosis, depending on your location and individual situation. These diseases are discussed in chapter 8.

Vaccinating Your Horse

You may choose to have the veterinarian vaccinate your horses or you may prefer to do it yourself. If you vaccinate your own horses, take proper care of vaccines: read directions for storage and handling and make sure all vaccines are fresh and potent, keeping none beyond their expiration dates. Also, learn how to give injections properly. Be aware of the possibility for reactions (some of which can be very serious) and what to do if you encounter this problem.

If you have never given an injection, have the veterinarian show you how. Observing the technique of an experienced person will help you learn to do it with less discomfort to the horse, and give you confidence in your practice of proper technique. A properly inserted needle causes little pain; the horse may not even notice.

HEALTH TIP

When giving an injection, make sure the site is very clean, free of mud and dirt. It should also be dry. Contamination is likely to be taken into the flesh with the needle if hair and skin are wet. Disinfecting the area with alcohol is not necessary if the hair and skin are clean and dry. If you do swab the spot with alcohol, wait for it to dry before inserting the needle; moisture increases risk of contamination of the tissues.

SHARP AND STERILE IS BEST

Use a sharp, new needle — a separate sterile needle for each horse. The sharper the needle, the more easily it goes in with the least pain to the horse.

Intramuscular Shots

Most injections are given intramuscularly — deep into muscle tissue. These are the easiest to give. Relatively few shots are given to horses subcutaneously — under the skin. Some medications (sedatives, painkillers, and some anti-inflammatory drugs) are given intravenously, but these are generally administered by a veterinarian. *Do not attempt to give an intravenous injection unless your veterinarian has instructed you on how to do it properly.*

Choosing the Proper Needle and Syringe Size

Most vaccines are given in small doses; use a small syringe for accuracy. Antibiotics and some other medications are often given in larger amounts and for these you'll need a larger syringe. It's best, however, not to inject more than 10 mL in any one site. A large injection can be split into two doses and injected in two places for better absorption and less muscle soreness afterward.

For intramuscular injections, an 18-gauge needle that's 1.5 inches (3.8 cm) long is about right, even for foals. A needle of larger gauge is too big — more likely to cause discomfort as well as leakage of injected material because it makes such a big hole in the skin. A needle of smaller diameter goes in nicely, but it's more difficult to push liquid through it; it takes longer and there's more chance of having syringe and needle "blow apart" (losing the vaccine or antibiotic) because of the extra pressure needed to push the liquid in. A needle longer than 1.5 inches (3.8 cm) is apt to penetrate too deeply and hit a bone or artery, and a needle shorter than that will not deposit the injection deep enough into the muscle for optimum absorption.

Filling a Syringe

Some vaccines come in convenient individual doses, already in small syringes. But, often, you have to fill a syringe from a bottle of vaccine or antibiotic. After inserting the sterile needle through the rubber top, tip the bottle up and slowly pull back the plunger of the syringe to create a vacuum in the syringe. It will be filled by liquid from the bottle.

FILLING A SYRINGE

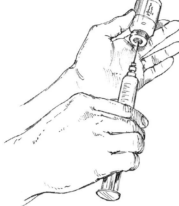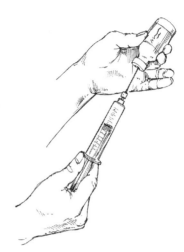

A. To fill a syringe, insert the needle through the rubber top of the bottle, then tip the bottle upside down so the needle tip is in the liquid.

B. With the needle tip inserted into the bottle, slowly pull back the plunger on the syringe to create a vacuum, allowing the fluid to flow into the syringe.

If you have a large bottle of vaccine and several horses to vaccinate, or a large bottle of antibiotic and are giving a large-dose injection, it will be harder to fill the syringe because of the vacuum created in the bottle when you withdraw some of the fluid. In these instances, fill the syringe with air by pulling back the plunger, and then insert the air into the bottle. The fluid will come out more easily this way. Hold the bottle at an angle so the needle injects air into the air space above the liquid. Injecting air into the liquid itself would create many tiny bubbles and make it harder to fill the syringe to accurate dosage.

BEFORE GIVING THE INJECTION

If it will be a moment before you give the injection, put the protective cover back on the needle so there is no chance of getting it dirty or contaminated if you set the syringe down for some reason. If the contents of the syringe are cold, hold the syringe in your hand for a moment to warm it before giving the injection to the horse. The exception would be a modified live-virus vaccine such as rhinopneumonitis. Once you "rehydrate" the rhinopneumonitis vaccine (by injecting a vial of sterile water into the dehydrated vaccine powder), give the injection as soon as possible — before the fragile contents are inactivated by sunlight or heat.

PROPER USE AND STORAGE OF VACCINE AND OTHER INJECTABLES

Note the expiration date on every bottle of vaccine or antibiotic, and don't buy any that is within a month or less of the expiration date; never buy any that's already expired. Read labels. Make sure a product that's supposed to be kept cool has been properly refrigerated. Try to buy vaccine in a bottle size that you can use up at one time. Modified live-virus vaccines must always be used within an hour of rehydration. Many antibiotics keep fairly well once opened, but a bottle of vaccine should be used right away. Don't save fractional contents. Follow label directions for storage, use, and administration.

Don't Have Air in the Syringe

Try not to get any air into the syringe when filling it. If this happens, hold it vertically with the air pocket at the top and gently push the plunger until all the air is expelled, making sure the proper dosage of liquid remains in the syringe. Squirt the air back into the vaccine or antibiotic bottle so no fluid gets wasted if you accidentally push too hard.

Injection Sites

Intramuscular injections should go into thick muscle. The best site is the buttocks. For shots of 2 mL or less, the side of the neck (a short distance above the shoulder) or the breast (chest) on either side of the breastbone can be used.

Horse people traditionally used the large rump muscle, but it's not recommended because it runs horizontally. This thick muscle is one of the best sites for absorbing a large shot, but if an injection should cause an abscess there, it would be difficult to establish good drainage. A properly administered injection carries little risk of contamination and infection, but there is always the possibility, so it's better not to take the chance. The neck, buttock, and breast muscles have the advantage of being more vertical, and an infection in that area would be able to drain.

The Neck

A safe and convenient place to give a small injection is the side of the neck, in the triangular area above the shoulder and below the crest. The horse's neck bones are low here; if you aim for this triangle, you will be in thick muscle.

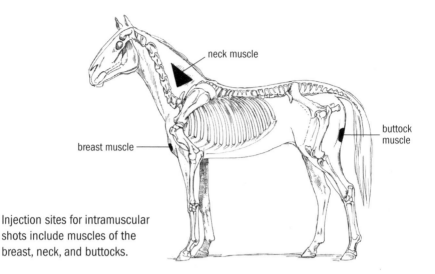

neck muscle

buttock muscle

breast muscle

Injection sites for intramuscular shots include muscles of the breast, neck, and buttocks.

The advantage of giving injections in neck or breast is that you are working at the front of the horse, where it is easier to control him and avoid being kicked. Don't put large shots into the neck because tissue swelling may make the neck sore; the horse may have trouble raising and lowering his head to eat and drink. If a horse gets a sore neck after being injected with a certain type of vaccine — a local reaction to that product — give it to him somewhere else in the future.

The Breast

For a horse who fidgets or resents shots, the best place to inject is the muscle on either side of the breastbone. The horse is less sensitive here and less apt to feel the needle. You can usually put it in without him noticing, especially if someone else holds him and gives him a bite of green grass or grain at the time you are doing it. For a large injection, however, this muscle is too small for proper absorption, more likely to cause tissue swelling. If giving a shot larger than 2 mL it's best to split it and put half in each breast muscle, or inject it in the buttocks.

The Buttocks

Large injections can be put into the horse's buttock muscle. This is a good location unless a horse is inclined to kick. When using this site, stand close to the horse.

1 cc = 1 mL

Syringe size is measured in cubic centimeters (cc); 1 cc is the same as 1 milliliter (mL).

SWELLINGS

Sometimes an injection site swells up for a day or two. Large shots given in a small muscle are more likely to cause swelling than if given in a larger muscle. Vaccines are more apt to cause swelling than are antibiotics. The amount of swelling from a vaccination depends on the type of shot and the individual horse's reaction to it. Some horses are more sensitive than others to certain vaccinations.

In most cases, the swelling is merely unsightly and might make the horse stiff and sore for a few days, but usually the site returns to normal within 24 to 48 hours. Sometimes, mild exercise after a vaccination helps prevent or reabsorb tissue swelling. If a horse swells alarmingly (more than 2 or 3 inches [5–7.5 cm] in diameter), it generally means he is quite sensitive to the injected substance. Discuss this with your veterinarian.

Giving the Injection

When giving the shot, have someone hold the horse for you if he is a nervous individual. Some old dependables can be tied, but if there's any chance the horse will react violently, it's better to have him held. If shots are given properly, the horse won't even flinch. Teamwork helps. It's not hard to give shots by yourself to a horse who knows and trusts you, but for a nervous or suspicious horse, it helps to have someone at his head holding him still and talking to him, distracting him so he won't move at the wrong time.

Make sure the injection site is very clean, then detach the needle from the filled syringe and, holding the needle in your hand, gently thump the area several times with that hand, using the side of your fist. Thumping tends to desensitize the area and mask the prick of the needle. If you try to jab it in with no previous preparation, the horse will jump. Thump quickly, only two or three times, and immediately put in the needle.

Some horses get nervous about the thumping because it's been done before and they know what's coming. For these individuals, it's better to just press the area firmly for a moment with your thumb or the side of your hand; this also desensitizes the skin. When giving an injection in the neck, you can desensitize the area by giving the skin a little pinch before inserting the needle.

GIVING AN INTRAMUSCULAR INJECTION

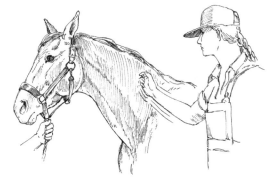

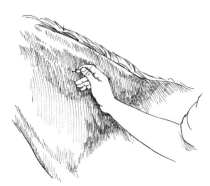

A. Before giving an injection, desensitize the target area by thumping it a few times with the side of the hand holding the needle.

B. Then thrust the needle into the muscle.

C. Next attach the syringe to the needle.

D. Check to make sure you have not hit a blood vessel by pulling back slightly on the plunger to see whether blood appears in the syringe.

E. If no blood appears, give the injection. If blood appears, withdraw the syringe and needle and start over.

F. After withdrawing the needle, rub or press the injection site to keep fluid from leaking out.

Putting in the Needle

After thumping, pressing, or pinching, stick the needle in with one swift motion. Most beginners make the mistake of not pushing forcefully enough to send it all the way in. After a few tries you will know how to do it. Remember that the faster the needle goes in, the less it will hurt.

Always take the needle off the syringe and put it in first. Then, if the horse flinches or moves around, you can wait until he settles down again before attempting to give the shot. Also, you can tell if you've hit a vein. If the inserted needle begins to ooze blood, take it out and start over in a slightly different spot. Never inject an intramuscular shot into a blood vessel.

Double-Checking

Once you attach the syringe to the inserted needle, double-check to make sure you haven't hit a blood vessel. Pull the plunger back just a little to create a vacuum. If blood appears in the syringe, remove it and start over. (A small amount of blood in the syringe is nothing to worry about when giving the injection.) If no blood appears, go ahead and give the injection. Push the plunger in steadily but not too quickly; if you create too much pressure for the needle, it will come apart off the syringe.

If you are using a screw-on needle and syringe, make sure it is adequately screwed together before pushing the plunger. Otherwise, it will come apart and spill the injection. If you are using a slip-on needle and syringe, make sure it's fully on; hold the syringe and needle firmly at their juncture with your other hand as you depress the plunger, just to make sure they do not come apart as you give the injection.

Removing the Syringe after Giving the Injection

Withdraw the syringe and needle together in one easy movement, pulling straight out. Then rub or press the injection site with your hand or finger. This not only keeps it desensitized for a moment but also prevents the fluid from leaking out before the hole constricts. If an injection starts to ooze out, press the site until it stops.

Adverse Reactions to Injections

Vaccinations offer tremendous protection against some of the most serious diseases of horses, but a few individuals react adversely to even the safest vaccines or to routine shots of antibiotics or other injections. Every horse owner should be aware of this and be prepared to act fast should a serious reaction ever occur. Even if your veterinarian gives the injection, the reaction may occur after he or she leaves and it will be up to you to treat the horse; in a very serious reaction, the horse might be dead by the time the veterinarian is able to return.

Despite the potential for adverse reactions, vaccination is very important in controlling the spread of disease and protecting horses from debilitating or fatal illness. If a horse is sensitive to a vaccine such as tetanus or sleeping sickness and you don't want to leave him unprotected, have the veterinarian try half doses or a half dose split into two injection sites. This will depend on the vaccine and the type of reaction your horse has. In some instances, you should never use that vaccine again. The veterinarian can advise you on this.

All injections should be given with caution, proper technique, and sterile equipment and at recommended sites. Only healthy horses should be vaccinated. Be aware that a horse coming down with strangles and given a strangles vaccine will have a serious reaction (see chapter 8). Even with proper procedure, problems sometimes occur — so be prepared.

Allergic Reactions

There are two kinds of allergic reactions although, in a sense, the difference is one of degree. A *mild* allergic reaction may manifest in excessive swelling of local tissue or hives all over the body. A severe allergic reaction may cause congestion of the lungs and shortness of breath, as an attack of asthma. Most reactions can be alleviated by giving antihistamines but, in extremely severe cases, the horse may go into shock and die unless promptly treated to reverse the shock.

Some horses react adversely to certain injectables or to the substance (carrier or adjuvant) with which a vaccine is mixed. If your horse has a history of such problems, you may have to change vaccines or even eliminate that kind from his annual vaccination program.

Anaphylactic Shock

Shock is a manifestation of hypersensitivity that affects the horse's entire body rather than just the local tissues. *Anaphylaxis*, a life-threatening reaction to an antigen, usually occurs after the second exposure to a protein substance entering the bloodstream — a substance the horse is already sensitive to (a particular disease organism or vaccine). Histamine is liberated in the affected tissues because of the reaction between the circulating antigen from the vaccine and the horse's own antibodies, which developed in his bloodstream from the previous vaccination or exposure to the disease. Histamine acts to dilate the capillaries, which start leaking fluid, producing a drop in blood pressure and causing smooth muscle, including the bronchial tissue in the lungs, to contract, making it harder to breathe.

Symptoms

Horses going into anaphylactic shock may react in various ways. Skin, lungs, feet, or blood circulation may be affected. The horse may have trouble breathing and suffocate. He may develop acute laminitis. He may collapse and die before treatment can be given (this sometimes occurs if an injection is put directly into a blood vessel instead of muscle).

Usually, the more serious the reaction, the sooner it occurs after the injection causing it. For example, horses who are allergic to certain drugs may collapse immediately after injection of that drug.

Treatment

Anaphylactic shock is usually reversible if treated in time. Steroids (dexamethasone) and epinephrine (adrenaline) reverse it if given immediately. Keep these drugs on hand in your barn or tack room for an emergency if you do your own vaccinating, especially if you live some distance from a veterinarian. Antihistamine works fine to reverse a mild allergic reaction but is inadequate if a horse goes into shock. In serious cases, the horse can die in a matter of minutes — delay in treatment can be fatal. The time it takes for a veterinarian to get to your place (or come back after giving your horses their vaccinations) may be too long, so keep these drugs on hand and know how to administer them. Discuss proper procedure and dosage with your veterinarian.

It's wise to observe horses for at least 30 minutes to an hour after vaccination, so keep them close by in a pen rather than turning them right back out to pasture.

Penicillin Reactions

Penicillin is one of the safest antibiotics, but one type of penicillin, procaine penicillin G, can cause an instant and alarming reaction if it is accidentally put into a blood vessel instead of the muscle. Once it gets into the bloodstream, it immediately affects the brain, causing seizures. The horse reacts violently, often rearing up and flipping over backward and thrashing. If this ever happens, the best thing to do is just get out of his way so you won't get hurt — because you can't help him. The horse may die during such an episode, but, often, the seizure will subside in a few moments and he will be fine, unless he injures himself in his struggles. Because of the possibility of this very serious type of reaction, take care to use proper injection technique if you ever have to give a horse penicillin.

GUIDELINES TO PREVENT AND MINIMIZE DISEASE

Protecting your horse from disease is a many-faceted challenge requiring conscientious management; if any one facet is ignored or neglected, you horse could be at risk. To minimize disease exposure:

- Vaccinate all horses against prevalent diseases in your area
- Keep horses in a healthy, safe, stress-free environment
- Keep feed and water supplies clean and uncontaminated
- Do not let your horse drink from a water tank used by other horses at a horse show, fair grounds, or trail ride
- Minimize exposure to strange horses
- Isolate any new horse in a separate pen for 2 weeks with no direct contact with your other horses until you are sure the new horse is healthy
- Quarantine any horse who becomes sick with a contagious disease (such as strangles, flu, or rhinopneumonitis) and take precautions not to spread infection to your healthy horses

Note: Specific precautions for certain diseases are given in chapter 8.

8

Infectious Diseases of Horses

THERE ARE MANY PATHOGENS THAT CAUSE DISEASE, including bacteria, viruses, fungi, protozoa, and tiny parasitic worms. All are opportunists that invade the body and multiply, causing illness. Vaccines are available to protect horses against many of the most dangerous diseases and some of these (specifically, tetanus, West Nile virus and equine encephalomyelitis) should be included in nearly every horse's annual vaccination program. The need for other vaccinations will depend on your locale and circumstances. Consult your veterinarian for information on regional vaccines and to discuss your personal situation.

The following diseases are discussed in order of seriousness and likelihood of being encountered. Some diseases near the end of the list may be deadly but are not a consistent problem in all regions of North America.

Tetanus

Tetanus is a highly fatal disease of all animals, but horses and human beings are especially vulnerable. This disease is caused by the toxin produced by *Clostridium tetani*, a spore-forming bacterium present in the digestive tract of many animals and in soils rich in humus and animal manure. Infections usually occur from wound contamination; deep puncture wounds in the feet are the most common portal of entry for the bacteria in horses. Clean wounds rarely result in tetanus; those that contain foreign matter such as soil — especially enclosed puncture wounds, where airless conditions are ideal for these anaerobic bacteria — are most dangerous.

TETANUS MAY BE DORMANT

Tetanus spores may lie dormant in body tissues after the wound that introduced them heals; disease may set in only when conditions turn favorable. The bacteria produce their deadly toxin if there is a lowering of local oxygen level in the tissues, as is the case with a wound that damages the tissues, disrupting the blood supply. This may happen immediately after the bacteria enter if there's moderate tissue damage, or it may be delayed until another injury at the same site causes additional tissue damage. The original wound might be completely healed by then.

The incubation period for tetanus is usually 1 to 3 weeks for a horse, with some cases occurring even later. After the toxin reaches the central nervous system by traveling along nerve trunks, muscle spasms begin. Mild stimulations to the nerve endings cause exaggerated responses by the animal. Death is usually by suffocation because the muscles used in breathing become paralyzed.

Symptoms

Early symptoms are muscle stiffness and tremor, followed by chronic muscle spasms. Facial and chewing muscles are affected, making it hard for the horse to open his mouth. The third eyelid membrane protrudes and the hind legs become stiff, making the horse's gait unsteady. He holds his tail out stiffly.

Other symptoms are an anxious and alert expression with ears up, eyes wide, nostrils flaring. The horse may eat and drink normally in early stages, but as the disease progresses, chewing becomes difficult and he can't swallow. Saliva may drool from his mouth. Treatment for tetanus is often unsuccessful; it is far better to prevent it.

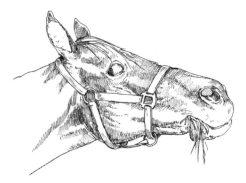

A horse with tetanus is characterized by a "sawhorse" appearance, with stiff head and neck, third eyelid protruding, flaring nostrils, inability to chew (lockjaw), and hypersensitivity to sounds, hence the cotton in his ears.

DEFINITIONS

- **Toxoid** consists of inactivated toxins used to immunize against tetanus, stimulating the body to build its own immunity.
- **Antitoxin** is a purified antiserum obtained from horses who have been immunized against tetanus. This product gives a passive (temporary) immunity, similar to that gained from the antibodies in colostrum.

Prevention

The best protection is vaccination with toxoid, starting the day the foal is born if his dam was not vaccinated during pregnancy. If she was vaccinated, producing antibodies in her colostrum to give the foal a temporary immunity, the foal can be given his first shot of toxoid at 3 to 4 months of age. The usual procedure is to give two shots 3 to 4 weeks apart, then an annual booster. Antitoxin is sometimes given if a wound occurs when a horse has not been vaccinated, but toxoid is sufficient if given within 24 hours of injury because horses build immunity quickly. Antitoxin should be used if a wound is not discovered promptly. It should also be given to foals with dams of uncertain vaccination history (the mare might not have had a toxoid shot) or with uncertain colostrum intake at birth. Antitoxin gives quick protection but doesn't last more than a few weeks. Annual vaccination is best.

Equine Encephalomyelitis

Called "brain fever" and "sleeping sickness," equine encephalomyelitis is often fatal. It's caused by a virus that is spread to horses from birds through mosquitoes. Cases occur when horse owners neglect to protect horses by vaccination. There are three kinds of encephalomyelitis in North America: Eastern, Western, and Venezuelan.

The Venezuelan strain (VEE) crossed into the United States from Mexico in 1971 after moving north from South America. U.S. horses had no immunity against this strain, and thousands died in Texas before a suitable vaccine was developed and the northward spread was halted. VEE has periodically moved north in some years but rarely gets very far into the U.S. Today, most horse owners don't need to vaccinate against VEE unless they plan to travel south of the border or live in southwestern regions where they might be at risk.

Eastern and Western strains have been in the United States for a long time. Horses vaccinated against one type are not immune to the other. The two types now overlap in territory; most horse owners use a combination vaccine that immunizes against both strains.

Symptoms

After being bitten by a virus-carrying mosquito, a horse usually shows symptoms within 3 weeks, often in 4 to 6 days. Early signs are fever, depression, and loss of appetite. The next stage involves invasion of the nervous system, with destruction of nerve tissue and brain swelling, but not in all cases. If the virus doesn't reach the nervous system, the horse recovers, developing antibodies that protect him for about 2 years.

If the virus attacks the nervous system, the horse becomes deranged by the time the fever reaches its peak. He acts restless, excited, and seemingly blind. He may grind his teeth, walk in circles, or stagger. He may be overly sensitive to touch and sound, crash through fences, or wander blindly into obstacles. Shoulder and facial muscles may tremble. Then, he may become severely depressed, with head hanging — perhaps resting his head on a fence or manger. In the final stages, he goes down and can't get up. Some horses don't develop paralysis and manage to survive, but often with permanent brain damage. Some horses die within a few hours after symptoms begin, but most deaths occur in 2 to 4 days.

Prevention

Every horse should be protected by vaccination in spring before mosquito season. In warm climates, horses should be vaccinated again in late summer. Foals and yearlings may need more protection; young horses have been known to die from the disease despite being vaccinated. They should be vaccinated at 3, 4, and 6 months of age, then every 6 months thereafter until adulthood (3 years of age). Because the source of the disease (birds, mosquitoes) cannot be controlled, the risk of sleeping sickness is always present, so vaccinate.

West Nile Virus

This bird disease sometimes spreads to horses and humans by mosquitoes. The mosquito bites a sick bird and then bites a horse or human. The disease itself, and mode of transmission, is very similar to EEE and WEE. West Nile virus is common in Africa, western Asia, Europe, and the Middle East, but is relatively new to this hemisphere. It first appeared in the United States in

New York in 1999 and then spread along the eastern coastal states and across the Midwest. It reached the West Coast by 2003.

Symptoms

Incubation time may be 3 to 15 days. Once the horse becomes sick, he may or may not have a fever but is usually depressed and lethargic. Usually, the signs are neurologic; the horse may have muscle twitches and become uncoordinated. Diagnosis can be made with a blood sample and lab test.

Treatment

The sick horse needs intravenous (IV) fluids and medications to reduce fever, inflammation, and spinal fluid pressure. Dexamethasone is often given, as well as nonsteroidal, anti-inflammatory drugs, dimethyl sulfoxide (DMSO), or mannitol (an alcohol given by IV). If the horse is too uncoordinated to stand up, he may need to be supported in a sling.

Prevention

There are several vaccines available. Mosquito control helps minimize risks for this disease, but vaccination is probably the most effective way to protect your horse. Vaccinations should be given before the peak of mosquito season to protect the horse when risk is highest. It takes a few weeks for the disease to become established and amplified in the mosquito population. If your mosquito season is long, the horse may need a booster shot before fall.

Strangles

Called "distemper" by old timers, strangles is caused by *Streptococcus equi*, a bacterium. Distemper is a misleading term; distemper of cats and dogs is a different type of illness, caused by a virus. Strangles in horses causes inflammation of the upper respiratory tract and swollen lymph nodes that often abscess and rupture.

Young horses are more susceptible than adults, but any healthy horse can contract strangles from a sick one or from contaminated feed, water, and surroundings (even a stall that housed a sick horse years earlier) if the horse has never been exposed and has no immunity. The disease begins with respiratory infection (cough, runny nose) and then produces swelling and abscesses in the salivary glands and lymph nodes of head and neck. Death may occur if an internal lymph gland abscesses and ruptures, especially if the infection has spread to other organs such as kidney, liver, or spleen.

> ## TO USE OR NOT TO USE PENICILLIN FOR STRANGLES
>
> Some veterinarians give penicillin in the early stages of strangles, but there is controversy over this practice; other veterinarians do not use penicillin in diseases that involve abscesses because it may temporarily delay their formation, only to have them develop later when the penicillin is discontinued.

Symptoms

After an incubation period of 3 to 8 days, the horse suddenly shows symptoms, with high fever (103 to 105°F [39.4–40.6°C]), loss of appetite, clear nasal discharge that soon becomes thick and profuse, and respiratory irritation. Infection in the throat may become so severe that the horse cannot swallow, and if he tries to eat or drink, he may regurgitate through his nose. The horse may also cough.

The fever goes down after a few days, then rises again as abscesses develop in the lymph nodes. These swellings, which are hot and painful, may begin to drain fluid approximately 10 days after symptoms appear. If the nodes rupture, they put forth a thick, creamy pus. If infection is severe, the abscesses may appear anywhere on the body, particularly on face and legs. Lower legs may swell to several times normal size. If the horse develops *purpura hemorrhagica* (characterized by swellings under the skin from damaged capillaries), he may die. Abscesses in the lungs cause pneumonia; an abscess in the brain causes meningitis — with paralysis and death. Abscesses in liver or spleen cause death if they rupture, which can happen weeks or months after the horse seems to have recovered from strangles.

How Strangles Is Spread

Bacteria are present in the nasal discharges of affected horses and can contaminate feed, water, bedding, and pasture. An infected horse can spread the disease for at least 4 weeks after contracting it. The bacteria may live on the ground or bedding and may infect other horses months or even years after the sick one has been removed. After a horse recovers from strangles, he'll have immunity for a while, but if the infection is persistent, horses in a group may suffer repeated attacks every 6 months.

Prevention and Sanitation

If a horse has been sick with strangles, the area in which he was kept should be thoroughly cleaned and disinfected. Old bedding should be burned. Any items used for that horse (brushes, tubs, blankets) should be carefully disinfected with chlorhexadine (Nolvasan) or a similar product from your veterinarian.

If other horses have been exposed, consult your veterinarian. Vaccination can be an effective means of prevention but should be given before a horse is exposed. A horse can have a serious reaction to the vaccine if he is already coming down with the disease. Vaccinating in the face of an outbreak is something you should discuss with your veterinarian. There is now an intranasal vaccine for strangles that is safer and less likely than the intramuscular vaccination to cause severe reactions. The intranasal vaccine is sprayed into one nostril using a long cannula inserted deeply up the nostril.

Influenza

This is an infectious disease of the upper respiratory tract caused by a virus. The term *influenza* has sometimes been applied to several respiratory diseases, and there are other viral infections (including rhinopneumonitis and equine viral arteritis) that cause similar symptoms. Influenza causes respiratory disease, with fever, and can become serious if a horse is worked hard or stressed during the early stages. It is very contagious and hard to treat because the virus does not respond to antibiotics. In fact, antibiotics given to a horse with influenza are actually intended to prevent or control secondary bacterial invasions that follow the initial virus attack. When a horse is weakened by the virus, bacterial complications can cause pneumonia and, sometimes, death.

The sick horse is often dull and depressed, with head hanging.

ILLNESS FOLLOWING FLU VACCINE

Some horses develop a mild fever and muscle stiffness for a couple of days following vaccination for influenza, and they may go off feed temporarily. The inactivated virus in the vaccine cannot actually cause influenza, but in some horses, an abnormal antibody response by the immune system makes them a little ill.

Symptoms

Influenza spreads rapidly among susceptible, unvaccinated horses — especially young animals who have not yet acquired immunity. The incubation period after contact is about 5 to 10 days. Symptoms appear suddenly. The horse is dull, depressed, standing with head down, and doesn't eat. He has a fever of 103 to 106°F (39.4–41.1°C) that lasts for about 3 days. In many cases, eyes are sensitive to light and eyelids may be swollen. The eyes water and the nose runs.

Prevention

Vaccination can prevent the disease, but this immunity is short-lived (3 to 4 months) compared to the natural immunity gained from having the disease itself; a recovered horse isn't susceptible to this virus again for at least a year. Horses at high risk for influenza, such as young individuals who have continued contact with other horses, should be vaccinated starting at 8 to 9 months of age, every 2 to 3 months after an initial series of two shots given 4 to 8 weeks apart. After 3 years of age, vaccination every 6 months usually gives adequate protection.

Rhinopneumonitis

"Rhino," caused by equine herpesvirus (EHV), is highly contagious and can affect horses in different ways: respiratory disease, abortion, and nervous system disorders. There are two strains of the virus: EHV-1, which causes abortion, and EHV-4, which causes respiratory problems.

Symptoms

Respiratory symptoms include runny nose, cough, and fever. These appear most frequently in young horses (they have low immunity) and in those who have not been previously exposed. Symptoms occur 2 to 10 days after exposure.

When broodmares are kept adjacent to infected horses, such as a sick group of weanlings or yearlings, they are at risk of aborting even though they may never show respiratory signs of this disease. A mare may abort as early as 2 weeks after infection or as late as several months after coming in contact with rhinopneumonitis. If she carries the infected fetus to term, it will be born sick and will not likely survive.

A less common form of the disease, paralytic rhinopneumonitis, can cause collapse. Early symptoms are weakness in the hind legs, floppy tail, and dribbling of urine. Horses suffering from a mild case of this form often recover without permanent damage, but severe cases do not survive unless given intensive care.

Signs of respiratory illness often include a runny nose.

Treatment

The main treatment for rhinopneumonitis is good care and lack of stress until the horse fully recovers. Most horses bounce back after 2 or 3 weeks. If a horse has a high fever and is off feed, call your veterinarian. The horse may need antibiotics to prevent secondary bacterial infection that might cause pneumonia.

Prevention

Several types of rhino vaccine give good protection against the respiratory form but not long-lasting protection against abortion. Breeders trying to guard against rhinopneumonitis abortions in broodmares generally vaccinate several times during pregnancy. The average horse owner may not need to vaccinate against rhinopneumonitis, depending on circumstances. If your horse lives where other horses come and go, or if young horses are present, it's wise to vaccinate. If your horse is kept with older horses who rarely leave the farm, and if he is rarely exposed to unknown horses, vaccination may not be needed — especially if there's not much incidence of rhinopneumonitis in your area. Check with your veterinarian to see whether your horse is at risk.

Pneumonia

Pneumonia is disease of the lungs. In the horse, it can be caused by a virus, bacteria, combination of the two, fungi, or migrating internal parasites. Bacteria in the lungs may be the primary invader, occur secondary to a virus, or develop as an aftereffect of some other disease, such as strangles. Any disease that lowers a horse's resistance and is accompanied by extra stress factors can open the way for pneumonia. The initial disease weakens the horse, then pneumonia sets in and becomes the actual killer.

Aspiration Pneumonia

Pneumonia can develop from moisture or foreign particles in the lungs, which cause irritation and infection. This may happen when a horse is given liquid medication or mineral oil and gets some in his windpipe that goes on into his lungs. It can also happen if a horse chokes on feed with blockage in the esophagus; fluid backs up and spills into windpipe and lungs. Anytime a horse can't lower his head to cough and get rid of a blockage (as when tied in a trailer), he is in danger of developing aspiration pneumonia if feed particles get into the windpipe.

Pneumonia from Septicemia

Most cases of pneumonia originate in the airways; pathogenic invaders come in through the windpipe; however, some lung infections are caused by blood-borne pathogens that enter the lungs through the bloodstream — such as in a newborn foal with septicemia from navel ill (infection that enters the moist umbilical stump) or some other serious infection.

Symptoms of Pneumonia

When a horse has pneumonia, his respiration rate increases (the body tries to bring in more oxygen), he breathes rapidly and shallowly, and he usually has a cough. The lungs make abnormal wheezing or rattling noises that you can hear if you listen to his chest with a stethoscope. In most bacterial pneumonias, there is toxemia — bacterial toxins in the bloodstream. In later stages of the disease, the horse has difficulty breathing because much of his lung tissue has been damaged and is useless.

There may be a discharge from the nostrils. The breath may smell foul. The horse may have a fever and be quite dull, with elevated pulse. He may not want to eat and might be reluctant to lie down; it is easier for him to breathe while standing up.

Treatment

With prompt and adequate treatment in the early stages, bacterial pneumonia usually responds quickly and completely. Viral pneumonia may not respond at all or just temporarily (antibiotics control secondary bacterial invaders), then the horse relapses. A horse with pneumonia should be kept isolated from other horses and in a warm barn, out of the weather. He should have fresh, clean water and light, nourishing feed. Your veterinarian can help with treatment — especially if the horse is not eating much and needs to be fed by stomach tube or IV — and determine which antibiotics would be most beneficial. The veterinarian may even give the horse other drugs to aid breathing and reduce pain and inflammation so the horse can better fight the illness and maybe feel well enough to eat and drink on his own.

Prevention

The best prevention against pneumonia is to keep a horse in good health and to minimize stress. This means vaccinating for diseases such as rhinopneumonitis, influenza, and strangles if your horse might be at risk because these diseases can weaken a horse and make him susceptible to pneumonia. Minimizing stress includes protecting your horse from extremes in bad weather.

Equine Viral Arteritis (EVA)

EVA is an acutely contagious viral disease that affects horses of all ages. It is spread by direct contact and breeding. Stallions can be carriers, shedding the virus and spreading the disease without ever showing symptoms themselves.

Symptoms

Symptoms are sometimes mistaken for those of a mild case of influenza: the horse may have elevated temperature and respiratory illness for a few days. Some horses experience swelling of the legs, muzzle, and eyelids. Stallions often have swelling in the sheath and scrotum. Other symptoms that sometimes occur are inflammation of membranes of the eye (conjunctivitis), skin rash, diarrhea, reddish gums, and nasal discharge. Fever may be mild or as high as 105°F (40.6°C), lasting for a day or two or as long as a week. Most cases are not life-threatening, but some horses become weak and unable to stand about a week after getting EVA; they may die unless given good care. The disease is most dangerous to foals and elderly horses.

The most serious complication of EVA infection is abortion of the fetus. At any stage of gestation, within 1 to 3 weeks of being exposed, pregnant mares may abort. A laboratory examination of the aborted fetus is necessary to determine the cause.

Prevention

EVA is generally not life-threatening but is serious because of the risk of abortion. A modified live-virus vaccine is effective in preventing outbreaks of EVA and can protect mares that will be bred to infected stallions. It is also wise to vaccinate disease-free stallions that breed other folks' mares because the disease is easily transmitted venereally. Horses can be blood-tested to see whether they are free of EVA.

Equine Infectious Anemia (EIA)

EIA ("swamp fever"), which is caused by a virus, occurs most often in horses pastured in wet places. Many cases occur in late summer and early fall, when bloodsucking insects spread EIA from horse to horse. It can also be spread from carrier horses to susceptible ones by use of dirty hypodermic needles and any other equipment that may transfer a tiny bit of blood from one animal to another.

Symptoms

EIA may appear in acute, subacute, or chronic form. The disease is usually characterized by high, undulating (up and down) fever; sweating; stiffness and weakness (especially in hindquarters); depression; anemia; jaundice; swelling of legs and lower body; and loss of weight, even if the horse has a good appetite.

When a horse suffers an acute attack, symptoms develop suddenly. He may have a fever (105 to 108°F [40.6–42.2°C]), appear dejected, refuse feed, become anemic, have jaundiced mucous membranes, and sweat profusely in warm weather. There may be watery discharge from the nostrils. Attacks may last 3 to 5 days or longer, then he may seem to recover. Other affected animals become progressively weak, dying within 2 to 4 weeks after several fever cycles. Some horses recover with a lingering type of anemia that leaves them weak and useless for the rest of their lives.

Chronic cases are most common. Horses with chronic EIA have intermittent attacks of fever—usually a fluctuating low fever—and suffer from anemia and weight loss. They become weak during attacks, sometimes with staggering or paralysis. Many chronic cases show no symptoms over long periods, but their blood contains the virus and is a source of infection for other horses.

In most chronic cases, anemia develops, heart action becomes irregular, body and leg swellings appear and disappear, and muscular weakness varies from being slight to so severe that the horse cannot stand or walk. Sometimes, the horse has diarrhea. Individuals that appear to recover from the first attack may be free of symptoms for days, weeks, or months, but sooner or later other attacks follow. The relapses often coincide with periods of stress, such as hard work or another disease. In any of these attacks, the horse may die. Horses who survive become carriers; after a few attacks they may develop a sort of immunity and seem to be healthy, but they still carry the virus. They usually die of the disease eventually.

Transmission

The disease is spread from a carrier horse to susceptible ones through transmission of tiny particles of blood carried by biting insects or on instruments that puncture the skin, such as hypodermic needles used on more than one horse. Incubation period is 12 to 15 days, sometimes shorter but often longer—even as long as 90 days.

EIA can also be spread by simple contact through infected secretions and from contaminated feed and water. Mares can be infected when bred by a stallion who carries the virus. Foals occasionally become infected by drinking milk from infected dams.

Prevention

Researchers have not yet been able to create a vaccine against EIA; therefore, other attempts are made to control it. Blood testing (Coggins test) of

---

EIA RESISTS DISINFECTING

EIA virus can withstand ordinary disinfecting methods and even short periods of boiling. Contaminated instruments and needles must be cleaned and boiled for at least 15 minutes to kill the virus — a good reason to never reuse needles. The virus is present in all tissues, secretions, and excretions of the infected horse.

---

all horses who travel from one state to another, quarantining or destroying horses who react positively, and following strict sanitation measures (including the use of disposable syringes and needles) have helped reduce the incidence of this disease. Horses who are raced, shown, sold, or transported across state lines must be blood tested. (The Coggins test is named after Dr. Leroy Coggins, who developed it.)

Blood is drawn from the jugular vein for a Coggins test.

Leptospirosis

"Lepto" is a disease of animals and humans caused by spiral-shaped bacteria. There are many kinds: some cause disease in dogs; others in cattle, pigs, horses, rodents, and humans. Leptospirosis can cause illness and abortions. A carrier animal often spreads the disease. This animal contracts the disease, recovers, but still harbors bacteria and sheds them in urine, saliva, and other bodily secretions or excretions. The most common carriers are rats, mice, and other rodents, although wildlife, pigs, cattle, and dogs may spread leptospirosis for a while after recovering from the disease. The bacteria live for many days in water or damp soil but die in a few hours when dried out; therefore, leptospirosis isn't highly contagious in dry climates or during dry seasons.

Symptoms

In horses, the disease is not usually life-threatening. A horse may have a fever, loss of appetite, dullness, or jaundice. Pregnant mares may abort. The bacteria may localize in the body and cause further problems in kidneys, joints, or eyes. In fact, leptospirosis is one of the main causes of moonblindness (periodic ophthalmia) in horses (see chapter 14). They usually recover easily enough from leptospirosis but later suffer recurring attacks of eye irritation that may eventually cause blindness. This is the most common cause of blindness in horses.

Prevention

There is no vaccine specifically for horses, but there is a cattle vaccine against five common types of leptospirosis. In areas where there is wet ground or ponds, horses can be at risk. Where leptospirosis causes abortion in mares or moonblindness in horses, some veterinarians recommend using the cattle vaccine to protect horses.

Potomac Horse Fever (PHF)

This is a serious disease of horses that first appeared in the Potomac River Valley in Maryland in 1979. Since then, it has been found in nearly all states. It's caused by a rickettsial organism, *Ehrlichia risticii*, belonging to the same family of organisms that cause Rocky Mountain spotted fever, anaplasmosis, and typhus — all spread by insects, fleas, and ticks. The larvae of flukes (parasitic flatworms) that live in freshwater snails transmit PHF. The disease occurs most commonly in summer and in horses who are pastured near streams or in irrigated fields.

Symptoms

The first sign is refusal of feed and water. There is usually fever up to 106°F (41.1°C). A thorough examination at that point would show low blood pressure and low blood count. Two to three days later, the typical symptom is severe, explosive diarrhea. The horse is in extreme pain with diarrhea and/or colic and becomes seriously dehydrated, requiring IV fluids. Some horses

---

SUSPECT PHF IN SUMMER

Many cases of PHF are never actually diagnosed because horses are not blood tested, but any horse with diarrhea and fever during summer is generally assumed to have PHF. At that time of year, most veterinarians promptly begin treatment — since early treatment is often crucial to survival of the horse.

founder, and some are euthanized to end their misery. Many horses test positive even though they have no history of illness; therefore, the disease apparently is sometimes very mild.

Acute cases have a high mortality rate (unless treated immediately) or develop founder, which can cause more damage than fever and diarrhea. Most horses respond to fluid and tetracyclines given by IV but may require constant care for a couple of weeks.

Prevention

Vaccination involves two doses the first year, several weeks apart, and an annual booster. Sometimes PHF still occurs in vaccinated horses but is milder because they have some immunity. The best immunity is gained if vaccination is given before summer and in proper sequence (two shots the first year). Incidence of the disease ceases after a hard frost. Whether you vaccinate will depend on where you live — find out whether cases have occurred in your area and determine the degree of risk. Ask your veterinarian to help with this. You can reduce risk of PHF by keeping horses off irrigated pastures and away from streams and ponds; provide a different water source.

Salmonellosis

Salmonella is a bacterium that causes disease in all species of animals, including humans. Several types affect horses, including *Salmonella typhimurium*, which can also cause illness in cattle, sheep, pigs, ducks, dogs, and small animals. The feces of all of these can serve as sources of infection for others. Farms with a lot of ducks or pigeons may have a problem with salmonella spreading from the birds to the horses.

Salmonella bacteria may live harmlessly in the horse's gut until stress enables them to cause disease, which can be mild or severe. Salmonellosis can be a primary disease — especially in foals — but, in most horses, it is often a secondary invader that attacks when the animal is stressed, suffering from some other disease or situation that lowers its resistance. The usual symptoms are inflammation of the intestine and diarrhea; mortality is high. The source of infection is a sick animal or carrier, wild or domestic, that contaminates feed, pasture, or drinking water with feces containing the bacteria.

Carriers

Salmonella bacteria often can't be eliminated from the horse's body even after he recovers; therefore, he becomes a carrier and spreads it to other horses.

This can become a chronic problem on a farm, with periodic outbreaks and high foal mortality. Young horses are most susceptible to fatal diarrhea, and adult horses are likely to become clinically normal carriers, breaking out with the disease only if stress or some other debilitating circumstance lowers their resistance and permits bacteria to multiply rapidly.

Symptoms

Infection results in rapid multiplication of bacteria in the intestine followed by invasion of the bloodstream, causing septicemia (bacteria or their toxins in the blood, poisoning the entire body), which can be rapidly fatal. The horse becomes restless and won't eat. He develops mild to moderate fever, and acute diarrhea may follow.

Newborn foals may develop severe septicemia, becoming dull and depressed with high fever and dying within 24 to 48 hours. Sometimes, the infection spreads so quickly the foal dies from the overwhelming septicemia before experiencing any diarrhea. Adults more commonly get extremely watery, sometimes bloody, diarrhea. There may be high fever at first and complete loss of appetite, but the animal may be very thirsty. Pulse and respiration rates are high. Pregnant mares may abort. The horse may have violent colic. Diarrhea causes severe dehydration and the horse loses condition rapidly. Some go into shock and die swiftly in spite of attempts to replace fluids. Others linger for weeks and then die or begin gradual recovery. Still others recover from the diarrhea within a few days, whereas other cases suffer chronic diarrhea for months.

Treatment

The most important treatment is fluids and electrolytes given orally and by IV. The veterinarian may administer an antibiotic, although this is controversial; several forms of the bacteria have developed resistance to commonly used antibiotics. The goal of treatment is to keep the horse alive and give him good care until he can develop resistance and avoid relapse.

Prevention

If salmonellosis is suspected, the horse should be kept in strict isolation from other animals and all equipment used on him kept separate. Even boots and clothing worn while treating him, feeding him, and so on should not be worn where other horses might come in contact with contamination. You don't want to track manure from a sick horse into a pen where there are healthy horses. Salmonella outbreaks can be prevented by not overgrazing pastures (an overused, overgrazed pasture is more apt to be contaminated), keeping healthy horses out of a contaminated area, and quarantining new horses until you are sure they are healthy. Isolating sick and recovering animals is extremely important. There is no effective vaccine for salmonella.

Botulism

Botulism is a highly fatal disease caused by toxins produced by a spore-forming bacterium, *Clostridium botulinum*, found in the environment and especially in soil. The bacteria live in the digestive tracts of plant-eating animals such as horses and cattle but do not usually cause disease. The bacteria multiply best in decaying plant or animal matter; under warm, moist conditions they create a lethal toxin. Botulism is responsible for "shaker foal syndrome," which causes affected foals to quiver, become weak, and collapse. Even with intensive care, most unvaccinated shaker foals eventually die of respiratory failure because of progressive paralysis.

In horses, there are two strains of botulism (B and C) that cause "forage poisoning." Adult horses usually become ill after eating contaminated feed but can also ingest botulism spores from soil. The bacteria then produce toxins while multiplying in the gut. Foals usually develop botulism by eating soil containing the spores. Although the spores are commonly found in soil in the Northeast, horses can develop botulism anywhere in the United States if they eat toxins in feed or water that has been contaminated. Dead rodents in haystacks or small animals baled in hay are common sources. The toxin can also be found in spoiled grain.

Symptoms

Symptoms of botulism show up 3 to 7 days after the horse has eaten contaminated feed. Incubation period can be shorter if large amounts of toxin are eaten. Most horses die unless treated quickly; even then, many do not survive. The toxin short-circuits the nerves controlling muscle function, causing paralysis and difficulty swallowing, standing, and breathing. The horse has progressive muscle paralysis and is soon unable to chew or swallow, drooling saliva. Paralysis of the chest muscles results in suffocation. In foals with shaker syndrome, usually the first sign is trouble swallowing; milk dribbles from the foal's mouth when he nurses. He becomes uncoordinated and lies down frequently. There is no fever.

Prevention

Vaccinated pregnant mares can transfer immunity to their otherwise defenseless foals through antibodies in colostrum. This gives foals temporary immunity for several weeks. Adult horses of any age can be given an initial series of three doses of toxoid at least a month apart, with an annual booster. Foals can begin their vaccination series as young as 2 to 4 weeks of age, with three inoculations at 2-week intervals. If an outbreak occurs, mares and foals can have an even tighter vaccination schedule, giving the foals' initial series a week apart; a fourth dose might then be necessary to provide adequate immunity. A mare's vaccinations can be spaced more closely also, to give adequate concentrations of antibody in colostrum.

Take care to make sure all feed is free of contamination. A dead bird, snake, or rodent baled in hay can be deadly, and toxins can contaminate the hay around it. Any feed found with traces of dead animals in it should always be discarded. Be sure to get rid of the hay surrounding the animal carcass as well. The best rule for feeding is "when in doubt, throw it out."

Rabies

Many people think of rabies as a disease that affects dogs, skunks, foxes, or bats; but horses and cattle get it, too. This fatal viral infection of the central nervous system can occur in any warm-blooded animal. It's transmitted by bites of affected animals or by contact with saliva from an infected animal. If the virus enters the body through a wound or break in the skin, it attacks the nerves. Unlike other viruses that travel through the bloodstream, the rabies virus affects nerve tissue, moving slowly through nerve cells and eventually reaching the brain.

Symptoms

When the virus enters the spinal cord and brain, it causes encephalitis (inflammation of the brain) with destruction of nerve cells. One of the first and most

---

INCUBATION OF RABIES

The incubation period depends on location of the bite and how long it takes the virus to reach the brain and spinal cord. A horse bitten on a hind foot may not show symptoms as soon as a horse bitten on the nose. Symptoms may not appear for weeks or even months after the bite. Once symptoms do show up, rabies is always fatal.

common symptoms is difficulty swallowing. The horse may try to eat or drink but can't. Saliva may drool from the mouth. Inability to swallow may be misdiagnosed as choking, which results in handlers being exposed to rabies as the horse owner or veterinarian examines the mouth and throat.

In the dumb or paralytic form of the disease, the animal is lethargic and depressed with head hanging, slack lip, dazed expression, and drooling. Temperature, pulse, and respiration are normal; there's no sign of shock or pain. The horse is uncoordinated; hind legs may knuckle under at the fetlock joints, and hindquarters may sag and sway if he tries to walk. The horse eventually becomes paralyzed, unable to get up.

In the furious form, the animal becomes aggressive, tense, alert, and hypersensitive to sounds and movement. He may pace about or run wildly, biting at anything. He may snap at his halter rope or the person handling him and become aggressive and vicious. Acute change in personality may be the first noticeable symptom.

Death from Rabies

Death from rabies is usually a result of respiratory paralysis of the horse; the muscles necessary for breathing no longer work.

No Set Patterns

Although symptoms vary, rabid animals often have fever, depression, loss of muscle tone in the tail area, muscle tremors, and paralysis of hindquarters. A veterinarian is often called because the horse shows lameness or hind-leg incoordination. Other symptoms may or may not appear:

- They may eat strange things such as straw or wood.
- Pulse, temperature, and respiration may be normal at first, then climb as the animal becomes frenzied.
- Some animals are sick 6 to 7 days before dying; others don't last more than 36 to 48 hours.
- They don't always display aggressive behavior or drooling.
- Some are colicky.

Rabies can be mistaken for many other illnesses or behavioral problems, and this makes it exceptionally dangerous; it may not be suspected soon enough to prevent human exposure. Precautions should be taken when handling a horse who might have rabies; wear protective clothing to avoid contact with his saliva.

HAS THE HORSE BEEN EXPOSED TO RABIES?

It's quite difficult to know if a horse has been exposed to rabies. The bite of a rabid animal is often small and goes unnoticed, healing in a few days and leaving no clue to the problem until the horse shows signs of infection. If you live where rabies occurs in wildlife, discuss vaccinating your horses with your veterinarian.

Diagnosis

Positive diagnosis can be made only through laboratory tests on the brain to confirm presence of the virus after the animal has died. If rabies is suspected, the veterinarian should submit the brain to a diagnostic lab for testing.

Treatment and Prevention

There is no treatment for rabies. Once symptoms begin, it is always fatal in human or animal. Because of the long incubation period (3 weeks to several months), humans can be vaccinated soon after exposure. But a horse usually can't receive the vaccine early because you may not know when or if the animal was bitten! The safest route is annual vaccination, rather than waiting until the animal is exposed. If vaccination is started early enough, the body builds immunity in time to prevent fatal inflammation of the brain.

To protect against rabies, all cats and dogs should be vaccinated; they are often more at risk than horses for coming in contact with wild animals. Horses can also be vaccinated. This is generally adequate protection, but any horse exposed to rabies (bitten by a rabid animal) should receive a booster shot immediately and be quarantined for 90 days. An unvaccinated horse exposed to rabies is more at risk. Post-exposure vaccination is routine in humans, but experts debate the effectiveness of this measure for animals. Some public health officials recommend immediate euthanasia or putting the horse into strict quarantine after post-exposure vaccination — for 6 to 9 months with minimal human contact.

Piroplasmosis

Piroplasmosis is caused by protozoa and spread by blood-sucking ticks. Incubation time is 7 to 20 days as protozoa multiply in the bloodstream. Death is usually from lack of oxygen in tissues, a result of severe anemia. If the horse survives, it becomes a carrier for a year (or longer if it becomes reinfected).

Symptoms

Intermittent fever, loss of appetite, depression, weakness, and increased pulse and respiration rates are typical. As the disease progresses, the horse shows evidence of jaundice, anemia, and dark-colored urine. Fever lasts a week; the total course of the disease is about 3 weeks. Abortion is common in pregnant mares. If the horse recovers, it takes a long time to overcome anemia and emaciation.

Treatment and Prevention

A few drugs such as imidocarb are helpful, but there are no vaccines. Prevention requires effective quarantine of horses and other measures to halt the introduction of the blood-sucking tick into new areas. Restricting imports or movement of animals from tick-infested regions has proved effective.

Anthrax

Anthrax is a serious disease that affects most warm-blooded animals. It comes on suddenly and runs its course quickly, ending in sudden death. The bacteria are present in the tissues of infected animals; if the carcass is scattered by birds or predators, soil and water may become contaminated. Spores are formed that can live a long time. The disease is often spread during floods, contaminating grass that grows in soil containing the bacteria. Grass grown in flooded lowlands may be infected, and animals fed hay grown on these lowlands may also get the disease. In hot climates, anthrax is sometimes spread by bloodsucking insects, particularly horseflies. Close grazing of scratchy stubble (causing injuries to mouth tissues) and confined grazing in contaminated areas (around waterholes, for example) can sometimes lead to spread of the disease.

Symptoms

Anthrax in horses is always acute. If a horse gets it from contaminated feed or water, he develops septicemia (bacterial toxins in the blood), enteritis (inflammation of the intestines), and colic. If infection is from an insect bite, the horse may get hot, painful swellings around the throat, neck, rib cage and abdomen, penile sheath, or mammary glands. There is high fever; the horse is depressed and may have difficulty breathing. Anthrax is usually fatal within 48 to 96 hours. Antibiotics and anti-anthrax serum are used in treatment; if given soon enough, the horse may recover.

Prevention

In areas where anthrax is a problem, horses should be vaccinated in spring. Occasionally, an animal develops severe or even fatal reactions to the vaccine. If an individual develops a reaction, he should be treated with antibiotics and antiserum, a blood serum derived from an immune animal that contains antibodies against anthrax.

Vesicular Stomatitis (VS)

This contagious viral disease occurs in warm regions of South and Central America and Mexico but sometimes spreads north into the United States, requiring quarantines and restricted movement of horses and livestock. VS is not a fatal disease, but health officials are diligent to prevent its spread because symptoms are so alarmingly similar to those of foot-and-mouth disease. Outbreaks must be accurately identified. Foot-and-mouth, a much more serious disease of cloven-hoofed animals, was eradicated from the United States in 1929.

Transmission

VS occurs during summer because of large insect populations and increased movement of animals to shows and competitions. The virus can be spread by insects and also by contact with saliva or fluid from the ruptured blisters of an infected animal. It is often spread by moving infected animals, failing to clean out vehicles used in transport, and being careless about human contact between infected and noninfected premises. Blister fluid or saliva on bits or in water buckets can infect susceptible animals. A farm that's had a sick animal cannot be considered free of VS until at least 30 days after the affected animal's lesions have healed. If a case occurs, the horse should be quarantined in a stall. Animals on pasture are more frequently infected. Isolating a sick one in a stall reduces the risk of spread.

Symptoms

The disease is rarely fatal but causes blisters in the mouth, tongue, lips, nostrils, teats, and feet. Blisters swell and break, and the skin sloughs away leaving painful ulcers; the animal stops eating and drinking (thereby losing weight) because of raw tissue and pain. Animals may die of starvation, dehydration, or secondary infection. Blisters on the coronary band cause lameness, founder, and, occasionally, loss of a hoof. The most common symptoms are drooling and high fever. Incubation time from exposure to first blisters is 2 to 8 days,

---

HEALTH TIP FOR VS

Use protective measures when handling VS-infected animals to prevent contact with body fluids. The virus can be transmitted to humans through skin or the respiratory system, causing flu-like symptoms. In humans, VS causes acute illness: fever, muscle aches, headache, general discomfort, with blister lesions only rarely. When handling a sick horse, wear gloves to protect hands from saliva and blister fluid; never let these fluids come in contact with your eyes, mouth, or any broken skin. If you come in contact with fluids, consult your doctor.

To prevent VS, keep horses in stalls or dry corrals away from insects. Avoid stress or medications that suppress the immune system. Use insecticide and repellent daily on horses. Isolate new horses for at least 21 days to make sure they don't have VS. This disease must be reported to authorities and, if it occurs, the premises must be quarantined.

---

sometimes longer. A close look at the mouth will reveal the blisters. In horses, blisters usually appear on the upper surface of the tongue. If there are no secondary infections, the horse usually recovers in 2 to 3 weeks.

Treatment and Prevention

There is no treatment for VS except to prevent secondary infection where the blisters have broken and raw tissue is exposed. Mild antiseptic mouthwashes and ointments help alleviate pain and speed recovery. Good sanitation and quarantine measures on affected farms can keep VS from spreading. Use disinfectants such as chlorine bleach or chlorhexadine (Nolvasan). A quarantine area should be as far away from other horses as possible. Avoid shared feeding equipment, or clean and disinfect it for at least 10 minutes between uses. If caring for sick horses, shower and change clothing and boots before handling healthy horses.

Lyme Disease

Lyme disease affects humans and horses and is spread by tiny deer ticks. It was first diagnosed in Lyme, Connecticut, in 1976 and has since been reported in almost every state. It is caused by the *Borrelia burgdorferi* bacterium.

> ### TICKS, MICE, AND LYME-DISEASE CONTROL
>
> Not all ticks are infected. They must first bite an infected white-footed mouse. The ticks can be controlled with a selective pesticide that is available in biodegradable tubes containing cotton balls impregnated with insecticide. When these tubes are scattered in a tick-infested area, mice, which account for spread of the disease in certain regions, gather the cotton balls for their nests and the insecticide spreads into their fur. When ticks make contact with the cotton or mice, they die — without causing any harm to the mice.

Symptoms

Symptoms include rash and swelling at the tick bite, fever, fatigue, and depression. Horses may also suffer laminitis, joint problems, and chronic lameness. Some are eventually euthanized because of chronic problems caused by the disease. Treatment with antibiotics is effective if begun early, but often, the disease is mistaken for something else. Diagnosis is usually based on whether the horse resides in a tick-infested region and has a positive blood test.

Prevention

The best ways to prevent Lyme disease are to keep you and your horse free of deer ticks, keep pastures mowed, and remove brush and wood piles from pasture areas to decrease rodent-nesting sites. Precautions during tick season include careful grooming after a ride and keeping your horse out of tick habitat, like brushy areas, as much as possible. If you find a tick, carefully remove it with tweezers, not your fingers. Grasp it as close to the skin as possible to be sure you get the head.

Equine Protozoal Myeloencephalitis (EPM)

This neurologic disease is one of the most common causes of neurologic problems in horses, affecting the spinal cord. It is caused by a protozoan, *Sarcocystis neurona*, that lives in opossums, with intermediate hosts that include raccoons, skunks, and house cats. Horses are accidental hosts that generally become infected by eating feed or water contaminated with feces from infected opossums. The one-celled organism then makes its way to the spinal cord, where it causes inflammation and nerve cell damage. This most often results in a vague lameness or weakness and incoordination.

Symptoms

The key to diagnosis is recognition of subtle changes in the horse's stance and way of traveling—little hints of weakness or incoordination such as toe dragging when tired, mild lameness in the hind legs, back soreness, choppiness of gait, or awkward stance. As the disease gets worse, symptoms become more obvious — muscle atrophy, seizures, facial paralysis (ears or lips drooping on one side, head tilting), or leaning to one side. Diagnosis can be aided with a blood test or spinal tap.

Treatment

Early diagnosis and treatment to rid the horse's body of the parasite can halt the infection in time to prevent permanent crippling, but if treatment is begun too late, there can be irreversible damage to the nervous system. Earlier treatments were long and complicated, requiring daily medicinal doses for 3 to 6 months. Now, there are several oral paste formulations of various anti-protozoal drugs and these treatments are safer and more reliable. Horses in early stages of the disease are more likely to respond well to treatment and recover more fully, but sometimes, all you can hope for is just to halt the process; the horse may not get much better, but won't get any worse either.

Prevention

Keep opossum droppings from contaminating feed and pasture. Keep grain covered, buy heat-processed feed (high temperatures kill the organism that causes EPM), and reduce opossum population on your property. Live trapping is time-consuming but effective. Some veterinarians recommend using vaccination if your horses are at high risk, if the vaccine is approved in your state.

Conscientious Care

There are other infectious diseases that can affect horses, but most are not common and this book can't cover all of them. The best way to protect your horse from illness is to conscientiously give the best care you can, vaccinate against diseases in your area (if there's a vaccine available), and consult your veterinarian if the horse ever develops signs of illness or displays a change from normal temperament or behavior. Proper diagnosis and prompt treatment make a critical difference in his recovery.

9

Noninfectious Diseases of Horses

THERE ARE A GREAT NUMBER OF HORSE PROBLEMS not caused by infectious organisms. Most of these diseases, such as "heaves," laminitis, navicular syndrome, and muscles tying up, can be prevented by proper care. Other problems, such as cancer or Cushing's disease, are not as preventable; they may occur despite the best care and responsible use of the horse.

Cancer

Cancer is not common in horses, but be aware of the possibility because many types are treatable in early stages. Benign growths enlarge slowly and rarely recur if removed. Malignant growths expand more rapidly and tend to spread to other parts of the body, infiltrating neighboring tissues or entering into the blood or lymph system.

A cancer can start from any type of cell (blood, bone, muscle), but more than half the cancers in horses originate in skin tissue. Most can be controlled if discovered in time and can usually be removed without major surgery. Some types of cancer are curable but only if they are detected early. Some kill the horse within a few weeks; others take years to cause death.

Melanoma

This skin cancer can occur on any dark-skinned horse, most commonly on grays. The tumors originate in cells that produce skin pigment (melanin). Melanomas eventually affect 80 percent of all gray horses over 15 years of age. Most gray individuals develop the black lumps in their old age. Because melanoma in horses is not as dangerous as in humans, only a few cases of these black lumps progress to the point of killing the horse. However, internal growths are fairly common in the intestinal tracts of older gray horses. You should suspect this if an older horse has chronic digestive or colic problems.

Look for Lumps

Most growths appear under the base of the tail; you are likely to find the bumps when grooming or taking the horse's temperature.

Location

In mares, the lumps (tumors) usually develop under the base of the tail or around the anus or vulva. The nodules, which contain an excess of melanin (black skin pigment), generally start out benign but may spread inward along the mare's genital or rectal area, and the expanding growths in the pelvic lymph nodes can interfere with breeding and obstruct foaling. In male horses, the melanomas may develop under the tail but are often found in the sheath, sometimes giving the whole sheath a swollen appearance.

The black lump can be felt as a hard nodule in the skin. Not all bumps are melanomas; have a vet take a look to be sure. Melanomas are usually smooth, raised areas without scabs or ulcers. Lumps may appear at any time after age four or five (but more frequently after age fifteen), and there may be few or many. Sometimes, a single large tumor appears on a young horse; this type of melanoma is likely to become malignant more quickly than those on older horses.

External tumors usually grow quickly and don't develop into malignancy, but internal melanomas involve the lymph system and are more dangerous, often spreading to other parts of the body rapidly. Sometimes, a fast-growing melanoma creates a sore that won't heal, pushing the skin apart. It may cause bleeding from the affected area.

WATCH MELANOMAS

A slow-growing melanoma may take several years to become very large but can become fast-growing, killing the horse in 6 months or less. A horse with melanoma lumps should be checked periodically to see if they're grown or spread. Habitually check every time you groom or use the horse. External melanomas may never spread, but the potential for spread should never be ruled out.

> ### SUNNY-DAY HEALTH TIP
>
> Protect horses with unpigmented eyelids from bright sunlight by keeping them stabled, providing shade during bright days, or painting a nontoxic dark substance such as black theatrical greasepaint around the eyes during bright, sunny weather.

Treatment

Slow-growing external lumps are best left alone. A biopsy to confirm the diagnosis may activate the melanoma and cause it to grow or spread, sending cancer cells through the blood or lymph system into the lungs, where they become deadly very quickly. Removal of a melanoma is usually a last resort, done only if the lump is growing rapidly or causing a problem such as blocking the rectum or interfering with use of a saddle. The veterinarian freezes or burns the area after removing the growth to kill any stray cells that might be left. There's been some success in treating melanomas with cimetidine (Tagamet), an ulcer medication for humans, which helps the immune system fight the cancer. Another treatment involves creating a vaccine from a portion of the growth, which also stimulates the body to combat the cancer.

Squamous-Cell Carcinoma

Another skin cancer, this may appear on the eyelid and/or around the vulva or sheath if the skin is unpigmented. It is slow to metastasize, or spread. Intense sunlight damages the skin and makes it more susceptible to cancer. Because skin pigment helps protect against ultraviolet rays, dark-skinned horses rarely get this cancer. It is more common in horses with pink eyelids (Paints, Appaloosas, Pintos, horses with white faces, those with white markings on the face with pink skin around an eye, or light-skinned horses). Tumors near the eyeball in light-skinned horses are common in sunny areas such as California, the southern United States, and south of the border.

Location

Because cancer of the eyelid can spread rapidly to nearby tissues, unless it is caught early you may need to remove the eye. Most squamous-cell carcinomas occur on inner parts of the lower lid or third eyelid and appear as a single raised bump or raw surface — a runny sore. The eye is irritated and waters. Not every bump or reddened area is a cancer, but if the abnormality becomes larger instead of healing and going away, or becomes redder or more irritated, the horse needs immediate veterinary attention.

Squamous-cell carcinoma can also occur on other unpigmented areas, especially those that are thinly haired, such as under the tail, around the mouth, or on the sheath. Check these areas periodically for any abnormalities.

Treatment

The most common treatment is surgical removal. Other methods include freezing with liquid nitrogen, immunotherapy (injecting vaccine into the tumor tissue to stimulate production of anticancer cells), radiation, and hyperthermia (burning off the growths). When the cancer occurs on the eyelid, however, some of these methods pose a risk to the eyeball. Your veterinarian will choose a method that minimizes the risks.

Granulosa-Cell Tumor

A granulosa-cell tumor in the ovary of a mare causes problems with reproduction and disposition but is usually not malignant (although it can rupture, spreading cancer cells to other parts of the body). The granulosa cell is part of the ovary and produces the female hormone, estrogen. If this cell becomes cancerous, the tumor produces a mix of hormones, altering the mare's hormonal balance and causing personality change and bizarre behavioral symptoms. She may become irritable, aggressive, or vicious. She may even show stallion-like behavior around other mares, especially if they're in heat. Some affected mares tease and mount other mares. As the tumor grows, the mare does not come into heat and she cannot become pregnant.

Treatment

Usually, the only effective treatment involves surgical removal of the tumor and ovary. Removal of one ovary does not interfere with the mare's ability to reproduce if the remaining ovary is healthy. Most mares return to normal

---

BENIGN, BUT FATAL, TUMORS

Granulosa-cell tumors sometimes cause death. Normally, each ovary is suspended from the roof of the abdominal cavity, but an ovary enveloped in a growing tumor may become so heavy that its attachment breaks and ruptures a blood vessel, causing fatal hemorrhaging.

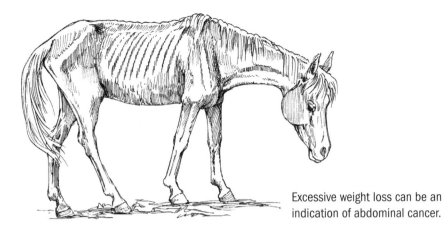

Excessive weight loss can be an indication of abdominal cancer.

after the removal and start cycling again. If a mare shows a drastic change in behavior, have her ovaries checked. If there's a tumor, it can be removed.

Other Cancers

Other cancers in horses include abdominal cancer and cancer of the blood. Some are rare and difficult to diagnose or treat. Signs of cancer include any unusual sore or abnormality: a growing lump or bump, a persistent sore, bleeding or discharge from any body opening, chronic cough (other than heaves), or persistent weight loss. Abdominal cancer in an old horse may cause dramatic weight loss.

Lipomas

Lipomas are not really cancer because they are almost always benign. Consisting of slow-growing balls of fat on thin stalks, these fatty tumors in the abdomen occur most commonly in older horses, especially fat individuals. Usually, the tumor poses no problem unless the thin stalk becomes wrapped around an intestine, strangling the gut. In these cases, the horse will suffer colic and a fatal blockage of the gut (the stricture causes death of a segment of intestine) unless the condition is promptly corrected by surgery. Many older horses have a number of these fatty tumors, but only rarely do they cause problems.

Heaves (Chronic Obstructive Pulmonary Disease)

The horse has a remarkable set of lungs that keep his blood supplied with oxygen. Anything that interferes with proper working of the lungs can limit his athletic or working career or compromise his health. A common problem that hinders a horse's ability to breathe is chronic obstructive pulmonary disease

HOW THE HORSE BREATHES

When a healthy horse inhales, his chest expands and draws air in with a vacuum effect, enlarging flexible airways. Inhaling is an active process and exhaling is passive; collapsing of the chest wall pushes air back out. In horses with COPD, air passages are constricted and the horse must work harder to force the air back out through the narrowed airways. This produces an audible wheeze.

(COPD), called *heaves* by horse people. It occurs when horses are confined indoors or in dusty pens or fed hay that is dusty or moldy; it can also be caused by allergies.

COPD develops gradually and becomes worse, usually the result of prolonged feeding of dusty hay or an allergic reaction to hay dust, pollens, molds, or bacteria. A horse in a barn is more susceptible to heaves than one kept outside; there's more dust in the barn air from hay and bedding. Even an outdoor horse may get heaves if fed dusty hay because he constantly breathes dust particles as he eats.

Once a horse is sensitive to dust or molds, he relapses every time he breathes dusty air. Even if you keep him outside, he'll cough if you bring him in overnight or to saddle him. If his problem is related to dust, he must be kept in a dust-free environment.

Symptoms

This chronic respiratory disorder is characterized by loss of ability to perform (exercise intolerance), constant or intermittent cough, and watery discharge from the nostrils. As COPD progresses, the horse loses weight and has labored breathing. Because the condition is not caused by infection, it doesn't respond to antibiotics. Similar to asthma in humans, breathing difficulty results from narrowed air passages and an increased effort for every breath.

Listen to the windpipe to check for wheezing sounds indicative of restricted airways.

HAY HEALTH TIP

Alfalfa is usually dustier than good grass hay, mainly because the leaves tend to shatter when dry. Horses who cough when fed hay usually do best with grass hay; alfalfa hay may make their allergies much worse.

There's a wheezing sound as the horse forces air out with two movements of the abdominal wall. Simple relaxation of the rib cage does not empty the lungs. The horse has to immediately follow it up with more effort, contracting his abdominal muscles and tensing his chest. In fact, many horses who suffer from COPD develop an enlarged ridge of muscle ("heave line") along the lower side of the abdomen from overworking these muscles while forcing air out of the lungs.

Treatment and Prevention

Horses with COPD can be treated with drugs that open the air passages as well as with antihistamines. Your veterinarian can prescribe medication. The best treatment is to keep the horse in a dust-free place and never feed dusty hay. If he coughs when fed hay, use pellets; nondusty pellets are available for horses with heaves.

You can prevent heaves if you avoid dusty hay and bedding and keep horses outdoors as much as possible (unless the problem is from pollens outside). Horses with allergies are especially susceptible to COPD. If a horse tends to cough when hay is dusty, shake it thoroughly to get rid of the dust (but don't shake it in the barn, as it will make the air dusty). Then, sprinkle it with water to settle any dust that is left. For a horse with a serious problem, dunk each flake of hay in water (then drain it) before feeding so it is absolutely dust free. Bedding material should be as dust-free as possible. Sometimes wood shavings are less dusty than straw.

Cushing's Disease (Pituitary Pars Intermedia Dysfunction)

Pituitary pars intermedia dysfunction, once called (and better known as) Cushing's disease, sometimes affects older horses in their late twenties or into their thirties. Young horses are rarely affected. This disease is caused by excessive output of adrenal hormones from overactivity of the adrenal glands (found in front of each kidney), often in response to a pituitary enlargement.

Hormones are the signals that communicate between various glands and organs. In a normal horse, the hypothalmas at the base of the brain constantly reads the levels of certain hormones in the blood and then tells the marble-sized pituitary gland next to it to secrete more or less of certain types of hormones to keep things functioning smoothly. For example, adrenocorticotropic hormone (ACTH, secreted by the pituitary) tells the adrenal glands to release cortisol. If the pituitary gland is secreting too much ACTH, the adrenal glands release more cortisol, which is detrimental to the horse's body.

The adrenal glands secrete several hormones known as corticosteroids that affect the body's use of electrolytes and glucose. These hormones also inhibit the immune system, a condition that slows the healing of wounds and results in lowered resistance to infection and disease.

Symptoms

The affected horse has an excess of hormones that raise the level of glucose in the bloodstream. This, along with the effect of the hormone on the kidneys, results in increased urine output; therefore, most of these horses drink a lot and urinate frequently. They may also have a shaggy coat that doesn't shed, suffer from weight loss, experience chronic infection, or develop laminitis. They may become swaybacked and potbellied from loss of muscle tone. Diagnosis can be aided with tests for certain hormone levels.

Treatment

There is no cure for Cushing's disease; however, if the condition is diagnosed early, the horse can be treated with drugs to alleviate symptoms. The drug most often used is pergolide, originally used for treating Parkinson's disease in humans. It replaces the missing dopamine that is not being supplied by the hypothalamus to regulate the pituitary gland. The horse must be treated for the rest of his life. The condition is eventually fatal, but treatment can help buy the horse more time and better quality of life.

Hepatitis

A healthy liver breaks down and filters out poisons, but occasionally in the process of protecting the body, it becomes damaged or infected, resulting in hepatitis, or inflammation of the liver. Both infectious and noninfectious, the disease can be caused by a virus, chemical or bacterial toxins, or poisonous plants. Primary liver disease is generally the result of poisoning. Secondary disease of the liver occurs as part of a generalized disease process in the body.

Noninfectious Hepatitis

Toxic (noninfectious) hepatitis, usually caused by a poison in the body, can be acute or chronic. Acute hepatitis comes on suddenly; chances of survival are poor because the liver damage is so severe.

Symptoms

Chronic hepatitis occurs from toxins or poisons in smaller doses; it comes on gradually and may not be noticed until symptoms worsen. This can happen if the horse is eating poisonous plants or contaminated feed over time or has a bacterial infection that creates toxins, which, in turn, damage the liver. Early signs are dullness and lack of appetite. Pulse, temperature, and respiration are usually normal, but the horse may have abdominal pain that may be mistaken for colic and yellow mucous membranes.

Toxic substances are processed by the liver and excreted, but in a damaged liver, they build up in the bloodstream and affect the nervous system. Ultimately, the affected horse may stagger or drag his feet. His mental condition deteriorates. He may stand with feet wide apart and head drooping, or he may have muscle tremors. He may become so oblivious to his surroundings that he walks into trees or fences.

In some cases, the horse becomes violent and unmanageable, a danger to himself and anyone trying to handle him. He may have diarrhea, colic, or founder; he may also suffer from photosensitization (skin inflammation and sloughing in unpigmented areas) because his liver cannot filter out photosensitizing agents from plant material he has eaten.

Infectious Hepatitis

Serum hepatitis (or "Theiler's disease") causes massive liver destruction. The virus is so hardy that chemical disinfectants cannot kill it.

Symptoms

Onset of symptoms is sudden, and the horse is almost always violent. He becomes mentally deranged and can't be caught; he runs wildly, crashing into fences, walls, or other obstacles, and soon dies. Hepatitis is easily misdiagnosed as some type of neurological disease.

Serum hepatitis and rapid death of liver tissue are often the result of vaccination, occurring 30 to 90 days after the injection. A horse injected with vaccine made from equine tissue or derived from horses may occasionally be at risk from serum hepatitis. Cases have occurred up to 6 months after injections of encephalomyelitis serum, tetanus antitoxin, pregnant-mare serum, and anthrax antiserum. About 90 percent of all serum hepatitis cases are fatal, with the horse dying within 12 to 48 hours.

Because of the risk for serum hepatitis, many veterinarians do not advise use of antitoxins, which have traditionally been used to treat or prevent a number of other serious conditions, except as a last resort. Several antitoxins are derived from horse serum (a possible source of the virus); the most widely used is tetanus antitoxin, often given for immediate protection against tetanus following surgery, foaling, or a wound.

Treatment

Treatment with antibiotics, glucose, IV electrolyte solutions, and B-complex vitamins may be helpful in some cases of serum hepatitis.

Prevention

Because it can be spread from horse to horse by use of contaminated needles, you should always use sterile, disposable needles and syringes. Discard each needle and syringe after one use. Keep tetanus vaccinations current so you never need to use antitoxin. Be careful with chemicals that contaminate feed or water. Keep horses well fed so they won't be tempted to eat poisonous plants. Check for poisonous weeds in pastures, feed, and hay. (See chapter 11 for information on poisonous plants.) Keep horses healthy with good care and a conscientious vaccination program against possible diseases. Most cases of hepatitis can be prevented with good management and forethought.

TOXOID VS. ANTITOXIN

Antitoxin is not the same as toxoid. Toxoid is a modified substance produced in a laboratory, not made with horse serum. There's no danger of hepatitis with toxoid injection. Antitoxin, by contrast, is derived from blood serum and contains antibodies to counteract tetanus toxins. Although produced under rigid safety standards — testing donor horses regularly for hepatitis — antitoxin always carries the remote chance for error; it's possible for contaminated serum to reach veterinary shelves. There is never a need to use tetanus antitoxin if you keep a horse's toxoid vaccinations up to date.

Laminitis (Founder)

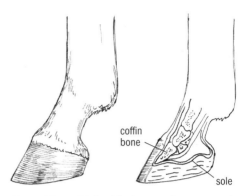

coffin bone

sole

Normal foot with healthy horn and proper angle.

Founder is a term for changes that take place in the feet as a result of laminitis — inflammation of the attachments of the hoof wall. Bones and tendons of the leg terminate inside the hoof in an area that serves as a cushion and attachment between the inner tissues and the outer horny shell. When a horse suffers from a disease or metabolic problem that upsets his body chemistry, many changes take place that sometimes affect his feet.

A condition affecting blood circulation may not cause permanent damage in other parts of the body, but in the feet, which are encased in solid walls, there can be terrible pain and damage to the sensitive laminae. The tiny capillaries feeding the sensitive laminae are actually affected before the horse shows pain. By the time the horse's feet become hot and tender, the damage is already done. The laminar tissues subsequently die, causing the hoof wall to separate from lack of support as the laminae give way and the coffin bone drops or tips downward at the front.

Pain makes it hard for the horse to walk; in severe cases, he reacts by lying down or rolling around. Lameness usually appears first in the front feet because they carry the most weight. Heat and pain occur in the hoof and around the coronary band. In acute cases of founder, there may be pain in all four feet, causing the horse great distress. He shivers, sweats, breathes rapidly and shallowly, and has a rapid pulse. He stands with all four feet bunched up under his body, back arched upward, and head hanging down. He doesn't want to move. If he does, his gait is stumbling and shuffling because of the pain. It is hard for him to get up and down, and he may lie flat on his side for long periods, sometimes hours at a time.

Several things can cause laminitis, including grain overload, lush pasture, concussion from fast speed on hard surfaces, increased weight on the feet, drastic changes such as drinking cold water when he is exhausted and hot and sweaty, infections, hypothyroidism, and complications (such as endotoxemia) from failure to shed the placenta after foaling.

Endotoxemia and Laminitis

Bacterial toxins in the bloodstream (which may occur with uterine infection from failure to shed the placenta after foaling, or when a horse overeats grain) are common causes of laminitis. When a horse eats too much grain, excessive carbohydrates in the hind-gut stimulate rapid multiplication of intestinal bacteria. The microbes that ferment the lactic acid proliferate, using up so much available oxygen that anaerobic bacteria (pathogens) begin to multiply rapidly. As they die, their endotoxins (poisonous substances in the bacteria) are released and seep through the intestinal wall into the bloodstream.

The Hooves Suffer First and Last

Once in the bloodstream, endotoxins damage the smallest blood vessels. Tiny arteries clamp shut to conserve blood pressure, but this deprives the tissues of nutrients and oxygen. The terminal ends of the hoof's arteries are especially vulnerable — the first victims of dying local tissue. As the horse's blood-clotting system malfunctions, small clots form in many small vessels. As blood supply to the hoof is impaired, oxygen starvation causes excruciating pain followed by congestion and swelling. The laminae begin to die.

Even after the original cause of the condition has been corrected, the horse's feet may become deformed if the laminae have become separated from the hoof wall and have allowed the sole to drop. The hoof wall spreads and develops rings and ridges, and the slope of the front of the hoof becomes concave. Sometimes, the separation of hoof wall is complete and the horny shell comes off. In this case, the horse may need special shoeing for the rest of his life.

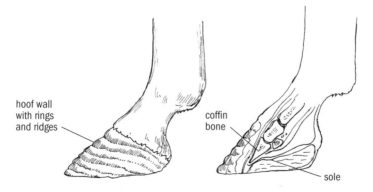

hoof wall with rings and ridges

coffin bone

sole

A foundered foot with rings and ridges and a dropped coffin bone. When a horse founders, the sensitive laminae, which bind the hoof wall to the coffin bone, lose their ability to hold the structures in place. With its anchor disintegrating, the coffin bone sinks downward, rotating toward — or even through — the sole of the foot.

Treatment and Prevention

If you suspect laminitis, call your veterinarian immediately. Prompt treatment can help relieve the condition before the feet suffer founder — the changes in the foot that may follow laminitis. The veterinarian will treat the primary condition (such as grain overload, uterine infection) and give medications to alleviate the circulatory problem and reduce pain and swelling in the feet.

Using and caring for a horse wisely is the best insurance against founder. Always feed a well-balanced ration and introduce new feeds gradually over several days; never overfeed or allow a horse to overeat on grain or lush pasture. Condition horses to their work gradually, never work them to the point of exhaustion, and always cool a horse out properly after a workout — gradually. Don't allow him to drink cold water until he is cooled out, and don't put him away hot. Keeping a horse fit and healthy and giving prompt attention to any problems such as illness or retained placenta will help prevent founder.

Navicular Syndrome

Navicular syndrome is a term that covers a multitude of problems within the horse's foot that were once all lumped together as navicular disease (problems within the navicular bone or bursa). This is a common cause of lameness in domestic horses but rarely affects the free-roaming equine. The practices of breeding certain horses with feet too small for body weight, using horses in inappropriate athletic activities, and keeping horses confined in unnatural conditions can lead to damage within the foot that may make the horse lame and lead to navicular syndrome. This is primarily a problem of front feet, because they are subjected to more weight and concussion than hind feet.

The small navicular bone lies deep inside the foot, above and behind the coffin bone, and serves as a pulley for the deep flexor tendon that glides along its underside and as a weight-bearing surface and shock absorber. It stabilizes the union of the coffin bone and short pastern bone, which are of dissimilar shape.

Anything that increases concussion to the foot, such as upright shoulders and pasterns, puts extra stress on the navicular bone and may lead to true navicular disease in which the bone itself (or its surrounding bursa) is injured. Jumping adds even more stress that squeezes the bone.

Often, the first step to diseased bone is poor circulation caused by clotting of tiny capillaries in the bone itself. Reduced blood flow weakens the structure; the bone begins to die. As it loses health and strength, it can't bear weight properly and the horse suffers pain and lameness. Poor circulation may result from increased pressure (concussion, jumping, or working immature horses too hard) or from inactivity such as confinement in a stall. Standing in a stall reduces circulation in the feet and puts constant pressure on joint cartilages that are dependent on motion for blood circulation.

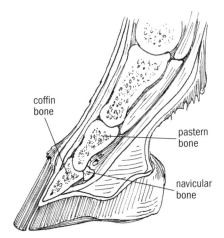

The navicular bone lies behind the coffin bone at the joint between the coffin bone and pastern bone.

Long toes and unbalanced feet also put more stress on the navicular bone, as does excessive body weight. A horse may have good foot and leg conformation and still have stress on that bone if he is kept in a stall with sloping sides, putting feet off level; the forces on the navicular bone as he stands on an upward slope are the same as if he had long toes and low heels. Navicular disease is often seen in feet that are too low at the heel, putting an extra squeeze on the bone between the deep flexor tendon and the lower end of the short pastern bone when the horse is standing. Pointing the foot forward relieves the pressure temporarily; a horse with a sore navicular bone often puts that foot forward.

Symptoms

Mild tenderfootedness is noticeable when the horse comes out of his stall. During early stages, the lameness diminishes with mild exercise, then gets worse if exercise is followed by confinement. Symptoms are usually not evident until the bone has already been seriously damaged. Lameness may

YOUNG AND SICK

Confinement and lack of exercise in a young growing horse, especially if he stands around on hard surfaces, can lead to navicular disease. Keeping youngsters in stalls and overfeeding them (causing them to become overweight and inactive) are also detrimental.

Overuse of young horses subjects their immature feet to excessive concussion. Navicular disease usually starts between ages 3 and 6, although some horses don't show symptoms until later.

appear in both front feet, but one
may be worse than the other.
Often, the first sign of lameness
is obvious when the horse is turning or walking on gravel or rocks.
He takes shorter, lighter steps,
trying to keep the weight off his
front feet. He gallops with a short,
choppy stride.

During early stages, the lameness may come and go but, eventually, it returns and gets worse.
The horse stumbles as he tries to
put the toe down first and not take
weight on the heel. Because of his
shortened stride, you might think

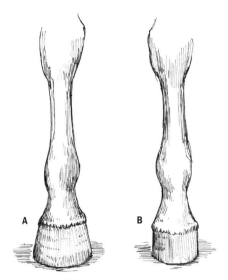

Normal foot (**A**) compared with a contracted
hoof (**B**) caused by navicular disease.

the lameness is in the shoulder instead of the foot. The horse may point one
front foot forward when standing still to relieve the pressure. He may also
shift his weight from one leg to the other. If the foot is unshod, it may be boxy,
with high contracted heels, a worn stubby toe, and a small frog that is well up
off the ground. If shod, the shoe's toe will be excessively worn.

Treatment and Prevention

The first step in treatment is proper diagnosis. Typical "navicular" foot pain
may be caused by injury to the deep digital flexor tendon within the hoof, one
of the supporting ligaments, or cartilage. Some of these injuries will heal with
rest and treatment. Until recently, there was no good way to actually tell what
was going on inside the hoof because nerve blocks, hoof testers, X-rays, and
ultrasound can sometimes be inconclusive. Now, a true picture of what's wrong
can be obtained using magnetic resonance imaging (MRI) and the horse can
then be treated appropriately — and you'll also know whether the horse has a
chance to recover from whatever damage has occurred. Some of these problems can be treated and some can't.

The best way to help a horse with actual navicular disease is to make
management changes to halt the deterioration of bone. Nothing will return
the damaged bone to its original shape and smooth surface, but changes in
stabling and shoeing and new medical treatments can help him. The horse
whose circulation and hoof balance is restored will be more comfortable and

---

HEALTH TIP FOR INACTIVE HORSES

Inactive horses have greatest damage to circulation in the feet. If a horse must be confined, give him a roomy box stall with a mound in the floor instead of a depression in the center. If he stands for hours on a floor that slopes downward to the center of the stall, his feet are in an unhealthy toes-up, heels-down position even if they are shod in perfect balance. The mound in the floor puts his feet at a better, less-damaging angle.

---

more sound. Treatment varies with the severity of the condition. Anti-inflammatory drugs and arthritis medication or injections into the navicular bursa or coffin bone/navicular bone joint can help halt joint degeneration if the disease involves the joint.

In a severe and chronic case, surgery is an option to block nerves so that the horse feels no pain. This treatment is not without risk or possible complication. Your veterinarian should assess the horse's condition and advise you on the best course of action. It is always better to prevent or halt navicular disease before it progresses.

Hoof care is important in preventing and controlling navicular disease and other injuries that can also cause foot pain. If shoes are left on too long, even a well-balanced foot will grow a long toe. For a horse with a serious problem, your farrier can use special shoes or wedge pads under the frog to give relief.

Muscles Tying Up (Exertional Rhabdomyolosis)

Muscle pain and cramping associated with exercise is fairly common in horses. The condition has had many names, including azoturia, Monday-morning disease, tying up, being corded, or being set fast. It involves painful cramping of the large muscles in the rump and, sometimes, the thigh and shoulders. Symptoms range from minor discomfort to collapse and death. The muscle cramping generally occurs during or after physical exercise such as a fast gallop or a long ride (or even running and bucking at liberty in deep mud or snow), but in some horses, there is a genetic defect in muscle metabolism and the cramping occurs even during mild exercise. It may also develop after stressful situations such as a fight with a farrier, frantic running during a thunderstorm, or a long trailer trip. Any muscle exertion beyond the horse's accustomed activity can trigger tying up.

There are two types of severe muscle cramping: chronic and sporadic.

Chronic tying up occurs early in a workout or soon after exercise begins and is caused by inherited muscle abnormalities. This category can be further broken down into two distinct types: polysaccharide storage myopathy (PSSM) in Quarter Horses, draft horses, and warmbloods (horses with heavy muscles) and recurrent exertional rhabdomyolysis (RER) in Thoroughbreds, Standardbreds, and Arabians.

Sporadic tying up can occasionally occur in any horse and occurs most often after many hours of steady work and muscle fatigue. The causes and treatment/prevention are different for each type.

Sporadic Exertional Rhabdomyolysis

In these cases, the horse merely did something out of the ordinary: He may have worked too hard for his fitness level, for example, or strained his muscles, creating soreness. Dehydration from a long day of hard work in hot weather may lead to inadequate circulation to exhausted muscles. Putting too much cold water over the large muscles when the horse has been working hard may also cause cramping. These horses usually are fine if they are rested, and this type of tying up can be prevented by not overworking the horse beyond his abilities. Treatment in some cases may involve fluids and electrolytes if the horse has been worked too long and hard.

PSSM (Polysaccharide Storage Myopathy)

This inherited defect is most common in heavily muscled horses. When draft horses were commonly used, this was called Monday-morning disease because it generally happened after a weekend — the draft horses were grained heavily and worked hard all week, rested on Sunday, and worked again on Monday.

Muscle biopsy research has now shown that horses with PSSM accumulate an abnormal amount of sugar in the muscles and generally tie up if not getting regular exercise. They develop this problem early in life and are very sensitive to insulin; they can't properly regulate energy metabolism in the muscles. This is because of a genetic mutation that probably occurred as long ago as the Middle Ages when people were breeding large, heavily muscled horses to carry knights in armor. These horses were the forerunners of today's draft breeds. The genetic defect has now been found in at least 17 breeds, including draft breeds and their derivatives, Quarter Horses, Paints, Morgans, Tennessee Walking Horses, Mustangs, and Haflingers. There is now a genetic test that can determine whether a horse has PSSM.

IS IT TYING UP OR COLIC?

Tying up is often mistaken for colic because the horse may paw and sweat, but if you try to walk him, as you might for colic, you will make his condition worse.

Symptoms

Symptoms develop 15 minutes to an hour after exercise begins. The attack is sudden. The horse comes to a stiff halt, begins to sweat, is reluctant to move, and may want to lie down. The rump muscles are tight, stiff, and sore — similar to sudden thigh or calf muscle cramping in your own leg when you sprint or do something you're not used to.

It may look like the horse has a urinary problem because of his stretched-out stance. If he can walk, his hind legs are stiff and dragging. The longer the cramp lasts, the more intense the pain because the muscle is oxygen deprived. The spasm squeezes the capillaries, hindering blood flow. There may be a peculiar odor to the horse's breath, urine, and sweat from wastes excreted through the lungs, kidneys, and sweat glands; his urine will be darker than normal.

In mild cases, the symptoms disappear within a few hours if the horse is given immediate and complete rest and not moved. In moderate cases, the horse is anxious, trembling, very stiff, and reluctant to move; these symptoms may last from 24 to 48 hours. If pain is extreme, the horse will go down.

Treatment

Prompt treatment usually relieves the muscle cramping. While waiting for the veterinarian, blanket the horse to keep him warm and relaxed, avoiding movement. Care for him where he is, or take him home in a trailer and keep him on his feet. The veterinarian may treat him with tranquilizers, pain relievers, and muscle relaxants to increase blood flow and relieve spasms in blood vessels. He or she may also give vitamin E and selenium to help the muscles return to normal quickly. If muscle cramps are relieved in an hour or less, chances are good the horse will recover swiftly.

Prevention

Horses with PSSM must be kept on a low-starch, high-fat diet. They should not be fed grain, sweet feeds, or molasses and should not be out on lush pasture. They need regular exercise, however, and should be kept in a large drylot with plenty of room, or a not-so-green pasture. These horses should be kept outside as much as possible rather than confined in a stall. Continual light exercise helps train the muscles to burn fat and also to access the excess glycogen to use as fuel.

Most horses with this problem are easy keepers and don't need grain. They should never be allowed to get fat. Today, there are several low-starch/high-fat/high-fiber feed products that are designed for horses with insulin-resistance problems and these products also work well for horses with PSSM.

Recurrent Exertional Rhabdomyolysis

Recurrent exertional rhabdomyolysis (RER) is another type of situation in which the horse ties up repeatedly, but it generally doesn't occur until the horse is quite fit (as in race training) and it's usually the nervous, young horses who are most severely affected. The tying-up episodes are often associated with excitement — young racehorses being galloped for exercise, fighting a rider who is trying to hold them back from full speed.

This is an inherited condition; certain family lines are more prone to this problem. The regulation of muscle contraction and relaxation is abnormal in these horses. Tying-up episodes can be minimized by careful management and allowing the horse regular exercise (turn out rather than confined in a stall). There is medication that can be fed daily to smooth out the erratic contraction of the muscles.

Diet management is helpful because many of these horses are more "hyper" and easily excited when fed a lot of grain. They tend to be very nervous horses and it's hard to keep weight on them; therefore, owners generally feed them a large grain ration. It's better to substitute fat such as vegetable oil or commercial high-fat supplements for part of the grain. This helps supply the energy they need without the extra grain. It also helps to keep the horse calm and reduce factors that may cause stress and excitement.

Other Diseases

There are other diseases of horses that you might experience; however, to cover all of them is beyond the scope of this book. Be alert to any unusual symptoms, and do not hesitate to contact your veterinarian if you think your horse has a problem. Some diseases occur only rarely, but the veterinarian may be able to help diagnose them and assist you with treatment.

10

Parasites

ALL ANIMALS HARBOR PARASITES, which are organisms that live on or within another animal. The small creatures feed on the host, sucking blood or absorbing nutrients. Responsible horsemen and horsewomen try to control these pests to keep them at a low level so they can't seriously damage the horse or interfere with his nutrition.

Internal Parasites

Most parasites are "host specific" (species specific); for example, horse parasites survive only in horses. Although no horse is free of internal parasites, horses have adapted over time to living with them — a young horse builds immunity to some worms after repeated infestations.

Most parasites need a host animal to complete their life cycle. An intestinal worm, for example, lays eggs that pass out of the body with the horse's manure. The eggs then hatch into larvae that migrate to grass and are eaten by a horse with his feed. Once inside the horse, the larvae migrate through body tissues and feed on the host, becoming adults in the intestine, only to lay eggs and begin the cycle again.

Immunity to Worms

The immune system not only fights bacterial and viral infections but also reacts to internal parasites because the invading worms release antigens into the bloodstream and body tissues. The antigens trigger the horse to produce immunity against the invader. Later, when more worms appear, the horse's immune system recognizes them and sets up a counterattack — reducing the females' egg-laying capacity and the males' fertility.

Young horses can build immunity to ascarids (roundworms) to keep larvae from maturing. Adult horses rarely carry large infestations of ascarids; these

---

PARASITES CAN BE DEADLY

Internal parasites were once a common cause of death in horses not regularly dewormed (for example, many colic cases were related to bloodworm infestation). A lifetime of carrying a heavy parasite load was a major factor in other causes of death: the horse lacked nutritional reserves and disease resistance because of the constant parasitic drain on his body. Today, however, we see less incidence of worm-related colic, because most people deworm their horses. And twice-a-year deworming with ivermectin or moxidectin often keeps bloodworms at bay.

---

damaging worms are primarily a problem of young horses who haven't yet developed immunity. Bloodworms are a different story; horses do not build an effective immunity against them. In earlier years, bloodworms were considered the most damaging parasites. But, with the advent of ivermectin use in the mid-1980s, bloodworms could be readily controlled and are no longer such a threat. In fact, they have all but disappeared from some farms.

How Horses Become Infested

Horses in small pastures are continuously exposed to their own droppings, grazing the same places over and over. They become heavily infested unless manure is picked up daily or a regular deworming program keeps the egg-laying adult parasites to a minimum.

Parasites thrive in a warm, moist environment. In springtime and early summer, the worm eggs passed in the manure hatch, and the larvae swarm onto pasture plants, waiting to be eaten. Sunlight, heat, and severe cold tend to kill larvae, but the eggs in manure are resistant to extremes in temperature.

Any previously grazed pasture is a continual source of infestation, but proper pasture management and a good parasite control program can reduce the problem. For example, harrowing a pasture during hot weather breaks up the manure piles and exposes the worm eggs, which are susceptible to sunlight, heat, and drying. In addition, alternating cattle and horses on a pasture can break the parasite cycle. The horse-worm larvae eaten by cattle will not complete their life cycle and, similarly, the cattle parasites will not survive in the horse.

Horses grazing irrigated or wet pastures where manure is not picked up are highly likely to be reinfested with worms. The most worm-free environments are dry corrals or paddocks where manure is picked up regularly and stalls are kept very clean.

Control of Worms

Control is aimed at preventing contamination of the horse's environment with worm eggs and larvae by means of pasture management, manure removal, proper feeding and watering, and deworming of horses at appropriate intervals to prevent reproduction of parasites. Newer dewormers, such as ivermectin and moxidectin (Quest), help to control worms and prevent damage to the individual horse by killing botfly larvae before they get to the stomach, bloodworm larvae before they damage the arteries, or small strongyles while still encysted in the intestinal lining. Before the introduction of ivermectin, worms were vulnerable to drugs only during their final stage in the digestive tract. This prevented further contamination of paddock and pasture by reducing the egg numbers in manure, but the larvae had already damaged the horse during migrations through the body and arteries.

There are many good dewormers that can be used in a control program. Knowing the parasites' life cycles and targeting them when most vulnerable will keep them under control. The most crucial time to diligently control roundworms, small strongyles, and bloodworms is spring and early summer, because larvae are most active in warm, moist weather. Deworming horses just before the parasites' "egg-laying spree" and the spring rise of larvae onto forage

---

IVERMECTIN

Since its introduction in the early 1980s, this drug (available in several different brands) has become widely used because of its effectiveness against all internal parasites except tapeworms, in larval as well as adult stages. It also kills blood-sucking skin parasites such as ticks, lice, and mites. Unlike other drugs that poison parasites, ivermectin works by paralyzing those having gamma-aminobutyric acid (GABA) in their nervous systems. Ivermectin stimulates release of this acid; the parasite is unable to move or eat, and it dies. The drug has no effect on tapeworms because they have no GABA. Ivermectin is not harmful to horses or humans. For small strongyles encysted and hibernating in the lining of the large intestine, however, moxidectin works a little better than ivermectin for killing this dormant phase of the parasite.

plants will halt a worm population explosion. Many veterinarians now recommend a simple twice-a-year deworming with either ivermectin or moxidectin — once in spring and again in late fall.

Botfly Larvae (Bots)

Most internal parasites of horses are worms, but the horse is also subject to infestations of fly larvae (grubs), especially the botfly, *Gasterophilus*. The larvae of this fly irritate the stomach and occasionally perforate the stomach wall. The adult fly is brown and hairy, the size of a small bee.

The botfly lays eggs on the horse's hairs, one to a hair, where they can be seen as yellow specks. One type of botfly lays its eggs on the front legs and flanks; the eggs hatch into tiny larvae that are transported to the mouth when the horse licks or nuzzles his legs. Other types of botfly lay eggs around the jaw or cheek, and these migrate directly into the mouth after hatching.

Once inside the mouth, the botfly larvae burrow into the gums. After about a month they emerge and are swallowed down to the stomach, where they attach, feed, and grow for 8 to 12 months. They pass out in the horse's manure in spring and early summer, migrate into the ground, and pupate, emerging as adult flies after 3 to 5 weeks to begin the cycle again.

Botflies in large numbers cause *unthriftiness* (thin body condition and poor hair coat), mild colic, and lack of appetite. Adult flies irritate horses; when attacked by egg-laying flies, they run frantically or strike out with a front foot. Horses can be dangerous when botflies are pestering them.

LIFE CYCLE OF THE BOTFLY

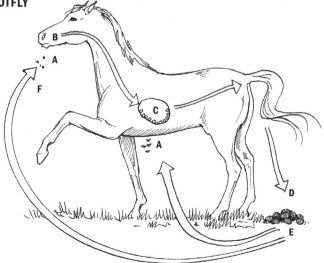

Flies lay eggs (**A**). Bots migrate to mouth and burrow into gums (**B**), pass to stomach (**C**), pass out with droppings (**D**), migrate into the ground to pupate (**E**), and become flies (**F**).

<table>
<tr><td>

ELIMINATING BOTS

Use of a good boticide such as ivermectin, combined with diligent egg removal from the hair coat, will eliminate a bot problem.

</td></tr>
</table>

Control

The crucial time for control is when botflies are most active: a 2- to 3-month late-summer season in northern climates or a year-round season in southern climates. Give a boticide 1 month after the first botflies appear and every 2 months thereafter, as long as botfly season lasts. Do a follow-up treatment late in the season after hard frosts have killed all flies to make sure that all the larvae within the horse are killed before winter. In northern climates, a boticide can be used once or twice a year in late fall and early spring (the latter to kill any botflies the fall deworming might have missed) to ensure that the horse will be free of bots.

Roundworms (Ascarids)

The ascarid, *Parascaris equorum*, is the largest intestinal parasite of horses. The white worms resemble an earthworm in shape and can grow up to 12 inches (30.5 cm) long. Adult worms live in the intestine, where a single female may lay more than a quarter of a million eggs per day. The eggs pass out with manure and, after 2 or 3 weeks, become infective if eaten by the horse, with the larvae migrating through the body tissues. They can be damaging to foals and young horses. Most pastures that have been previously used by horses, especially young horses, are heavily infested with roundworm eggs. These remain viable for several years under the right conditions — mild weather and lack of pasture management, such as harrowing.

Ascarids are most common in foals. By the time a horse is 2 years old, his immune system can hinder development of larvae. The ascarid robs the foal of nutrients and may cause mechanical blockage of the small intestine — and subsequent colic or death — if worms become too numerous. The most common and extensive damage, however, occurs during larval migration through the lungs. Ascarids are the main cause of lung damage and cough in foals under 6 months of age.

After ascarid eggs are eaten, newly hatched larvae penetrate the intestinal wall and enter the bloodstream. Within 48 hours, they reach the liver, where they remain for a week; then they travel through the blood to the lung, where they pass through small blood vessels, through tiny air sacs, and into the bronchial tubes. Respiratory problems develop 3 to 4 weeks after the foal eats ascarid eggs; symptoms include nasal discharge and, sometimes, a cough. The lung damage in a heavily infested foal can be permanent.

After spending time in the lungs and bronchial tubes, the larvae are coughed up into the mouth and are swallowed, returning to the digestive tract. There they continue to mature and become egg-laying adults. The interval between ingestion of an egg and development of an egg-laying adult is 2 to 2.5 months; therefore, fecal examinations for parasite eggs are usually negative until the foal is almost 3 months old.

Control

Don't wait long before deworming the foal; ascarids have already damaged his lungs and other body tissues by the time they return to the intestine. Some veterinarians recommend deworming the mare with ivermectin right after foaling, and making sure her udder, flanks, and belly are free of manure. Pyrantel tartrate (Strongid C) given daily in grain is effective against ascarids; the foal can be started on it as soon as he eats grain, and mares can be started on it a few weeks before foaling.

LIFE CYCLE OF THE ASCARID

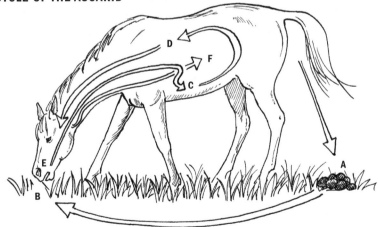

Eggs are passed in manure (**A**); horse eats eggs with grass or hay (**B**); eggs hatch and penetrate intestine, and larvae travel through bloodstream and liver (**C**); larvae enter lungs (**D**); larvae are coughed up and reswallowed (**E**); larvae return to digestive tract to mature and lay eggs (**F**).

DEWORM FOALS ROUTINELY

Foals should be routinely dewormed for ascarids even if they don't show symptoms. Most veterinarians recommend you first deworm when the foal is about 2 months old. By the time symptoms appear, significant damage has occurred. A heavy infestation may stunt the foal's growth and make him rough-coated and potbellied; irritation from the worms may also give him diarrhea or colic. An extensive infestation can be fatal, as can deworming when the foal has a heavy load of worms — the bulk of dead and dying worms causes intestinal impaction and blockage. When deworming a foal with a heavy load of ascarids, it's safest to use a drug with a slow "kill rate," such as ivermectin, to prevent sudden blockage.

Bloodworms (Large Strongyles)

Large strongyles, called "bloodworms," "red worms," or "palisade worms," are active blood suckers that cause debility and anemia. Three species affect horses, varying in length from less than 0.5 inch (1 cm) to about 2 inches (5 cm). The most serious problems are caused by the smallest, *Strongylus vulgaris*. Some of these lodge in the walls of the blood vessels that supply the intestine and in the large arteries leading away from the heart.

Eggs are passed in the feces of the host horse and hatch to produce larvae that are resistant to drying and low temperatures, living about 3 months or

**LIFE CYCLE OF
THE BLOODWORM**

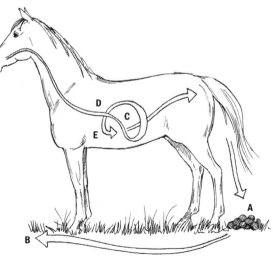

Eggs are passed in manure (**A**); larvae migrate onto forage plants (**B**); when eaten, larvae penetrate the intestine, enter the bloodstream, and live in major arteries (**C**); larvae travel to the liver (**D**); larvae return to intestine and lay eggs (**E**).

even a year or more. In warm, wet weather, they migrate onto forage plants. When swallowed by a horse, they penetrate the intestine and begin extensive migrations throughout the body before coming back to the intestine to attach and lay eggs.

Damage to Blood Vessels

Too many bloodworms in an artery may block it or cause it to rupture. They cause inflammation, scarring, and roughening of the inside wall, as well as weak spots. This damage can lead to clots or aneurysms. If a clot breaks loose (becoming an embolus) and travels down the vessel, it eventually comes to a place too narrow to pass through and promptly blocks it, shutting off blood flow to the tissues supplied by that artery. Aneurysms (dilated sacs) bulge out from the vessel wall in a weak spot. If an aneurysm gets large or the blood vessel is weak, it may suddenly rupture, causing sudden death of the horse from internal hemorrhage.

Although any artery can be damaged by bloodworm larvae, *S. vulgaris* most often damages the mesenteric arteries that supply the intestine. When this happens, a section of intestine suffers decreased blood supply, resulting in pain and colic. Severe blockage leads to death of a section of intestine. Another common site for blockage is where the aorta (the large artery leaving the heart) branches into arteries supplying the hind legs. If blockage occurs here, the horse becomes lame in one or both legs and the affected limbs feel cold. If blockage is permanent, the leg tissue dies from lack of blood, killing the horse.

Control

Bloodworms can be readily controlled by periodic deworming with ivermectin. This dewormer kills the immature forms wherever they are in the body,

SYMPTOMS OF BLOODWORM INFESTATION

Mature bloodworms fastened to the intestinal wall may cause diarrhea, anemia, weakness, and unthriftiness. Other symptoms of infestation are unexplained bouts of colic and lameness in the hind legs. Bloodworms deprive the horse of nutrients and blood, causing young individuals to grow slowly and working individuals to be inefficient.

as well as the egg-laying adults in the intestine. Veterinarians once recommended deworming every 2 to 3 months, with frequency dependant on rate of reinfestation. Today, however, on farms where diligent deworming with ivermectin has been used, bloodworms are no longer much of a problem. Twice-a-year deworming is generally adequate.

Small Strongyles

These worms, which cause debility and anemia, are blood suckers about 0.25 to 1 inch (0.5–2.5 cm) long. They live in the intestine (mainly the cecum), ulcerating the mucous lining. Heavy infestations may leave little normal area in the intestinal lining. Ulcers made by these small worms are deeper than those made by large strongyles and occasionally cause fatal hemorrhage. All horses are infested to some degree with small strongyles, and today, in the relative absence of bloodworms, the small strongyles have taken their place as the most damaging of all internal parasites.

The seasonal mass emergence of small strongyle larvae from the intestinal wall causes the most problems. Affected horses show a subtle decrease in performance; are slow to shed; and have rapid weight loss, diarrhea, mild colic, or severe inflammation of the intestine (enteritis), which can be fatal.

Control

The presence of immature worms embedded in the gut wall is hard to detect because they're not yet laying eggs (which show in fecal counts) and are beyond the reach of many dewormer drugs except moxidectin. Seasonal emergence from the intestinal wall takes place in late winter and early spring in northern climates, and in fall and winter in the South. For best control, time your deworming program to coincide with certain stages of the life cycle,

---

YOUNG HORSE HEALTH TIP

Weanlings and yearlings need frequent and continual deworming because they are more vulnerable and adversely affected by parasites than are older horses; the youngsters have not yet developed much immune resistance to worms. Some adult horses, however, also have poor immunity to worms and carry heavier infestation than other horses on the farm. Fecal egg counts can help you determine which individuals need deworming more frequently than others.

hitting them when most vulnerable. Most veterinarians recommend targeting small strongyles diligently from April to August if you live in the northern part of the country where winters are cold, and as early as February or March in southern climates.

Ivermectin is effective against small strongyles, but not encysted ones, if given in three treatments at 8-week intervals. This kills bloodworms at the same time, as well as botfly larvae that might have been missed in the previous fall deworming. The newer dewormer moxidectin is effective against encysted small strongyles in the intestinal wall, and one treatment is sufficient. Most veterinarians now recommend using it instead of ivermectin because ivermectin gets only the adult worms and any larvae that are active, and not the inactive larvae hiding in the intestinal wall.

Pyrantel pamoate, oxibendazole, and dichlorvos are effective against small strongyle larvae only after they emerge from the gut lining. Fenbendazole is sometimes used; if given in a high dose and repeated over several days' time, it kills the adults with the first dose, signaling to the immature worms to come up and take up the space vacated by the eliminated adults. The repeated doses then kill the younger worms as well; however, many types of small strongyles have now developed resistance to these commonly used dewormers. Because worm populations are constantly adapting and changing, a parasite control program that worked a couple of decades ago may not be effective today. As a result, the conscientious horseman must work with his or her well-informed veterinarian to change, adapt, and update the deworming strategy as needed.

Pinworms

Pinworms are bothersome to horses but not as damaging as roundworms or strongyles. The parasite, *Oxyuris equi*, most commonly affects horses confined in small areas; it is not much problem in big pastures with lots of room. Pinworms inhabit the large intestine and cause intense irritation and subsequent disfigurement as the horse rubs his tail trying to relieve the itching. Many infested horses break off most of their tail hairs.

The irritation is caused by worms depositing eggs on the horse's anus and in the rectum. In some cases, the parasites cause inflammation of the cecum and large intestine. Sometimes a secondary infection affects the raw sores created by constant rubbing. Irritation may cause weight loss because the horse spends more time rubbing the itch than eating.

The mature pinworm is gray, white, or yellow. The female grows up to 6 inches long (15.2 cm) and has a threadlike tail. Mature females crawl out of the horse's large intestine and lay eggs by the anus in yellow clusters. The eggs hatch in about 3 days. The horse becomes reinfested by biting at areas on the body where eggs were laid or by eating feed contaminated with eggs and larvae from manure.

Control

A horse with pinworms can be treated with any good deworming drug. You should also thoroughly clean his stable area or pen because eggs may stick to stall walls, floors, and bedding.

Tapeworms

Horses in many regions often become host to tapeworms, whose life cycle depends on orbatid mites, the intermediate host. Horses infested with tapeworms spread eggs in their manure. In areas where winters are mild and temperatures are rarely below freezing, the eggs hatch nearly year round (because warmth and moisture promote hatching) and the host mites are active in pastures all year. The mites take in the tiny immature tapeworms, which undergo several changes inside the mites. Any horse who eats grass infested with the tapeworm-bearing mites becomes infested. The young worms continue to develop inside the horse.

Tapeworms cause problems if large numbers collect at the end of the small intestine, leading to colic from partial blockage of the cecum. Tapeworms also rob the horse of nutrients, just as other internal parasites do.

Control

Tapeworms are resistant to many dewormers and are not affected by ivermectin, but pyrantel pamoate kills them if given at two to four times the dosage required for other worms. Today, however, most horsewomen use the combination dewormers containing praziquantel — ivermectin or moxidectin with praziquantel added (such as Equimax, Zimectrin Gold, and Quest Plus).

Lungworms

Lungworms may be a problem if horses are pastured with donkeys or burros. Donkeys are natural hosts for this worm, *Dictyocaulus arnfieldi*, but can pass it to horses. Symptoms of lungworms in horses are coughing (especially during exercise), occasional difficulty in breathing, and lung irritation. Some heavily infested horses have no symptoms. Donkeys rarely show symptoms because they are usually not asked to exert enough to trigger coughing.

Control

If a horse develops a cough with no other symptoms and there is a donkey on your place, suspect lungworms. Deworm every equine, including the donkey, with ivermectin. If all equines are kept dewormed on a regular basis and ivermectin is used several times during the year, lungworms won't be a problem.

Warbles

Occasionally, cattle grubs develop in horses instead of cattle, but they rarely do much damage other than causing irritation under a saddle. The late larval stages are sometimes found in the backs of horses, creating cysts (warbles). The grub comes out when mature and ready to become a fly. Fully developed larvae are about 0.33 inch (8.5 mm) thick and 1.25 inches (3 cm) long. They perforate the skin and fall to the ground to pupate before becoming flies. Adult flies are active during the heat of summer, pestering cattle, and occasionally laying eggs on the hairs of horses. When the eggs hatch, the larvae burrow through the skin and migrate to the horse's back, coming to rest under the skin, where they make a breathing hole to the outside and continue to grow.

Control

Warbles can be eliminated if nearby cattle are treated for grubs every fall and if horses are routinely dewormed with ivermectin; this drug will kill the immature stages before they arrive at the horse's back.

The Battle against Internal Parasites

A regular deworming schedule and diligent pasture management will keep a horse relatively free of internal parasites. Usually, only 1 to 5 percent of the worm population is actually inside the horse; the majority of eggs and larvae are in the pasture. In small areas where horses are confined, cleaning up manure at least twice a week greatly reduces the chances of reinfestation.

Deworming Drugs

Some dewormers are used in purge doses on a periodic schedule to kill adult worms in the gut but do not kill the larval stages migrating through the tissues; after a time, the worms become resistant to most of these drugs. Ivermectin seems to be an exception because it affects not only mature worms in the gut but also the immature migratory larvae. It paralyzes rather than kills, but the worms eventually die because they cannot move or eat. Because of this mode of action, however, ivermectins are ineffective if you underdose the horse.

To prevent worm resistance, make sure drugs are given in adequate dosage. This is one reason it's better to give paste or tube application of purge doses rather than putting the dewormer in feed. Unless the horse eats the medication readily, he may get only a partial dose, which does more harm than good: surviv-

---

WORM RESISTANCE TO DRUGS

Some drugs that were traditionally effective against bloodworms and roundworms create worm resistance if used repeatedly, especially if given in doses that are not quite adequate to kill all adult worms. Because those that survive are the most resistant to the drug, their offspring are not particularly susceptible to it. The benzimidazole family of dewormers is a very safe and effective group of drugs, but it has the disadvantage of promoting worm resistance after being used repeatedly for several years. These drugs include cambendazole, oxibendazole, fenbendazole, oxfendazole, mebendazole, and thiabendazole. This family of drugs is not effective against botflies.

On farms where benzimidazoles and febantel have been used, large and small strongyles may have developed resistance. Horses may carry a heavy load of parasites despite regular deworming. There can also be a problem if horses with resistant worms have been added to a herd. When a worm population becomes resistant to one benzimidazole drug, it is also resistant to other drugs in that class. If there's a chance that worm populations on your place are resistant to these dewormers, change to ivermectin, pyrantel pamoate, or moxidectin (which is related to ivermectin but gives longer lasting protection).

Many veterinarians no longer recommend the old deworming schedules we once followed meticulously (since frequent deworming of all horses on the farm invites resistance). They suggest we concentrate more on strategic deworming of the individuals who need it. For example, all foals, weanlings and yearlings probably need regular deworming, but for adult horses, selective deworming may be more practical — and not as apt to aid the development of resistant worms as much as when you continually deworm the whole herd.

ing worms develop resistance to the drug. Most dewormers are safe for horses even in fairly large overdoses, so it's better to err on the side of overdose rather than to underdose when estimating dosage. The one exception to this rule is moxidectin; be careful never to overdose with this drug.

To combat drug resistance, rotate classes of drugs. If the benzimidazole drugs are used alternatively with an unrelated drug such as febantel, pyrantel pamoate, or ivermectin, worm resistance won't become a problem so quickly.

Using Paste Dewormers

Many horses are fussy about eating purge doses of dewormers in feed. Pastes have made deworming much simpler but must be given correctly; otherwise, the horse may spit out some or have a reaction to the dewormer. Always deposit the paste far back in the mouth, on the back of the tongue, so the horse can't spit it out.

A few horses are sensitive to the ingredients in dewormers. A local reaction in the mouth can occur in certain individuals if paste ends up under the tongue, between the tongue and cheek, or in some area other than the back of the tongue. Prolonged contact with mouth tissues sometimes causes problems, including excessive salivation (drooling) and swelling of the tongue and lips.

If a horse protests, tossing his head or trying to avoid the paste, train him to be more tolerant by giving him pleasant-tasting treats for a while until he looks forward to having a syringe put into his mouth. Use an empty, well-washed deworming syringe and give him applesauce or something else he likes.

HOW TO GIVE A PASTE DEWORMER

1. Stand beside the horse — not in front — and hold the halter so he can't toss his head or move around.
2. Put the syringe between the lips at the corner of the mouth, aimed far up toward the back of the tongue.
3. In one quick motion, insert the syringe far back into the mouth and depress the plunger, depositing the paste before the horse can toss his head. If he moves, don't depress the plunger; start over.
4. As soon as you've squirted in the paste, raise his head and hold it high until you're sure he has swallowed. You can tell when he swallows by the movement of his throat.

When it's time to deworm him, first give him the treat. Molasses and water works well for this because it's liquid and won't stay in the mouth to interfere with the dewormer, which must be administered into an empty mouth so the horse can't spit it out. Then slip him the dewormer; afterward, reward him for his good behavior with another syringe full of the treat. With this type of program, very few problem horses continue to resist deworming.

After squirting in the dewormer, hold the horse's head up until he swallows so he can't spit out the dewormer.

Daily Deworming

One deworming option is pyrantel tartrate given daily in the horse's grain to keep him continually dewormed. It is available in a pelleted alfalfa–molasses feed additive. The strategy of this deworming program includes first giving the horse a purge dose of ivermectin to clean out all major worms (larval stages and all), then keeping him on a daily dose of pyrantel tartrate to keep him free of the parasites. The drug works well for broodmares and young horses, but some athletic horses — primarily endurance horses — have trouble with occasional instances of colic or anemia after administration. In these cases, something else should be used. Because pyrantel tartrate is not effective against botflies, seasonal treatment with ivermectin is still necessary.

PLAN A GOOD DEWORMING PROGRAM

Work with your veterinarian to plan a deworming program that will be effective for your horses in your climate and situation. Remember that young horses need more diligent protection than older ones. Warm climates present year-round problems with many parasites that are only a seasonal problem in colder climates. Your veterinarian can advise you on new drugs being developed. The battle against parasites is constant, but an understanding of their life cycles and the weapons provided by modern research can help you keep these pests from becoming a health hazard.

External Parasites

There are several parasites that feed on the horse externally, causing irritation, discomfort, weight loss, and, at times, disease. Flies and mosquitoes are the most common during warmer months. Lice are a problem during colder months. Ticks are an occasional problem in some locations in spring and early summer. Also, some types of mites are more active in winter than in summer.

Flies

Flies cause annoyance and blood loss, and they spread disease. Their bites can result in skin allergies and hypersensitivity reactions. The annoyance may disrupt grazing as horses run frantically to get away from biting flies or seek shady areas in which to swish and stomp. An understanding of flies' life cycles and habits can help you control them. Flies go through four stages: egg, larva (maggot), pupa, and adult. The adult fly is the most troublesome, feeding relentlessly on the horse. The female fly must ingest blood before she can produce eggs, hence the biting.

Stable Fly

This fly is similar in size and appearance to the house-fly. It has a long, slender proboscis that sticks up in front of its head. The stable fly rests on vertical surfaces such as fences, walls, trees, or structures near horses, and lands on horses only long enough to get a blood meal. Usually feeding on the lower legs, flank, and belly in early morning or late evening, stable flies cause great annoyance. Their bites are quite painful,

Stable fly

making the horse not only stomp and kick to get rid of them, but also lick at the bite wounds. The bites often bleed freely. In large numbers, these flies can cause considerable blood loss.

The stable fly lays eggs in rotting hay and straw and in horse manure. Several generations develop during summer. In areas with large numbers of flies, there may be more than 25 flies per horse. This doesn't seem serious until you realize that the number observed on a horse at any given time may be only 2 to 3 percent of the total feeding on the horse that day. Large numbers cause anemia and loss of grazing time. Stable flies spread diseases, including anthrax and EIA (see chapter 8). They also act as an intermediate host for *Habronema microstoma*, a stomach worm that causes habronemiasis (summer

TABANIDS CARRY DISEASE

Tabanids are often mechanical vectors, carrying bloodborne disease from one horse to another because of their habit of feeding on one horse and then immediately attacking another. EIA is often spread this way, and, at times, so is encephalomyelitis.

sores). Stable flies are mechanical vectors (spreaders) for strangles and other diseases as well, physically carrying the germs from one horse to another.

Control. Remove rotting organic matter that the flies use as breeding sites. Clean out manure and soiled bedding daily from stalls, along with piles of grass clippings or wet hay around the barnyard. Keep flies from multiplying in compost piles by spraying with insecticide or larvicide or by covering with black plastic. Use repellents to keep the flies away from horses. Spraying barns, stables, and foliage where the flies rest also reduces their numbers.

Horseflies and Deer Flies

These flies belong to the Tabanidae family, the most annoying group of biting flies that attack horses. The flies may be black, brown, yellow, or gray, varying in size from 0.38 inch (1.0 cm) to more than 1 inch (2.5 cm) long.

Horsefly

When tabanids slash the skin, blood flows freely from the bite; the fly laps up the blood with a lobe at the tip of its proboscis. Only the female takes a blood meal. Males feed on vegetable sap or the juices of soft-bodied insects.

Tabanids are numerous in areas of permanently wet ground. Females lay eggs on plants growing near water or on objects projecting over water or marshes. The eggs hatch within 5 to 7 days. The larvae drop into the mud, where they burrow in and feed on organic matter or the juices of other insect larvae or earthworms. They remain in the larval stage during summer, fall, and winter and then migrate in spring to drier areas to pupate for 2 to 3 weeks, after which the adult flies emerge. Soon after emergence, the female fly needs a blood meal. Some horses have local reactions to the bites, developing lumps in the skin. Large numbers of flies may leave the hair crusted with blood and the bites dripping blood. Horseflies drink more than their own weight in blood at each feeding.

Control. Control of horseflies and deer flies is difficult unless swampy areas are drained or horses kept away from them when flies are most active. Some repellents are helpful. Sprays or wipe-ons may protect horses for 1 to 2 days. A shed (situated as far as possible from marshes) that horses can go into during the warmer part of the day when flies are active can help, as most tabanids prefer to be in the sunlight. These flies are most active on hot, sunny days and less active in cool or overcast weather.

Horn Flies

The horn fly is a cattle parasite that can also be a nuisance to horses living nearby. These blood suckers practically live on the host animal, taking 20 to 30 blood meals per day. Constant irritation from the bites may cause skin problems such as abdominal midline dermatitis.

Horn fly

Horn-fly bites often leave the skin raw and crusted. These flies generally do not transmit disease but cause skin irritation and interfere with grazing. They sometimes crawl into the ears and cause severe annoyance. Heavy infestations can lead to weight loss or death in cattle, but extensive blood loss is not common in horses because they attract fewer horn flies.

Adult horn flies stay on the host most of the time, leaving only to lay eggs in fresh cattle manure. The eggs hatch within 24 hours and the larvae burrow into manure, completing development in 4 to 5 days. Then, they crawl to a drier part of the manure or into the ground to pupate. Adults emerge and seek a host to drink blood, then lay eggs. In cold climates, they spend winter as

HORN FLIES ON CATTLE AND HORSES

Adult horn flies are half the size of a housefly and gather on withers, back, neck, shoulders, underline, and around the eyes of the host. When disturbed by a swish of the tail or toss of the head, they swarm up, only to immediately settle back down onto the animal. Cattle may have hundreds at once. Horses are much more sensitive; even as few as 20 horn flies may seriously annoy a horse.

pupae. Although the mature flies can live equally well on horses or cattle, the eggs are generally laid in cattle manure.

Control. Apply insecticides directly to the host animal because the flies spend almost all their time on the host. Some people fasten cattle insecticide ear tags to horses' manes or halters, but this is not a good idea because the insecticide dose in cattle tags is too strong for horses and is not recommended. There are some effective repellents and insecticides that can be applied to the horse's legs, belly, and flanks to get rid of horn flies. Check labels or ask your veterinarian.

Black Flies, Buffalo Gnats, Sand Flies

These small (0.19 inch; 4.8 mm) gray or black flies are often present in large numbers after periods of flooding or in areas with many streams. Larval stages are passed in flowing streams. The flies congregate in swarms and attack all animals, biting the legs, belly, and head. Irritation is so great that many animals stampede or mill about nervously, injuring one another or trampling young ones. Hypersensitive reactions to the bites can be a serious problem.

Black flies cause annoyance and itching. They feed inside the ears; on the chest, udder, and scrotum; inside the thighs; and under the belly. Their bites cause swelling, oozing blood, and bloody crusts. Swarms of these flies can travel 3 to 5 miles (5–8 km) to seek a host.

Control. Put horses in a barn or shed on mornings and evenings when flies are most active. Apply fly repellents and wipe-on products frequently, two to three times a day if necessary. Try ear nets. Apply petroleum jelly to protect the inside of the ears; the flies will not bite through it.

Midges

Tiny biting midges (also called "no-see-ums") sometimes attack in swarms. Thousands may feed on the horse at night. Their bites cause irritation and hypersensitive reactions, and the horse may damage his skin by constantly rubbing and scratching. The itching dermatitis (inflammation of the skin) causes the horse to rub his mane, tail, and withers; his face, chest, and belly may also be affected.

Control. Put horses indoors before dusk and keep them in until after dawn. If stabling is not possible, try frequent use of repellent sprays or wipe-ons in the late afternoon. Be sure that a horse showing an allergic reaction to bites, such as hives or swelling, is treated by a veterinarian.

Face Flies

Face flies are similar to houseflies in size and color. They feed on secretions from the eyes and nostrils — and on blood from the bites of other flies. They stay on the horse only a short time; the rest of the day, they sit on vegetation, fence posts, and other nearby structures. They lay their eggs in cattle manure.

Control. Apply insecticide to the horse's head and neck. Also try fly masks or face screens.

Preventing Flies

You can help your horses in their fight against flies through diligent management and careful use of insecticides. Provide shelter during peak fly season. Install screens on stalls and coat the screens with insecticide. Run a fan in the stall area. Attach fringe to the halter or bridle browband to dislodge flies around the head when the horse shakes his head. Use mesh fly masks to keep flies from landing around the eyes. Obtain summer flysheets (open cool-weave blankets) to keep flies from biting the horse's body.

Good manure management and control of wet areas can limit fly breeding. Remove manure and wasted feed daily from stalls and pens; spread it thinly for quick drying or compost it in a covered pile. Make sure there is good drainage in the barnyard and no moist areas or leaky faucets or waterers.

Insecticides and repellents can be purchased in several forms. Be sure to use only preparations meant for horses because other types may be harmful. Always follow label directions. Be aware that if a horse sweats much after an insecticide or repellent has been applied, the effects won't last long and you'll need to apply more.

Topical spray repellents usually contain a small amount of insecticide and some sunscreen, coat conditioner, or other ingredients to help the repellent

BENEFICIAL PREDATOR WASPS

One way to break flies' life cycle is to introduce predator wasps. These tiny nocturnal wasps lay eggs in the pupae of houseflies, stable flies, and some other flies that lay eggs in manure and rotting bedding. The wasps use the fly pupae as food, eating them before they can fully develop. The tiny wasps are harmless to people and animals. If you use this biological form of fly control, be careful when using insecticides because they also kill the beneficial wasps.

VINEGAR: A NATURAL FLY REPELLENT

Apple cider vinegar can be added to a horse's feed or water as a natural fly repellent. It can be mixed with the daily grain, using ⅛ to ¼ cup (30–59 mL). The vinegar seems to change the horse's smell; some of the acid quality may be excreted through the skin pores. The flies still land on the horse but may not bite.

To start a horse on vinegar, put a few drops on the grain at first, then gradually increase the amount. Use only pure apple cider vinegar; if a horse doesn't like the flavor, add a little molasses to disguise the vinegar taste. Most horses don't mind it. Gradually increase the amount to the point at which it begins to work, usually about ½ to 1 cup (118–237 mL). The minimum workable amount must be determined for each horse, and it may not work for some horses.

Horseflies, mosquitoes, and even botflies dislike the smell or taste of vinegar. It can be used externally — rubbed over the horse's coat — as well as internally in the feed. The mild acid doesn't seem to bother most horses' fly bites or raw spots in the skin.

stay on the hair longer. Read labels and find out if the product is to be diluted or used at full strength. In addition to the standard sprays, wipes, and sticks (rub-on repellents), strips impregnated with insecticides can be attached to a horse's halter.

Methods of Fly Eradication

Some forms of insecticide have long-lasting, residual action for up to 6 weeks; these are designed for application on fly resting areas such as bushes or barn rafters. Fogs and mists can be used daily—expelled into the barn air or sprayed on the horse's body with a hand sprayer. Fly-killing strips work in enclosed areas such as tack rooms and feed rooms. Sticky baits and fly traps are effective in areas with large numbers of flies, but must be cleaned or emptied periodically. Electric fly attractors also kill flies. There are chemical larvicides that you can feed to your horses; the chemical goes through the digestive system in small pellets, with no effect on the horse, and out with the manure—where it kills any larvae that hatch in the manure from flies laying eggs there. Another

KNOW EACH TYPE OF FLY

In a fly control program, become familiar with the life cycle and habits of the type of fly you're trying to eradicate. That way, you can attack it at a stage when it's most susceptible to the control measure you're using.

type of feed-through contains an insect hormone that hinders maturation of the larvae, which causes their death.

If a certain type of fly is a problem all summer, the best control is to eradicate it early in the season before it appears in large numbers. This is especially true of stable flies. If you are using an insecticide, be sure to repeat treatment before the flies become numerous again so there won't be a chance for a large breeding population to develop. For information on fly control programs, contact your veterinarian or county Extension agent. With proper care, you can keep the fly problem to a minimum.

Mosquitoes

Although they are not flies, these insects are a serious nuisance, especially in swampy areas where water or frequent flooding creates breeding sites. Mosquitoes also spread disease, such as encephalomyelitis and West Nile virus, and some types of tiny filarid worms.

In most species, the female mosquito must have a blood meal before she can lay eggs. Males feed on plant juices. Eggs are laid on the water surface or on low ground that floods. Several batches hatch each summer. Strong winds can carry mosquitoes many miles. Snow mosquitoes appear in swarms in mountain and northern regions in spring; they overwinter as eggs and develop in pools of water when snow melts.

Control

Eliminate breeding sites, destroy larvae by treating breeding grounds with the proper insecticides, and chemically destroy adults. Some insecticides kill mosquitoes on contact, but there are no preparations that can be applied directly to horses for long-lasting relief. Daily application of certain products gives partial protection; for example, pyrethroid insecticides reduce mosquito numbers by about 75 percent. Observing horses just at dusk will give you a clue as to how many mosquitoes are bothering them.

SUMMER SORE STRESSERS

A horse with a summer sore on the penile sheath may spray urine instead of producing a normal stream because a lesion is obstructing normal flow. A summer sore on the leg may grow quite large, interfering with use of the horse until the wound is treated or until the lesion regresses.

Summer Sore Larvae

Summer sores (habronemiasis) are caused by the larvae of a stomach worm trying to develop in the horse's skin instead of the stomach after being brought to the skin by flies. Adult worms lay eggs in the horse's stomach; the eggs are passed in manure. The larval stage is eaten by larvae of the stable fly and horsefly; the tiny worms infect the fly. When infected flies feed on skin wounds, eye secretions, or other moist tissue such as the sheath, the worms are transmitted to the skin. This problem is most prevalent in warm climates. Skin lesions occur in summer after infected flies deposit the worms, but the worm larvae cannot mature and finish their life cycle in the skin. They can, however, live up to 2 years in the skin, causing problems for the horse.

Summer sores appear on any part of the body but most often on the lower legs, at the corner of the eye, and on the sheath. At first they look like infected wounds, but they don't heal. The area enlarges with granulation tissue, such as that created when a wound repairs itself, protruding from the skin surface — reddish brown (bloody sometimes) and often covered with coagulated wound drainage. After the wound is 2 or 3 weeks old, it becomes circular and may have areas of dead tissue. There is usually drainage and severe itching.

If a wound is untreated in summer, it may develop into a summer sore and not heal until cold weather sets in. The skin lesions usually heal in fall but often reappear the next spring because larvae are still present. Some horses develop a exaggerated immune response to the worms, resulting in swelling, hives, or even asthmalike respiratory symptoms. Treatment for a hypersensitive reaction generally includes antihistamines and steroids.

Control

Take proper care of wounds so they stay clean, and keep flies away from them. Give ivermectin (two doses 3 to 6 weeks apart) to kill most of the worm larvae in the lesions. Horses who are routinely wormed with ivermectin may not have as many problems with summer sores. Practice fly control measures to reduce the population of horseflies and stable flies that spread the worms.

Lice

Lice are sometimes a problem on horses, especially in winter when hair is long. These tiny insects spend their entire life cycle on the host, although some species can survive for up to 2 weeks away from the horse. Eggs are laid on the hairs next to the skin. The life cycle varies from 2 to 4 weeks.

Lice are host specific; horse lice are not the same as cattle lice or human lice. You can't get lice from your horse. A horse can't get lice from cattle or chickens — only from another horse. Two kinds of lice infest horses: sucking lice and biting lice. Sucking lice feed on blood and severe infestation can cause anemia from blood loss. Biting lice are smaller and more active, causing skin irritation and itching. The horse may rub and bite affected areas, and large patches of skin may become hairless or raw. Lice are found primarily under the mane or on buttocks, rump, and tail.

Although lice are very small, you can see them on affected areas with close observation, especially if you raise the mane and rub your hand along the neck opposite the lay of the hair. Lice will be evident on the skin between the hairs. You can also see the eggs if you pull out a few tufts of hair (it comes out easily where the lice have been feeding) and hold the hair up to the light.

Most active in winter, lice multiply rapidly on thin animals. Some horses are more susceptible to heavy infestations than are others; individuals who are young, old, or have a suppressed immune system usually are most vulnerable. Lice rarely cause a problem during warm months because they are inhibited by sunshine and high temperatures.

BUG-FREE GROOMING

Remember that lice are transferred from one horse to another by direct contact; anything that comes in contact with an infested animal can carry lice to another horse. Blankets, grooming tools, harnesses, and saddle pads may remain infective for several days after being used on a lice-infested horse. Be sure to treat these with insecticide before using them again.

Control

Treat affected horses with insecticides; several sprays and powders kill lice on horses. If weather is cold, use a dust rather than a spray so as not to chill the horse. When using any insecticide, be sure to follow label directions and treat the horse in an open area where there's no danger of contaminating feed, water, or feeding areas. Clipping the horse makes lice more accessible to treatment but is not recommended in cold weather unless you keep the horse indoors.

Administer a second treatment 2 weeks after the first one to kill lice that have just hatched or that weren't accessible to the first application (dusts and sprays don't kill the eggs attached to the hairs). Also, be sure to treat any brushes, blankets, saddle pads, or other items used on the infested horse.

Ticks

Ticks are sometimes found on horses, especially those pastured in brushy areas. Ticks carry several diseases and cause anemia in severe infestations. They are the main vectors of many protozoal diseases; the protozoa survive from one generation to another in the ticks by infecting their eggs. Ticks can also spread diseases caused by bacteria, viruses, and rickettsia.

Rocky Mountain tick

All ticks found on your horses should be removed and destroyed. Be careful handling them; the tiny deer ticks can carry Lyme disease, and the larger hard-bodied ticks can carry Rocky Mountain spotted fever. Both diseases are dangerous to humans.

Control

Keep horses out of tick-infested pastures in spring. If horses get heavy infestations, spray with insecticide, preferably a product against which the ticks in your area are not resistant. Be sure the product you use is safe for horses. Amitraz, a chemical that is highly effective for cattle and pigs, is not safe for horses; it causes colic and sometimes death. Deworming with ivermectin every 6 weeks during spring and summer also helps.

Ear Ticks

Ear ticks are irritating to horses because they live deep inside the ear. Horses with ear ticks may become ear shy and head shy—hard to bridle because they resist any handling of the ears. They may shake their heads, rub their ears, or cock the head to one side (see also chapter 14).

LIFE CYCLE OF TICKS

Some tick species spend their entire life on the host, while others pass different stages on successive hosts. Still others are parasitic only in certain stages. Eggs are laid in the soil, and young ticks crawl onto bushes and attach themselves to a passing host. Adult females suck blood or lymph from the host then drop to the ground to lay eggs. Complete eradication of ticks is often difficult because many species live on several hosts, including wild animals, and adult ticks can live for weeks or months apart from a host.

Control. You can kill ear ticks by putting a half-and-half mix of mineral oil and insecticide, such as lindane, into the ear. Commercial mixtures are available from your veterinarian. The horse may have to be restrained and twitched, or you may have to use a Stableizer (see chapter 5), before he will stand still for you to put anything into his ears because the ticks have made them sensitive. If ticks have been present in the ear for some time, there may be a great deal of debris and secondary infection in the ear, requiring further medication.

Mites

Mites cause dermatitis (inflammation of skin). Trombidiform mites are harvest or grain mites that primarily attack harvested grain and infest animals only secondarily. However, they can transmit disease. Natural hosts for trombidiform mites are small rodents. Larvae drop off the rodents and develop into a nonparasitic stage that lives in grain and hay. They cause dermatitis in animals grazing in pastures or eating newly harvested grain. Horses are affected around the face and lips; the affected areas become itchy and scaly.

Mange mites infest the hair follicles of horses, causing chronic inflammation; loss of hair; and, sometimes, secondary infections such as pustules or small abscesses. Some mange mites live deep in the skin, creating small nodules that feel like birdshot in the hide. These mites are most active in winter and early spring, usually affecting the area around the horse's face and eyes. This problem is sometimes confused with ringworm.

Sarcoptic Mange

One type of mite causes a severe, itching dermatitis (also called "barn itch" or "red mange") that makes affected animals too restless to graze and, therefore, causes them to lose weight. These mites are most active during cold, wet weather. Eggs are laid in tunnels in the skin made by adult mites; the hatching nymphs burrow into the skin and mature in approximately 17 days. Mite infestation can be spread from horse to horse or by materials such as bedding, saddle pads, grooming tools, and clothing. In moist, protected sites, the adult mites can live as long as 3 weeks away from the horse.

When a horse has sarcoptic mange, affected areas of skin contain small red elevated patches. The area is intensely itchy and may be injured by constant scratching and biting. Hair is lost and thick brown scabs form over raw surfaces with thickening and wrinkling of surrounding skin. The most common site on horses is the neck.

Psoroptic Mange
Another type of mite migrates to all parts of the skin, preferring areas covered with hair. The adults puncture the skin to feed on lymph and cause local irritation and itchiness. The skin weeps and forms crusts that spread because the mites are most active around the crusts. In horses, the large, thick crusts are found on parts of the body with long hair, such as the base of the mane and root of the tail, and on hairless areas such as the udder.

Chiroptic Mange
Also called "leg mange," this causes great annoyance to horses. The mite is most active in winter, spread by direct contact from horse to horse or by infested bedding or equipment. In horses, the parasites occur almost entirely in long hair on the lower legs, rarely on other parts of the body. They cause severe irritation and itchiness; the first sign is usually stomping of the feet and rubbing of the back of the hind pasterns on anything available. The horse may resent having his feet handled. In longstanding cases, the skin may be swollen, scabby, cracked, and greasy. Scrapings from affected areas contain numerous mites.

Treatment and Control
Any itchy rashes on a horse should be examined by your veterinarian. He or she can usually differentiate among the various skin diseases by making a close examination and by taking skin scrapings to observe under a microscope. Several types of sprays and insecticides can be applied to the horse to eliminate mites, and the veterinarian can advise you on an effective treatment. Quarantine affected horses so the problem does not spread.

11

Toxic Plants
and Poisons

MOST POISONS THAT AFFECT HORSES are found in toxic plants containing dangerous alkaloids. Other agents that are harmful to horses include chemicals in contaminated feed and lead-based paint on fences or barns where horses chew or lick. Some poisons are so deadly that they kill a horse quickly, even if he ingests very small amounts, whereas others are lethal only after many small doses accumulate to dangerous levels in the body.

Toxic Vegetation

Most poisonous plants, trees, shrubs, and grasses vary in level of toxicity and conditions during which poisoning occurs, but some are extremely poisonous at all times. Others are harmless under normal conditions — in small amounts or at certain seasons — but become dangerous if eaten in large amounts or if chemically altered by freezing or enzyme activity (as when cyanogenic glycosides in chokecherry leaves turn into hydrogen cyanide). Certain plants absorb substances from the soil that accumulate to toxic levels (as selenium does in locoweed).

Horses usually avoid poisonous plants, but hunger, curiosity, or boredom may lead them to eat things they shouldn't. They may nibble shrubs through the paddock fence or plants along the trail when you ride if you aren't paying attention. They may also eat harmful plants inadvertently baled in hay or raked into lawn or hedge clippings. Even some plants and shrubs used for landscaping (in the yard or garden, barnyard or fairgrounds) or along a fence row, road, or trail may be poisonous. Most poisonous plants that might be eaten by horses are "forbs" — broad-leafed flowering plants. Only a few grasses are dangerous to horses.

HOW POISONOUS IS IT AND
HOW LIKELY TO POISON MY HORSE?

Poisoning depends on many factors, including:
- Palatability (some plants are readily eaten; others aren't unless a pasture is overgrazed and grass is short or the horse is bored)
- Stage of development
- Growing conditions (drought may increase toxicity levels)
- Soil type
- Moisture content of the plant
- Portion eaten (some have highest concentration of the poison in the roots, seeds, or new sprouts)

If you suspect a horse may have eaten something toxic or is showing behavioral changes or digestive upset, get veterinary help immediately. Some types of poisoning are irreversible and always fatal, but others can be resolved with prompt emergency treatment.

Plant poisoning should be suspected if horses experience sudden illness or death after being moved to a new pasture or fed different hay. Poisonous weeds in hay can be a serious threat because the hay-fed horse has no other food options, unlike a horse at pasture. Some horses sort out strange weeds, but greedy eaters may eat them despite their taste or smell. Even a not-so-hungry horse may eat harmful hay because the strange and unpleasant taste or smell of a living poisonous plant may be diminished in the drying process.

There is not space here to discuss all vegetation that is poisonous to horses. More than 500 plants in the eastern United States alone are toxic enough to kill a horse, and toxicologists are discovering more. This chapter touches on some of the common offenders.

Common Poisonous Plants

Following is a partial list of forbs (herbaceous plants that are not grasses) that are known to cause problems for horses.

- **Garden flowers and plants** poisonous to horses include buttercup, narcissus (daffodil), lily of the valley, and delphinium. Wild delphinium is called "larkspur"; tall larkspur is highly poisonous to cattle and not as deadly to horses; however, if a horse exerts after eating delphinium, he may die from impaired muscle activity and breathing. Rhubarb leaves contain soluble oxylates that crystallize in the kidneys, causing kidney failure and death. Don't let a horse eat in your yard or garden; never let him eat any plant that is not a part of his regular diet if you are not familiar with it.

Tall larkspur (delphinium) flowers and seedpods

- **Brackenfern** is part of the fern family. This large perennial herb grows in burned-over areas, woods, and shady places, as well as on dry, sandy, or gravelly soil. Horses eat it when other feed is scarce, when it is mixed in hay, or when allowed to grab a snack along the trail. The fern contains an enzyme that inhibits absorption of thiamine (vitamin B_1) in the gut. Because the affected horse cannot absorb it, symptoms are those of vitamin B_1 deficiency: his nerve function deteriorates and he loses weight and becomes progressively uncoordinated and depressed. He stands with back arched, legs apart, and muscles twitching. Symptoms develop slowly over days or weeks after eating a large amount of fern, but if affected horses are not treated, they eventually die. If the horse is taken off the offending feed and treated with IV injections of thiamine hydrochlorate when symptoms first appear, he will recover.

- **Castor beans** are grown as a crop (for castor oil) in California and the Southwest and as an ornamental plant in other regions. The seeds are the most toxic part, causing mild colic, diarrhea, sweating, and a thundering heartbeat. Only a few ounces of castor beans can kill a horse.

- **Cocklebur** is an annual plant that grows in lakebeds, lowlands along rivers, and disturbed soils of fields, pastures, and roadways. Poisoning occurs mainly when horses eat the leaves in early spring because seedlings are more toxic than mature plants. Symptoms include rapid and weak pulse, labored breathing, and spasmodic contractions of the leg and neck muscles.

- **Crotalaria,** a member of the Pea family, contains pyrrolizidine alkaloids that destroy the liver. In the Midwest, the plant is called "rattleweed";

IDENTIFY AND ERADICATE TOXIC PLANTS

Walk through your horse's pen or pasture and find out what's growing there. If you cannot identify certain plants or trees, take leaf samples to your county Extension agent. Some toxic plants are easy to eradicate by taking out individual offenders growing along a fence or ditchbank, but others may require the use of herbicides. Your county Extension agent can advise you on how best to get rid of them.

in Florida it may be called "showy crotalaria." The effects of liver damage may not show up right away because the liver is a large organ that continues to function until about 60 percent has been destroyed. Then, liver failure is quite sudden, with the horse dying within 3 to 5 days.

- **Death camus** grows on foothills and pastures in western and plains states. It takes less than 10 pounds (4.5 kg) of it to kill a horse. Symptoms are excessive salivation, weakness, difficulty breathing, and coma. Death occurs soon, within a few hours to 2 days after ingestion.

- **Equisetum** is called "horsetail," "scouring rush," or "jointfir"; several species grow in meadows and floodplains or wherever the water table is high. These plants may end up

Stalk-of-death camus with flower (cluster of tiny blossoms) at top

in meadow hay. Poisoning is cumulative and may occur after horses eat hay containing large amounts of horsetail or graze pastures short. Symptoms include excitement, muscle weakness, trembling, staggering, diarrhea, and loss of condition. As in brackenfern, the poisonous element in horsetail is the enzyme that destroys vitamin B_1. If diagnosis is made early, treatment with massive injections of thiamine can save the horse.

- **Fiddleneck, fireweed, tar weed, buckthorn,** and **yellow burr weed** are common weeds in wheat and grain crops in the West and Midwest, causing liver failure if eaten. Fiddleneck may grow in pastures or hay fields. A single, large dose, or small amounts eaten over days or weeks, causes poisoning from pyrrolizidine alkaloids. These weeds aren't very palatable, but a horse may be poisoned by eating small amounts at pasture or in hay or grain over an extended period.

- **Lantana** is an ornamental flower commonly found in yards and along roadways, especially in the Southeast, West, and Southwest. Horses eat it if other forage is poor or unavailable; 20 to 30 pounds (9–14 kg) will kill a horse. Symptoms of poisoning are respiratory distress, diarrhea, and low blood sugar; the horse is dull and weak.

- **Locoweed,** called "crazyweed," "milkvetch," or "poisonvetch," is native to arid and semi-arid rangelands; there are more than 50 kinds. It is a legume, part of the Pea family with alternate leaves and pea-like blossoms. Its seeds resemble garden peas in a dry, leathery pod. Most locoweeds and vetches grow in selenium-rich soils; therefore, they have concentrated selenium in their tissues. Some types are unpalatable; horses eat them only when they are short on better feed, as in dry years or during a late spring when pasture grass is slow starting. Locoweed comes up early, ahead of the grass.

Locoweed blossoms

SIGNS OF LOCOWEED POISONING

- Nervousness
- Weakness
- Emaciation
- Impaired vision (because the optic nerve is affected, small objects appear large and spatial relationships are distorted — the horse may walk into fences or over cliffs)
- Depraved appetite, eating unusual things
- Incoordination
- Difficulty breathing
- Heart failure
- Respiratory paralysis
- Seizures

Other types are quite palatable and vary in toxicity, ranging from good high-protein forage plants to extremely poisonous ones. Some are addictive; once a horse starts eating these, he seeks them out, even if better feed is available. In moderately toxic plants, the poison is cumulative, building up to dangerous levels over a few days or weeks, but the most toxic plants cause death within just a few hours. Nerve cells in the brain are destroyed and even if a horse stops eating the plants, he can never completely recover.

- **The Nightshade family** includes black nightshade (also called "horse nettle"), jimsonweed or thornapple, ground cherry, and cultivated varieties — potato, tomato, and eggplant. Fruits of the cultivated plants are not poisonous, but potato skins are toxic to horses. The leaves and stems of potato plants and other members of the nightshade family contain the alkaloid "solanine," which causes nervous system impairment (weakness and incoordination), digestive problems, and abdominal pain. Green parts of the garden plants are toxic, and all parts of the wild nightshades are poisonous.

Deadly nightshade: leaves, blossoms, and buds

  One to 10 pounds (0.5–4.5 kg) of black nightshade (horse nettle) will kill a horse. Deadly nightshade is a vinelike weed that looks somewhat like a tomato or potato plant with tiny purple blossoms and small red berries. Horses usually don't eat it but might nibble it in a pen where there is nothing else to eat.

- **Poison hemlock** is part of the Carrot family and should not be confused with water hemlock, which is actually more poisonous but less often eaten. Poison hemlock grows 2 to 10 feet (0.5–3 m) tall with hollow, purple-spotted stems and a solid taproot. The leaves are finely divided, like carrot leaves. They smell like parsnip when bruised and have a nauseating taste. Poisoning is most common in spring when the tender, succulent, highly toxic new leaves appear; horses generally don't eat hemlock unless they are short on other feed.

  Poison hemlock is found most often in pastures and along roadsides; less often near creeks and irrigation ditches. Once eaten, it causes incoordination, staggering, drooling, and slow heart rate. Death is usually from respiratory failure, and treatment is generally of no help. Whether the animal lives or dies depends on how much he has eaten: it takes less than a pound (0.5 kg) to kill a horse.

- **Russian knapweed** is a member of the same family as yellow starthistle and causes similar symptoms if eaten over several weeks. Brain damage from this type of poisoning is irreversible, and affected horses starve from the inability to eat. They should be euthanized to spare them this slow death.

- **Senecio** and other species of senecio (groundsel, stinking willie, ragwort,

Leaves and seed heads of poison hemlock

and tansy ragwort) contain pyrrolizidine alkaloids that destroy the liver. A horse who eats these plants or contaminated hay over a period of several weeks will eventually show signs of liver failure. Because the plants grow in early spring, they are often found in first-cutting hay. A horse is more apt to eat them in hay than at pasture unless good forage is lacking.

- **Saint John's wort**, called "Klamath weed" or "goat weed," causes reaction to sunlight (photosensitization) and severe lesions in unpigmented areas of the body (areas with white markings). Saint John's wort is toxic in all stages of growth but is most likely to be eaten in spring when young and tender. Hay containing this plant can also cause problems. The toxic compound is absorbed into the bloodstream from the gut; when it reaches the skin it sensitizes unpigmented areas to sunlight. Light-skinned horses or those with white markings develop red, swollen, sore, and itching skin that peels or comes off in sheets. Affected animals must be moved out of the sun and treated. Photosensitization is discussed in chapter 12.

- **White snakeroot** grows in sandy areas of the Midwest. Because it often stays green in early fall after other plants become dry, horses may be tempted to eat it. Symptoms of poisoning are weakness, lethargy, stiffness, and incoordination. The plant contains the alcohol trematol; a lethal dose is 2 to 10 pounds (1–4.5 kg) of plant material. Enzymes in the plant damage the heart muscle, after which stress can cause heart failure. Trematol also damages the liver. It can be excreted in a mare's milk and poison a foal if a lactating mare eats the plants.

- **Yellow starthistle** is common along the Pacific Coast as well as in several western, plains, and eastern states — and it is spreading. Some horses develop a taste for it and eat enough to permanently damage the brain. Signs of poisoning include twitching lips, flicking tongue, involuntary chewing movements (ranchers call it "chewing disease"), and drowsiness. Animals have trouble eating and drinking but are able to swallow. The horse may hold his mouth open as if something is stuck in his throat. The muscles of jaw and lips become rigid, and the horse has trouble grasping or chewing food.

Poisonous Grasses

Make sure your pastures and hay do not contain any grasses that may be toxic to horses.

- **Tall fescue** causes several problems in mares, including abortion, retained placenta, and failure to produce milk. The fescue grass itself is not toxic, but the fungus (*Acremonium coenophialum*) that infests many stands of it is.
- **Sorghum and Sudan grasses**, common in pastures in the Southwest and Southeast, can be harmful. Their toxic effects are intensified during a drought or after a frost because the glycosides in the plants are changed to cyanide, a potent poison. The affected horse is starved for oxygen, breathes heavily, grows weak, falls into a coma, and dies within a few hours.
- **Torpedo grass** is a lush pasture grass in Florida and the South that is toxic to horses. Animals who eat this grass often develop severe anemia and die.

GRASSES AND PHOTOSENSITIZATION

Some grasses, such as perennial rye grass and dried buckwheat, can cause photosensitization, as can certain clovers.

Poisonous Trees and Shrubs

Many trees and shrubs cause problems for horses — even apple trees. If a horse eats too many apples, he may develop colic. When horses have access to fallen apples in a pasture or paddock, they may overeat, especially if the fruit is ripe. Some crab apples are especially troublesome. The horse may be fine until he exerts; then, the combination of stress and apples in his digestive tract leads to an acute case of colic.

- **Wild cherry (chokecherry)** grows in most areas of the United States. It has dark green leaves and clusters of white or cream-colored blossoms that produce dark red or black fruit. The leaves contain a substance that produces cyanide and hydrocyanic acid poisoning when damaged. The chokecherry is most deadly when the leaves are damaged.

    Because the poison interferes with oxygen transfer to the cells, the affected animal suffers from acute oxygen starvation; the horse breathes heavily, grows excited and agitated, then becomes weak and may suffer convulsions before going into a coma. Symptoms appear soon after he eats the leaves, and unless treatment is immediate, he will die. Cherry, plum, peach, and apricot leaves are also toxic if wilted or otherwise damaged, so horses should be kept out of orchards. Cherry laurel (often planted in yards and gardens) is another potent source of hydrocyanic acid, as is crab apple.

    The chokecherry grows wild in most parts of North America and is often found in pastures, lining fence rows, or growing along streams in arid regions. It may be merely a very tall shrub in dry places, but it grows into a tree where there is plenty of water.

Chokecherry leaves and fruit

TREE HEALTH TIP

Many ornamental trees and shrubs, such as oleander or yew, are highly toxic. Make sure horses never have access to them.

- **Rayless goldenrod,** a bushy, unbranched perennial shrub, is also called "rosea" or "jimmy weed." Most common in the Southwest, it grows 2 to 4 feet (0.6–1.2 m) tall with yellow flowers. The toxic substance (trematol) causes muscle tremors, depression, stiff gait, and weakness. It can poison foals nursing mares who eat the shrub.

- **Black locust** is a legume with short thorns, creamy white blossoms, and thin brown seed pods. It grows east of the Mississippi and is used as an ornamental tree in other regions. Its leaves, twigs, and bark are poisonous to horses, causing colic, diarrhea, dilated pupils, weak and irregular heartbeat, depression, and death.

- **Red maple,** also called "swamp maple," is native to the eastern half of the United States. It can be poisonous to horses, killing them within a day or two after they've eaten small amounts of wilting leaves in fall or nibbled them off fallen or trimmed branches. The wilted leaves contain

Red maple leaves

an unidentified substance that causes breakdown of red blood cells. Symptoms of red maple poisoning include rapid breathing, weakness, depression, brown urine, and dark blood.

- **Oleander** is a common ornamental shrub in warmer regions that is deadly if eaten. One-fourth pound (113 g) or less will kill a horse; often just a few leaves can be deadly. Horses rarely eat leaves from the shrub itself, but if wilted or dried, they are more palatable because they don't retain their bitter taste (but do retain their toxicity). Oleander contains two toxins, oleandrin and nerioside, that affect the heart much like digitalis, producing erratic heartbeat and diarrhea. Death may occur within a few minutes or hours. It's best to assume that any ornamental tree or shrub is unsafe for horses to nibble on. Magnolia, azalea, privet, boxwood, and rhododendron cause illness in horses.

- **Black walnut** trees are often grown in the eastern part of the United States and in California. Wood shavings from these trees cause colic and laminitis in horses who are bedded on them. In addition, horses at pasture can be seriously affected if they have access to black walnut trees. Pollen in spring causes allergic respiratory reactions, and chewing of bark or branches leads to laminitis and colic at any time of year.

- **Yew** is a tree or shrub — an ornamental evergreen with red berries (arils). Its leaves are stiff and needlelike, dark glossy green on top and yellow-green underneath. There are several species — English yew, Japanese yew, and American or Canadian yew. Yew contains taxine, a highly poisonous alkaloid, even when dried. It has a depressing effect on the heart. About 4 ounces (113 g) (one mouthful) of yew needles can kill a horse. The animal dies of heart failure, often within minutes of nibbling on yew.

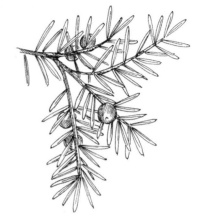

Yew leaves (needles) and arils (red berries)

BEWARE OF YEWS

Yew is poisonous to people and livestock. Symptoms include shortness of breath, muscle tremor, and sudden collapse. With horses, death may be so sudden that some of the plant will still be in the mouth. Horse owners should avoid these shrubs in landscaping. Check your farm and neighboring fence lines to make sure there's not a potentially lethal hazard waiting around the corner for an unsuspecting horse.

- **Other trees** that are mildly toxic if chewed by horses are oak, Ohio buckeye, and the Kentucky coffee tree. The responsible horse person should look at all the plants in a pasture or paddock before putting any horses in. Check all trees and shrubs — even if they are on the outside of the fence line but within reach of the horses, or where branches or leaves can blow into the pens.

PREVENT ACCIDENTAL PLANT POISONINGS

Conscientious attention to details can usually reduce the risk for poisoning. To keep your horses safe, be sure to:

- Never dump trimmings, pruned branches, or discarded garden plants near horse pens or pastures. Bored or curious horses may nibble on strange plants, even reaching under or through a fence to taste and eat them.
- Never let horses near ornamental shrubbery, plants, or flowers.
- Remove fallen tree limbs or leaves from pens and pastures immediately after a storm.
- Carefully inspect bedding straw when you put it in stalls to make sure there are no weeds in it.
- Always take time to look at and smell hay when feeding hay; inspect it for strange plants and discard anything you aren't sure about.

Poisonous Fungi

Fungi, a group of primitive plants that includes molds, are microorganisms that live almost everywhere. Most prefer a moist environment and multiply rapidly in warm, humid areas or during wet seasons. Many are harmless, but some cause trouble when ingested.

Hay or grain can become infested by fungi while growing or may spoil and be contaminated during storage. Factors that affect severity of the problem include weather conditions during the growing season and exposure to moisture during or after the harvest. An animal may eat the same kind of hay or grain for years with no problem and then suddenly be poisoned as a result of unusual growing conditions that cause fungi to flourish in the feed.

- **Black patch disease** occurs on clovers and other legumes that are sometimes host to the toxic fungus *Rhizoctonia leguminicola*, which causes "slobbers" in horses. The fungus appears as spots on leaves and stems. Red clover, alsike, white clover, soybeans, and alfalfa are commonly affected. Because the fungus overwinters, it can persist in a field or pasture once established. When a horse eats affected plants, his salivary glands are irritated, resulting in slobbering and drooling. The toxic alkaloids also cause fetal malformation or abortion in pregnant mares, along with diarrhea, frequent urination, weight loss, and even death, depending on how much is eaten.

- **Ergot** is a fungus that sometimes occurs on grasses, infecting the seed heads. Susceptible plants include wheat, rye, barley, bluegrass, red top, brome, and reed canarygrass. Ingesting small amounts of ergot daily can lead to circulation problems in horses and cattle. Blood vessels constrict and shut off blood supply to feet, ears, and tail. The horse may lose ear tips or ears and may develop gangrene in feet and legs, which may lead to death.

- **Moldy corn poisoning** is caused by *Fusarium moniliforme*, which produces deadly toxins. This fungus lives in soil. It is commonly found on moldy corn stalks or corn that has been rained on in the stalk or stored wet. Symptoms are often mistaken for those of encephalomyelitis. Poisoning can occur in horses who eat commercial feeds, including pellets. Horses ingesting a small amount of the toxin may suffer depression,

---

BEWARE OF CORN

Corn is usually a good feed for horses, but only if it is of high quality and mold free. Corn that has not been adequately dried in the field, or in the drier after harvest, may spoil and mold when stored. Even fully dried grain may mold if it is stored where it can draw moisture, and condensation inside a metal bin can cause trouble unless all old grain is cleaned out before a new batch is put in. Corn is more subject to mold during growing and storage than are less starchy grains. Some of the visible molds are not as deadly as the aflatoxins, or poisons, produced by *Aspergillus* or other molds that may develop during a wet growing season. Often the deadly corn looks perfectly normal.

Even in shelled corn, bad kernels sometimes get mixed with good ones. Usually, only a small percentage of individual kernels in a moldy batch contain high levels of toxin. This is why some horses develop fatal poisoning even though others on the same feed are unaffected. It takes only a few kernels of bad corn to produce problems.

> ### FUNGAL TOXINS AND PHOTOSENSITIZATION
>
> Fungal toxins, called mycotoxins, can be deadly — damaging liver and kidneys, destroying red blood cells, or producing other serious effects. A damaged liver can lead to photosensitization.

weight loss, lack of appetite, and bleeding disorders. Larger doses are usually fatal. The horse circles, runs into fences, or stands with head pressed against a fence or wall. He may cross his legs, walk sideways, or circle aimlessly, losing control of his movements as his brain deteriorates. He may die within a few hours or linger for several days.

- **Rusts and smut** are other fungi that may contaminate grain and pastures. Rusts are reddish brown patches that infect the leaves and stalks of grasses, and smuts are black powdery clumps that appear to replace the grass heads or grains. Eating smutty grain may cause convulsions, paralysis, and death within a few hours. Rusts and smuts also cause colic — and respiratory problems if inhaled.

- **Mowed hay that gets rained on before being baled** may develop molds and other fungi. For this same reason, piled lawn clippings are not safe for horses because they have a high water content and spoil quickly. Within 3 or 4 hours, the grass undergoes chemical changes in piles or clumps that can't dry quickly. It then heats and ferments, producing mold. The fermentation process may also produce nitrate, which can poison a horse, causing respiratory problems. A horse should not be allowed to eat grass clippings or graze a pasture that has been clipped by a rotary mower that tends to wad up the grass. Not only may clumps start to mold, but they may cause a horse to choke.

Blister Beetle Poisoning

Many species of blister beetle — the most common being the black and yellow striped variety — can cause poisoning. Blister beetles are small flying insects that are attracted to the blossoms of tomatoes, potatoes, soybeans, and alfalfa. The beetles travel in swarms and are baled up in alfalfa hay if it is cut while they are feeding on it. Their bodies contain a toxin, cantharidin, which causes severe blistering of the digestive tract. Ingestion of only 3 to 10 beetles will kill a horse. Cantharidin is just as toxic in a dead beetle as in a live one.

Blister beetle poisoning has been diagnosed in nearly every state, but hay grown west of the Mississippi is more frequently contaminated than hay grown in the East. Horse poisonings are most common in the Southwest. There are blister beetles wherever alfalfa is grown, but some species of the beetles are more toxic than others.

Blister beetle

Symptoms

Cantharidin causes extreme tissue damage in the gut. The affected horse goes off feed and has intense abdominal pain. Most horse owners treat for colic, not suspecting blister beetle poisoning unless several horses become sick from eating the same hay. The horse may survive if he eats only one or two beetles, but if several are ingested, his chances of survival are diminished. The tissue damage causes shock and dehydration. Another sign of poisoning is frequent urination caused by inflammation of the urinary ducts. A burning sensation in the mouth from the blistering makes the horse want to cool his mouth with water, although he can't swallow.

Treatment

Treatment can sometimes save a horse if it is begun immediately and the animal hasn't ingested a large number of beetles, but often the true cause of his colic is not recognized in time. Treatment for blister beetle ingestion includes drugs to minimize the pain and IV fluids to keep the horse from going into shock.

Prevention

First-cutting alfalfa is safer than later cuttings because adult blister beetles may not emerge until midsummer. Use only alfalfa cut before bloom stage; it is less likely to contain beetles that feed on pollen in the blooms. Beetle larvae eat grasshopper eggs, so there is usually a large beetle population following a dry year with lots of grasshoppers. Be cautious about buying alfalfa hay grown in areas that were dry during the previous year.

Chemical Poisons

Toxic materials commonly found around barns and stables include rodent poisons, insecticides, and lead-based paint on fences or buildings. Horses can also be poisoned by accidental overdose on medications, certain dewormers or vitamins, accidental consumption of seed grain or feeds not meant for horses, and ingestion of toxic substances such as antifreeze.

Antifreeze

Because antifreeze has a sweet taste, many animals sample it. When you drain it from a vehicle, be careful not to spill it and to rinse out the pan you drain it into. Even dried antifreeze crystals contain the poison, ethylene glycol, which is also used in brake fluid, windshield de-icers and motor coolants.

Pesticides

When using pesticides, never apply or spray them near feeds, and make sure the wind isn't blowing the spray into areas that shouldn't get contaminated (such as mangers, feed buckets, and water sources). Keep all pesticides far from horses and grain bins; a single bottle of pesticide can cause trouble if it gets spilled in the granary or where a horse might later clean up hay or grain off the barn floor.

Treated Grain

Almost all seed grain (for planting) and most other seeds are treated with fungicides to protect against fungi that damage or destroy seeds during storage or after sowing. The fungicide may be poisonous to horses.

Overdose of Medication

Horses can be poisoned by accidental or inappropriate use of certain medications, dewormers, or vitamins, especially those prepared in a palatable base. If such a preparation is left where a horse can find it, he might eat the whole container and poison himself. Vitamins A and D are toxic in quantities well above the recommended dosage, as are some dewormers. Other medications, such as wound treatments and liniments, are potentially dangerous if eaten

---

GRAIN SAFETY TIP

It is extremely important to know the origin of all grain. Usually the feed from a commercial dealer is safe, but beware of using grain that comes in unmarked sacks or from an unknown source. Never keep any seed grain in the barn. That way, it won't get fed to horses accidentally.

or spilled in feed. Keep all medications labeled and out of reach of horses. Spilled medicine might entice a curious animal to smell and taste it.

Lead Poisoning

Lead is especially dangerous and can accumulate in the body over time. Even small doses, if repeated (as when a horse chews or licks a painted wall or fence), can prove fatal. The most common sources of lead poisoning are lead-based paint on old buildings or other structures (painted before the advent of lead-free paint in the late 1970s) and old batteries. Don't let horses have access to old dumps that might contain batteries and other harmful materials. Fumes from wet paint or certain solvents are also toxic.

Prevention of lead poisoning is by far the best course. Make sure your barn and stable area contain no questionable paint chips, painted surfaces, or batteries that might poison a horse.

Monensin (Rumensin) Poisoning

Many horses have died from eating cattle feeds containing monensin (Rumensin). This drug is often added to alfalfa pellets for ruminant livestock such as cattle and goats. It is fatal when eaten by horses. It causes damage to muscles, especially the heart muscle, often leading to heart failure. Symptoms of monensin poisoning in horses include rapid heart rate, weak and irregular pulse, blue mucous membranes, and abnormal lung sounds. Only 0.04 or 0.07 ounce (1 or 2 g) of the drug can kill an adult horse. Some additives in poultry feed are also unhealthy for horses.

Prevention

To avoid monensin poisoning, never feed grain mixes intended for other livestock. Also, keep horse feeds completely separate from other feeds and be careful to feed in separate areas. Even the fine particles of dust left at the bottom of a trough or bin may contain enough monensin to kill a horse. Don't put horses in pastures where beef cattle might be fed mineral blocks or protein supplements containing monensin.

12

Skin Problems

THE HORSE'S SKIN IS HIS FIRST LINE OF DEFENSE. It is subject to problems caused by bacteria, fungi, biting flies, allergies, and other things that range from being mere annoyances to serious conditions. Chapter 10 discusses skin problems related to external parasites such as flies, lice, and mites. This chapter focuses on dermatitis related to allergies and hives, photosensitization, fungus infections, warts, and sarcoid tumors.

Allergies and Hives

An allergy is a reaction to a substance that would normally cause little or no response by the body. Some horses are allergic (hypersensitive) to insect bites, certain plants, drugs, vaccinations, or medications. If the horse's defense system works overtime, an inflammatory response may become worse with each encounter. Sometimes, an allergic reaction creates raised areas in the skin. This condition is called "urticaria"; the bumps are called "hives" or "wheals." They are caused when small blood vessels near the skin surface dilate and

---

DERMATITIS

Dermatitis means "inflammation of the skin." Many things, including external irritants, burns, allergy, or infection, can cause dermatitis. It can also result from systemic disease. The affected horse's skin may become itchy (and he rubs against fences or walls) or show redness, swelling, or lumps. There may be oozing or scaliness as well.

Skin problems in horses can be minor or serious. When in doubt as to the identity or cause of any abnormality, have your veterinarian examine the horse.

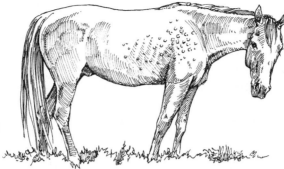

Horse with hives

blood seeps from the damaged capillaries. The hives develop suddenly after the horse comes in contact with the allergen. If you are with the horse when it happens, you can see the bumps growing as you watch. They usually pop up on both sides of the body and sometimes on the head and neck.

Insect bites or stings, contact with certain plants and/or pollen, and ingestion of certain weeds or plants can cause hives. Moreover, some drugs or vaccines produce allergic reactions. For example, penicillin causes hives (or some other manifestation of allergy) in some horses.

Treatment of Hives

In mild cases, the lumps subside without treatment within a few hours or by the next day. In severe cases, they persist for several days and may be accompanied by diarrhea, fever, or breathing difficulties. The affected horse may need an injection of antihistamine or corticosteroids.

Hives may recur if the allergen is not removed, the diet is not changed, exposure to the causative insect bites or plants is not prevented, or the same offending drug or vaccine is used again. Many times, the cause of hives is readily obvious, as when a horse reacts to a certain injection, drug, or feed change; walks through a patch of stinging nettle; or suffers insect bites. At other times it may be difficult to discover the cause. If hives recur or do not disappear within a few days, have your veterinarian do sensitivity skin tests to determine the cause of your horse's allergy.

Photosensitization

This is a skin disease in which unpigmented skin is damaged by a reaction to something the horse has eaten, and occurs when the skin is exposed to sunlight. Photosensitization is most common during summer (long days and stronger sunlight) when horses eat pasture plants that may contain certain photodynamic agents — chemicals that cause a toxic reaction when exposed

LIGHT SKIN AND PHOTOSENSITIZATION

Areas most affected by photosensitization are pink or unpigmented skin, such as white face markings and white legs. Pigmented skin is more resistant to the sun's rays.

to sunlight. A horse with long winter hair is not as susceptible because the hair shields the skin from the sun.

If a plant containing photodynamic agents is eaten, these chemicals are absorbed into the bloodstream and travel to all parts of the body. If they accumulate in the skin, the sun's rays cause a reaction, with inflammation and cell death. This problem is triggered by a number of plants — especially fescue grasses and some species of clover — and usually occurs when they are lush, green, and growing rapidly. The disease can also occur when a horse's liver is impaired and unable to filter out the by-products of chlorophyll breakdown in the digestive tract. When these by-products accumulate in the skin, the reaction kills skin cells.

Symptoms

The ultraviolet rays penetrating the sensitized skin tissues trigger release of histamine, which causes dilation of capillaries and swelling of tissue. Irritation and itching may be intense because of painful swelling deep within the layers of skin. The condition may be mistaken for sunburn at first because the area turns red and starts to peel. Then, it becomes more swollen, breaking out in blisters and cracks, and the skin develops thick crusts and painful scabs. In severe cases, the horse becomes quite ill, with patches of skin sloughing off. The horse may go into shock.

Treatment

Remove the affected horse from the pasture and put him in a barn or covered area out of direct sunlight. In serious cases, your veterinarian may give laxatives such as mineral oil (a quart [1 L] or more by stomach tube) to help get rid of toxic materials still in the digestive tract. Antibiotics and antihistamines may be necessary as well. Corticosteroids reduce soreness and swelling but should be used only on the advice of your veterinarian. You may, however, apply soothing ointments or creams to raw areas to help alleviate the pain and itching.

Fungal Infections

Many types of fungal infections lead to skin problems in horses, including the fungi that cause ringworm, girth itch, rainrot, and scratches. Some fungal infections, such as ringworm and rainrot, are most common in winter when daylight hours are less (sunlight tends to inhibit them), and several can be spread from horse to horse through tack or grooming tools.

Ringworm

Ringworm is caused by several types of fungi that send out spores to start new infections when rubbed into skin. Ringworm is spread from horse to horse by infected brushes, curry combs, cinches, saddle pads, blankets, and the like. Young horses, older horses, those deficient in vitamin A, zinc, or selenium, and any horse with an immature or compromised immune system may be quite susceptible. Some families of horses have less resistance than others. Occasionally, the horses who get ringworm most often on a farm are related.

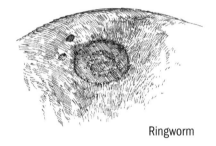

Ringworm

After the fungus spores enter the skin, circular areas in which hair falls out or breaks off become noticeable about a week to a month later.

Treatment

A horse with ringworm should be treated to prevent spread. It is also important to isolate the horse and disinfect all grooming tools and other equipment used on him. Several topical medications kill ringworm fungi. You can obtain them from your veterinarian.

Prevention

Ringworm will crop up sporadically if the fungal spores are present in the environment, but incidence of cases can be minimized by keeping horses healthy (adequate vitamin A and selenium in the diet) and not sharing grooming tools and tack among horses.

Girth Itch

Girth itch is a fungal skin problem that occurs in the girth area, generally after the horse has been ridden with a saddle. When tender skin at the girth (especially behind the elbows and in the armpits) becomes irritated by the cinch or

---

DON'T LET GIRTH ITCH SPREAD

Girth itch can spread from horse to horse if you use the same cinch, girth, or grooming equipment on different horses. The fungal spores are durable and can survive in the environment and on brushes and tack. If there is one infected horse on the farm, the problem can spread unless different equipment is used on each horse.

---

girth rubbing the skin, the fungus gets a start in the broken skin and then the skin begins to peel. Unless it is halted, the infection leads to ever-widening raw spots, and the horse may become so sore that he can't wear a saddle until the infection is halted and the skin heals.

Treatment

Most good fungicides halt girth itch, but take care to use one that will not be irritating to tender girth skin. Tincture of iodine is too harsh, but "tamed" iodine (Betadine) is fine. Another product that works well for halting serious cases of girth itch is the garden fungicide, captan (Orthocide), used on fruits, vegetables, and flowers. It is a wettable powder. Follow label directions, read label precautions, avoid getting it in eyes or inhaling it, and wear rubber gloves when applying it to horses. The solution can be sprayed or wiped on. It can also be used as a body wash if a horse has extensive fungal lesions or a skin problem such as rainrot.

Prevention

Don't use the same equipment on more than one horse (each should have his own brush). If you must use the same saddle, be sure to change cinches. If you use the same cinch or girth on more than one horse, wash it with a fungicide or bleach before using it on another horse. A neoprene girth is handy; you can wash it with warm water and chlorhexadine

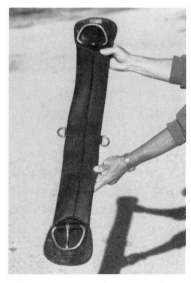

Use a washable neoprene girth if equipment must be worn by different horses.

(Nolvasan) and have it dry and ready to reuse in just a few minutes because the rubbery material does not absorb moisture.

Rainrot

Rainrot, also called "rain scald," often affects horses in wet weather. It's common in fall or in a wet climate. Patches of hair may come out in clumps; the horse may have red, raw spots on his back where the hair comes out. There may be random crusty patches in the skin that are sensitive when you touch or brush him.

Rainrot is caused by a funguslike organism that lives in the soil. It is often present in dust particles. The combination of water and dirt, as when a dusty horse gets wet, makes an ideal environment on the skin for this invader. Dusty conditions before a rain can fill the hair coat with dirt, especially if the horse rolls a lot; then rain (especially prolonged wet weather) provides the moisture for the organism to multiply and infect the skin. Hair becomes matted in clumps. Patches of irregular scabs, often oozing serum from beneath, may form along the horse's back, rump, shoulders, or fronts of the cannon bones — anywhere the horse's body was continually wet. The problem often lingers into winter, especially if the weather stays wet, and the skin may become raw and sore. A thick winter coat provides ideal conditions for the fungus to multiply.

In early stages of this skin disease, it is easier to feel the emerging bumps than to see them; they have not yet developed crusts that affect the hair. The bumps may be hot and tender, and the horse may be sensitive when you brush him. If your horse reacts adversely to being groomed, look more closely at his skin.

Rainrot is not contagious from horse to horse, but it may appear in several horses at once if they are kept in the same dusty pen and then experience a long period of wet weather. It can also be spread by dust and dirt on grooming tools and saddle

A dirty horse is more at risk for rainrot when wet.

pads, so be sure to keep equipment clean. Rainrot is most common in humid regions with periods of prolonged wet weather, such as the northern Pacific Coast, Southeast, and Gulf states.

Treatment

Treatment consists of cleaning up the skin and hair, using an iodine shampoo (or a human dandruff shampoo or antiseptic pet shampoo), and then applying medication recommended by your veterinarian. Daily bathing for several days may be necessary to get rid of fungus-laden dirt in the hair coat. Clean the skin thoroughly. Use a soft brush or your fingers to massage the skin as you wash it, gently working the scabs loose so you can pick them off. The horse may be a little sensitive to having the hairy scabs picked off, but the skin underneath is not so sensitive.

After the horse is clean and dry, you can use an ichthammol salve (ointment made from coal tar), or whatever your vet recommends, on affected areas. Mineral oil mixed with tamed iodine was a traditional treatment, but most vets don't recommend this anymore because the fungus grows best in the absence of oxygen, and oil seals it away from the air. The horse will get over the infection rapidly if you keep him out of the rain during treatment or use a blanket to keep water off his back while the skin is healing. Treatment is important because neglected cases may develop bacterial infection in the raw areas and the resultant staph or strep infection may be harder to treat.

Prevention

Rainrot is rarely a problem in dry weather because the organism needs moisture to multiply and cause sores; therefore, the best prevention is to keep the horse clean and dry. In climates where rainrot is a recurring problem, you should follow a program of regular grooming with periodic vacuuming to get as much dust and dirt as possible out of the hair, and a bath twice a month with a medicated pet shampoo.

Horse vacuums can be purchased through many mail order supply companies. People who show horses often use them, to help get all the dirt and dust out of the hair coat. Vacuuming is sometimes healthier for the horse than constant washing.

Scratches

Scratches is a skin problem on the lower legs that affects horses most often when they walk through wet areas; the fungus (and sometimes bacterium) is present in boggy mud or in the dust of a paddock. If the horse has a nick or scrape on the skin, the fungus can enter. The infected leg area becomes crusty, scabby, and thickened. In severe cases, the skin may ooze or the lower leg may swell, and the horse may become lame. This condition affects horses with unpigmented (white) skin more readily than dark-skinned ones.

Treatment

Traditional treatments for scratches comprise astringents such as methylene blue, iodine and glycerine mix, ointments made with zinc oxide, nitrofurazone, and steroids. A better treatment involves thoroughly scrubbing affected areas and applying a mix of one part nitrofurazone ointment (antibiotic), one part DMSO, and one part dewormer paste containing a benzimidazole (thiabendazole, fenbendazole, cambendazole, oxibendazole, oxfendizole, or mebendazole — good fungicides as well as dewormers).

DMSO reduces swelling and inflammation and also helps the fungicide penetrate the area deeply, taking the medication into the underlying tissues. The nitrofurazone combats bacterial infection and buffers the DMSO so it won't burn or irritate the skin. The wormer paste kills the fungus. Applied daily, this mixture clears up most cases of scratches quickly. Bandaging is not necessary. In fact, it's actually detrimental; you don't want to hold the moisture in.

Prevention

Prevent scratches by keeping the horse in a drier environment so he doesn't have to walk through mud. Prevent serious cases by treating small lesions when they first appear, never allowing them to become extensive and deep.

Warts

Warts are a common problem in horses, especially young ones (yearlings and 2-year-olds), and are caused by a virus that lives at the surface of the growth. They spread from animal to animal by direct contact. On horses, they are often spread by grooming equipment or tack used on more than one horse, or by a person who touches an infected horse and then a susceptible one. Small cuts and scrapes in the skin can lead to wart infection if the virus is present. Once established on the animal, warts can last for several months up to year. After exposure, a susceptible animal may break out with warts a few weeks later. Infected horses often pass the virus to other horses before their own warts are evident because of the long incubation period.

WARTS AND IMMUNITY

Once warts start to appear, they usually develop rapidly over several days and continue to grow and spread for several weeks. Warts affect young animals because they have not yet developed immunity to the virus, and they occur in older animals who have not been previously exposed and have no immunity.

Warts are unsightly but not a serious problem for the animal unless they grow large enough to interfere with breathing. In horses, they are usually found in clusters around the muzzle but, occasionally, they appear on other parts of the body or inside the ear. They are distinguished from other skin growths by their dry, rough appearance. They may be pale or gray.

Treatment

Warts usually run their course and then disappear after the animal builds immunity. When left alone, they often go away after a while. If they must be removed for health reasons, have your veterinarian remove them surgically. You can hasten their disappearance by crushing them with forceps or twisting off the larger ones; in fact, the veterinarian may choose this method. It encourages the immune system to fight them because some of the wart tissue makes contact with the bloodstream, stimulating the animal to make antibodies.

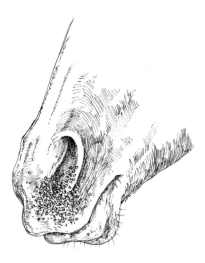

Horse's muzzle covered with small warts

Sarcoid Tumors

Sarcoids are also caused by a virus. Some horses develop them following an injury or break in the skin; others for no apparent reason. The tumor first looks like proud flesh (excessive granulation tissue that protrudes from a healing wound; see chapter 16) but continues to grow, producing a problem that may ultimately threaten the horse's usefulness.

Sarcoid tumors do not travel through the bloodstream or lymph system to other parts of the body, but they do continue to spread and grow locally. They are moderately malignant in that they usually grow back if removed — often in a larger form. If not removed early, a tumor may become so large that it disables the horse. Because removal often stimulates regrowth, some veterinarians leave small, nongrowing sarcoids alone.

At first, a sarcoid is hard to differentiate from proud flesh. The virus attacks young fibrous cells in the proud flesh and causes them to multiply rapidly. This excessive growth cannot be controlled by medications used to control proud flesh. Wounds in areas where there is motion (joints or moving parts on the legs) seem most susceptible to sarcoid tumors. The growth is usually mushroom shaped; the base of the attachment is generally smaller than the outer surface. Wounds that are not properly treated to prevent contamination, irritation, and delayed healing are quite susceptible to sarcoid.

Treatment

Sarcoids can sometimes be successfully removed with surgery, followed first by cauterization of the area to control bleeding and kill stray tumor cells, and then by radiation treatment. The area is kept clean and bandaged while it heals. Other removal techniques include freezing and burning. About 50 percent of them will grow back, however, so other treatments are often tried.

A vaccine made from the tumors has also been effective in stimulating the immune system to build its own defense. The treated tumors usually shrink and eventually dry up and fall off. The vaccine doesn't always work, but it is one of the most effective treatments available and rarely leaves blemishes.

Yet another treatment involves use of Aldara cream that contains a drug (imiquimod) that fights the virus by stimulating the body's immune system. This is a human prescription drug that your veterinarian can prescribe for your horse, to be applied three times a week for several months until the tumor shrinks and disappears.

13

Dental Care and Mouth Problems

MOST PROBLEMS IN A HORSE'S MOUTH are related to his teeth and can be resolved by periodic dental care such as filing off sharp edges. Sometimes, a horse gets a sore mouth because of problems in the soft tissues of the gums or roof of the mouth. If you have your veterinarian check a horse's mouth and teeth once or twice a year and address any problems promptly, small problems will not become serious and he will be healthier and more agreeable when you handle him.

Sore Mouth

There are a number of problems that can affect the soft tissues of the mouth and cause a horse discomfort. As a result, he may have trouble eating or resent the use of a bit.

Gums

Irritation to gums can be caused by rough feed, tartar buildup, gaps between the teeth where food catches and exerts pressure, swellings from injuries and bruises, weak gum tissue in older horses, irritation from equipment such as an ill-fitting bit, harsh use of any bit, or lip chains and twitches. Gum injury sometimes leads to gum disease. The bruising may result in gingivitis (an irritation of the gums), which can lead to tooth or bone problems if not resolved.

Tartar Buildup

Often, tartar buildup starts as gum deterioration, putting pressure on surrounding gums. The most common place for buildup is on the canine teeth

— the small, sharp teeth that appear a short distance behind the front teeth and are present in most male horses and a few females. Horses at pasture rarely develop tartar. Finely ground feeds such as pellets and mashes tend to cause more buildup on the teeth than do hay and roughages; the latter actually clean the teeth and gums.

Lampas

Lampas is a swelling of the fleshy lining of the horse's hard palate — the roof of his mouth just behind his top front teeth. You may not notice the swelling until you examine the horse's mouth. The condition is often first discovered by a veterinarian who looks at the teeth to determine why the horse is not eating well or spilling feed out of his mouth.

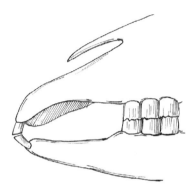

Lampas; swelling of the roof of the mouth behind the front teeth

Lampas commonly occurs in young horses when they are introduced to hard feeds such as grain or pellets. Because they have not yet learned how to handle it in their mouths, they bump and bruise the tender roof of the mouth right behind the front teeth, which causes swelling and pain. As a result they may go off feed and lose weight.

Tongue Injuries

Occasionally, a horse bites his tongue (as when hitting his head while falling) or has it injured by a bit. If a horse steps on his reins when wearing a bridle and jerks his head up, he can seriously injure the tongue. The horse may bleed profusely for a while from a cut tongue, but it usually heals. Sometimes, the

SOFT FEED SOOTHES LAMPAS

If a young horse develops lampas as a result of eating hard or abrasive feed, try switching to a very soft pellet or softening his grain by soaking. Good pasture grass is the most ideal and natural feed for a young horse; turnout on pasture — with less dependence on grain or hay — may quickly solve the problem.

injury makes it difficult for him to eat or interferes with future use of a bit. Have your veterinarian check the horse.

Embedded Seeds

Sharp seeds, such as those of cheatgrass and foxtail, can become embedded in a horse's mouth and create a sore or abscess. If the horse has a sore mouth, he moves the food around more than usual while chewing or even spits it out. A sore mouth may make him uncomfortable when wearing a bridle. If the horse has any obvious swellings at the mouth or cheeks, or has trouble eating, call your veterinarian. The horse may have an embedded seed or a tooth problem.

The Teeth

Horses have two kinds of teeth — incisors in front (used for biting off grass), and premolars and molars in back (used for grinding feed). Between the incisors and the premolars is the interdental space, a gap of several inches.

HOW TEETH "GROW"

Sometimes horses have tooth problems because of the way their teeth develop. Whereas the teeth of human beings and other carnivores grow rapidly and then stop once the permanent teeth are in, the horse's teeth keep erupting slowly through the gums throughout his life to compensate for the constant wear of chewing feed. The teeth do not actually become longer but are gradually pushed through the jaw bone out of the sockets. At first, the crown of the tooth wears down, then the neck, and then the root. The very old horse may end up with short stubs that eventually fall out.

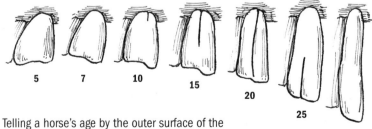

Telling a horse's age by the outer surface of the corner incisor: Galvayne's groove appears at age 10, reaches the lower edge of the tooth by age 20, and gradually disappears, from the top first, by age 30

A foal's teeth emerge soon after birth. First come the central incisors and initial set of premolars. By the end of his second week, the next two sets of premolars appear. The second set of incisors comes in between 4 and 6 weeks of age, then the third set between 6 and 9 months of age. The first set of wolf teeth may come in at about 5 or 6 months of age — they are technically first premolars, just ahead of the large second premolars — and are often shed around the same time as the temporary first premolars. Some horses have wolf teeth and others don't. Time of eruption is variable.

The first permanent molars appear at the back of the mouth behind the baby teeth, between the ages of 9 and 12 months, and the second set comes in (replacing baby teeth) at around 2 years of age. The baby teeth begin to shed at about 2½ years of age, replaced by permanent teeth in the same order they appeared. All the baby teeth are replaced by the time the horse is about 4½ years old. At that time, he has 36 teeth (12 incisors and 24 molars). In addition, he may have as many as four wolf teeth and four canines.

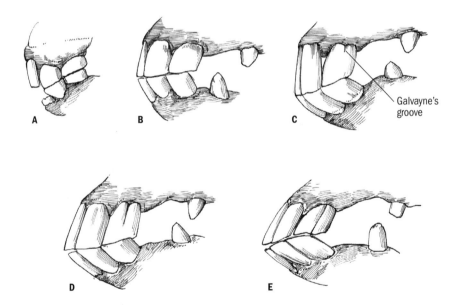

Incisors of horses at different ages; notice how the angle of the teeth changes over time: **A.** Teeth at 2½ years **B.** Teeth at 5 years (full mouth) **C.** Teeth at 10 years; notice Galvayne's groove beginning at top corner incisor **D.** Teeth at 15 years; notice Galvayne's groove extending halfway down tooth and incisors angling forward **E.** Teeth at 20 years; notice Galvayne's groove extending down length of corner incisor and incisors protruding farther forward

Canine Teeth

Not all horses have canine teeth. Male horses have them and so do a few mares, but theirs are much smaller. Canine teeth, if they come in, emerge behind the incisors at about 4 years of age. A horse cutting canine teeth may be irritable and go off feed or salivate a lot. The lower canines are usually larger than the upper ones and may be more than an inch long and fairly sharp. When putting your fingers in the interdental space to paste worm a horse or to encourage him to open for the bit, be careful to avoid the canines.

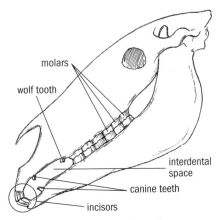

Canine teeth appear a short distance behind the incisors, top and bottom. Wolf teeth are located just in front of the back teeth, often just in the upper jaw.

Canine teeth appear early and are shed with the baby teeth. Permanent canines appear when the horse is between 4 and 5 years old. Canines rarely cause problems and are best left alone unless too sharp, in which case they can be smoothed with a dental rasp by your veterinarian.

Wolf Teeth

There is often confusion between wolf teeth and canine teeth. Wolf teeth sit farther back, next to the first premolars, and are harder to see. You must open the mouth and look toward the molars to find them. Most wolf teeth appear in the upper jaw and come in when the horse is 5 or 6 months old. Wolf teeth may be large and well formed, sometimes as large as a molar, but they are usually small and shallow rooted. They can cause problems with using a bit and many people have them removed when they start a young horse in training, before putting a bit in his mouth. If they cause pain and head tossing, or a bad attitude about the bit, it's best to have them taken out.

Tooth Problems

Good teeth are crucial to the horse's health and comfort. If he has a bad tooth, it may cause pain that interferes with chewing. Indeed, the condition of the teeth affects the horse's entire performance. A horse with a bad tooth may resent the bridle, travel with his head crooked, hold his head stiff, or become clumsy because of the awkward way he travels. Horses rarely suffer from tooth decay or cavities but are plagued by a number of other tooth problems.

TEETH TELL THE HORSE'S AGE

To keep the grinding surface rough, teeth are made up of three materials that vary in hardness and wear away at different rates. It is this rate of wear that enables horse people and veterinarians to tell a horse's age by looking at the teeth; different areas of the surface wear away at different stages in the horse's life. In a young horse, the age can also be determined by noting which teeth are present.

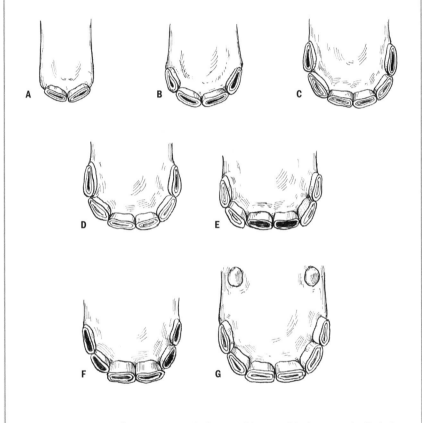

How teeth wear at different ages: **A.** At 8 days old, central incisors are in. **B.** At 2 to 3 months, central and lateral incisors are in. **C.** At 8 to 10 months, corner incisors appear. **D.** At 2 years, centrals and laterals are wearing. **E.** At 2½ to 3 years, permanent centrals are coming in and show no wear. **F.** At 3½ to 4 years, centrals are fully in, showing wear, and the permanent laterals are coming in. (Note that the corner teeth are still baby teeth.) **G.** At 4½ to 5 years, all permanent incisors are in; also, male horses will have their canine teeth by this time.

Retained Caps

Human baby teeth come loose as the permanent teeth come in. In horses, as a permanent tooth erupts, the baby tooth deteriorates and grows hollow, becoming a cap. Sometimes, these do not detach from the gums as they should when permanent teeth push them out. This condition, known as a "retained cap," can cause inflammation of gums, a painful mouth, and sometimes a sinus problem if it is in the upper jaw. If not removed, the cap may cause the new tooth to grow in at an improper angle or become impacted.

Sometimes, caps are retained when there's not enough room in the jaw for cheek teeth to expand. Signs that a horse has retained caps are bad breath, drooling, tilting the head back while eating, and dribbling feed from the mouth. Loose caps may move and irritate the horse and provide entry for infection. Bacteria from the infection may find their way into the bloodstream and cause serious trouble, so don't neglect any tooth problem.

Impacted Teeth

A young horse (about age 3) may develop bony protrusions on the lower jaw as the permanent cheek teeth come in. First and fourth molars may pinch the second and third ones as they emerge, temporarily inhibiting their upward growth, especially if the horse has a narrow jaw. This impaction creates bony lumps on the bottom of the lower jaw.

"Tooth bumps" on the lower jaw of a young horse indicate that the permanent molars are coming in.

Most of the time, impacted teeth gradually force their way up, and the problem eventually corrects itself. The second molar comes in at age 3; the third emerges at age 4. After the teeth come into their proper places, the bony protrusions on the jaw smooth out. The lumps are gone by the time the horse is 5, 6, or 7 years old. Sometimes, the impaction doesn't correct itself and the teeth must be surgically removed. If a young horse has "tooth bumps" that rapidly become larger or are quite tender when touched, have your vet examine the jaw and mouth. X-rays may be necessary to see whether the teeth must be removed.

Uneven Wear

Often, sharp points develop on the molars from uneven wear. Remember that the horse's incisors and molars continue to emerge upward, with the rate of growth balanced by an equal rate of wear. If lower and upper teeth are not in perfect opposition, uneven wear creates sharp points — usually on the inside of the lower cheek teeth and on the outside of the upper cheek teeth.

> ### SIGNS OF TOOTH TROUBLE
>
> - Difficulty chewing
> - Quidding (dropping wads of chewed food)
> - Drooling
> - Weight loss
> - Pain in the mouth or sides of the face

Because the upper jaw is wider than the lower jaw, the upper molars are slightly farther apart than the lower ones. When the molars do not meet evenly, they don't wear properly, so sharp points are a common problem in older horses.

Sharp Ridges

Sharp ridges on teeth interfere with the horse's ability to grind his food, and they may lacerate the tongue and sides of the mouth. The sharp edges can cause great discomfort, making cuts along the tongue or inner surface of the mouth. A horse with a jagged tooth may not chew his food completely; it passes through his digestive tract partly undigested, leaving him prone to colic and impaction. In serious cases, the horse may starve because chewing is so painful that he refuses to eat adequately.

Malocclusion: Parrot Mouth and Sow Mouth

Improper jaw and mouth conformation contribute to the likelihood of a horse having tooth problems. A common abnormality is "parrot mouth," or overbite — the top teeth (molars as well as incisors) are not aligned with the bottom ones and protrude over the lower teeth. As a result, the last lower molar also protrudes up behind the last upper molar. The overlapping tooth grows long because it has nothing to grind against, creating a "hook."

The opposite condition, "sow mouth," occurs when bottom teeth project farther forward than top ones, but this is not as common as parrot mouth. In a horse with sow mouth, the last upper molar protrudes down behind the lower one, eventually creating a hook. These sharp hooks sometimes become so long that they puncture the opposite gum. Your veterinarian or equine dentist can file them down or cut them off with nippers or shears. Mismatched jaw

A. Parrot mouth, or overbite — the top incisors protrude past the bottom ones
B. Sow mouth, or underbite — the lower jaw and teeth extend beyond the top ones

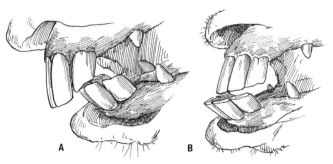

A   B

length is an inherited trait; therefore, do not use horses with parrot mouth or sow mouth in a breeding program.

Shear Mouth

Another conformation abnormality in some horses is an exceptionally narrow lower jaw. In these horses, the molars do not meet squarely and the grinding action eventually wears the teeth at an angle. This makes the grinding surfaces so steep that the teeth begin to slide past one another — a condition known as "shear mouth." If this is discovered before the teeth start sliding past one another, regular dental care to make sure the grinding surfaces remain normal will keep the problem in check.

Lost Tooth

A horse may occasionally lose a tooth. When this happens, the opposing one has nothing to grind against and keeps growing. It may get so long that it grows into the opposite gum unless it is cut off or periodically rasped down. A horse may also lose the central core of a molar, which creates a channel down into the root, opening the way for infection. When this happens, the tooth becomes abscessed and infection may break into the sinus cavities, creating foul pus that discharges through the horse's nostrils. A veterinarian must remove the abscessed molar or drill the central core. Sometimes, a horse's teeth become damaged or broken, in which case the broken pieces must be removed.

HEALTH TIP

Whether a horse will have tooth problems depends on the individual and also on the type of feed he has been eating over the years. In general, horses have fewer tooth problems if they've been on pasture most of their lives rather than fed hay and grain. The teeth wear more normally when the horse is on pasture, grazing.

OLD HORSES NEED DENTAL CARE

As a horse becomes elderly, his teeth may start to fall out or wear down to the gums. It may become difficult for him to eat coarse roughage; green grass and soft feeds are easier to chew than hay. It is a good idea to feed older horses by themselves so they don't have to compete for feed with younger horses who eat more rapidly. If your older horse eats too slowly or begins to lose weight, it is time to have his teeth checked again.

Dental Checkups

As a horse gets older, dental problems are more likely to occur — so you should have his teeth checked. Horses between the ages of two and four may also have problems. During this period, it's wise to have the mouth examined two or three times a year. If necessary, premolar caps or wolf teeth can be removed and sharp edges filed. Most horses, however, get by with a dental examination just once or twice a year.

Floating

If there are any sharp edges, they can be rasped smooth. This filing action is called "floating"; the rasp is called a float. The chewing surface itself is not rasped, just the rough uneven spots and sharp hooks that develop along the edges of the molars. The veterinarian or equine dentist puts the float into the mouth with the rasping surface next to the tooth that needs filing. He or she then moves it firmly back and forth, slowly at first, then more rapidly as irregularities are rasped down. If the file is kept in close contact with the tooth edges, it will not bump or injure the tongue or cheek. Most horses tolerate the procedure since it is not painful if done properly. There are nerves in the horse's teeth, but they are deep below the grinding surface, so filing off the sharp edges causes no pain.

The veterinarian or equine dentist may examine the mouth by grasping the horse's tongue (through the space between incisors and molars) and holding it outside the mouth, which forces the horse to keep his mouth open. The veterinarian then checks the teeth while feeling for sharp tooth edges. Most veterinarians use a gag or mouth speculum to hold the mouth open, but nervous horses resent this and tolerate the procedure better if the veterinarian merely holds on to the tongue. A veterinarian who uses a speculum typically gives the horse a mild sedative or tranquilizer during a dental exam.

14

Eyes and Ears

THE HORSE DEPENDS ON KEEN EYESIGHT and hearing to become aware of predators and danger. His eyes are located at the widest part of his head to give a good view of everything around him and most things behind him. In this position, the eyes are vulnerable to injury and sometimes get bumped if he tosses his head when startled in close quarters. Flicking branches may also injure the eyes. A blow to the eye rarely ruptures it, however, because of its design; there's a fatty cavity behind it, into which the eye can retreat without bursting.

Ears are sensitive too, but highly mobile. The horse can move them in any direction and flatten them back against his neck to protect them from an aggressive, biting herdmate.

THE NORMAL EYE

To detect an eye problem, you must first know what a normal eye looks like. Then you can notice abnormalities. The normal pupil is a horizontal oval, large in dim light, and constricting to about half its original size in bright light. The pupil of a blind eye may fail to constrict at all. The upper edge of the pupil has projections from the iris (the colored portion of the eye) that make a tree-like pattern in the pupil — like an awning over it. This keeps out bright light. The cornea has a smooth, clear surface.

PARTS OF THE EQUINE EYE

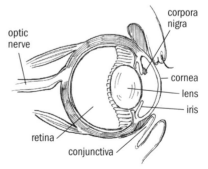

optic nerve · corpora nigra · cornea · lens · iris · retina · conjunctiva

---

WATCH FOR SIGNS OF TROUBLE

Some of the simplest eye problems often produce the most obvious symptoms, whereas more serious problems may be harder to notice. Early recognition and proper treatment usually make the difference in outcome. Check your horse's eyes closely whenever something seems amiss, and get professional help for diagnosis and treatment if you are unsure of the cause of the problem or how best to treat it.

Every visible eye disorder shows at least one of several possible signs:

1. Squinting or keeping the eye completely closed
2. Tears or pus
3. Constriction of the pupil
4. Discoloration behind the cornea (the transparent membrane covering the visible part of the eye)
5. Damage to the cornea

Signs 1 and 2 above are most common but often the least serious. If either or both of these signs are combined with any of the others, however, consult your veterinarian promptly.

Eyes and ears are vulnerable to problems, and the horse owner should be aware of these threats to his animal's health.

Eye Problems

A scrape or blow may result in pain and watering or a blue spot on the eye. Cataracts make the lens cloudy. Tiny slivers or grass seeds can become embedded under the lid and scrape the eye, causing an ulcer. Infections sometimes damage inner parts of the eye and lead to impaired vision or blindness. You should never ignore a horse's eye irritation. Minor problems often become serious if they are neglected.

Squinting and Tears

If these are the only signs of trouble, the horse probably has "conjunctivitis," that is, local irritation of the normally pinkish lining of the eyelid (the conjunctiva). Dust, smog, or some other irritant usually causes this most common and least serious eye problem. If irritation leads to infection, the eyelids may become swollen and there may be yellow mucus or pus at the corners of the eye. Conjunctivitis can be treated with eye ointment or eye drops.

Sometimes, a horse has a watery eye from a blocked tear duct — clogged by mucus or swollen shut by inflammation. If the eye is constantly running or seeping, your veterinarian may have to flush out the blocked tear duct from the nose end, using antibiotic solution and sometimes steroids.

Corneal Injury

Another common eye problem is corneal injury from a tree branch, a slash across the eye by another horse's tail, a blow from a whip, or anything that rubs or scrapes the outer covering of the eye. These injuries cause severe pain. If the eye becomes infected, the horse will need an antibiotic ointment. If it does not become infected, the scratched eyeball may heal within 24 to 36 hours. Large injuries and those that are slow to heal should have daily treatment for a while, with guidance from your veterinarian, to help prevent or eliminate infections that might damage inner tissues.

Lacerated Eyelids

Nails, barbed wire, or sharp protrusions in a stall can catch and tear a horse's eyelid. Any eyelid laceration should be given immediate attention by your veterinarian. Most of these rips can be repaired. The veterinarian will tranquilize the horse, wash the torn skin, and suture it back together. Most eyelid injuries heal well unless a chunk of eyelid has actually been torn away.

Cataract

Loss of transparency behind the pupil is a cataract, which can be caused by a low-grade infection or old age. Sometimes, a horse is born with cataracts. There is no pain, but haziness dims the animal's clarity of vision. By the time the older horse's lens becomes cloudy, the problem is already well advanced. Cataracts generally result in blindness but, sometimes, they can be removed — especially in young horses whose lenses are still soft.

Embedded Seeds and Slivers

Foreign bodies that get caught in the conjunctiva (the delicate membrane that lines the eyelids and covers the exposed surface of the white of the eye) are usually seed awns or other bits of plant material from bedding, feed, or tall pasture. Large seeds and pieces of hay or straw in the eye usually travel to the corner, where tears wash them out. Sometimes, however, they get trapped under the eyelid and cause direct irritation and scraping of the cornea, which leads to erosion or ulceration of that transparent covering. During an eye examination, the veterinarian can usually see large foreign particles and flush them free or remove them.

Microscopic Slivers

Tiny particles less than 0.12 inch (3 mm) long are very hard to locate and remove. They easily become embedded in folds of the eyelid lining. When caught there, they create a sore by scraping against the eyeball each time the horse blinks. Symptoms include sensitivity to light, holding the eye closed, discharge and weeping, and constriction of the pupil — all of which may lead the veterinarian to think the horse has a corneal injury, until conventional treatment (administering a topical antibiotic) fails to work. See the Burdock section below for information on veterinary eye exams.

Burdock

Like cocklebur, the burdock produces prickly burrs that adhere to hair or clothing. When ripe, they release hundreds of microscopic barbed slivers that are difficult to remove from clothing, skin, or an eye. The slivers are most likely to get into the eye in fall or winter, after the seeds ripen. The inflammation persists and does not respond to treatment. The cornea may become inflamed and ulcerated; the eye may turn cloudy and have a white spot or bulge on it.

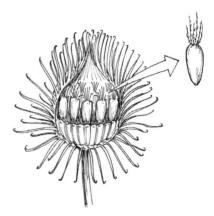

Cross-section of a burr (burdock plant) showing enlargement of tiny seed head and its microscopic barbed slivers

BEWARE OF BURDOCK

Fall and winter sometimes bring unexplained eye problems in horses and cattle: irritation, inflammation, and corneal ulcers. These are generally caused by microscopic barbed slivers embedded in the eyelid — bristles from the seed heads of the burdock plant.

If a horse has an inflamed eye that is not responding to treatment, you should suspect a burr fragment, especially if there are burdock plants on your property or you have noticed burrs in bedding or hay. Most injured or inflamed eyes respond to topical antibiotics and show improvement within 3 or 4 days of medication, but if an eye does not improve, you should take the horse to a clinic that has specialized equipment for eye examinations. Technicians will use a special dye to stain the eye, making it easier to locate the sliver with a powerful magnifying lens. Then, it can be removed with small tweezers.

Backyard horses, horses at pasture where burdock grows in wet or shady areas, and draft horses with lots of hair at the fetlocks, which may collect burrs, are at risk for burdock slivers. If there are any burrs caught in your horse's mane, tail, or fetlocks, suspect a burr sliver as the cause of his eye problem. Because burrs hang on the dead plant all winter and into spring, horses may come in contact with them even before the new plants have put forth seeds. A horse can also get a burr sliver at any time of year if his bedding contains shredded burrs that were baled with the straw.

Moonblindness (Recurrent Uveitis)

This chronic, painful eye disease is the most common cause of blindness in horses. Long ago, people thought the recurring attacks were related to phases of the moon, hence the term "moonblindness." The main causative factors were thought to be heredity; dark, damp stables; unwholesome food; and marshy pastures. The actual cause, however, is an infection.

Moonblindness (also called "periodic ophthalmia") is inflammation of the inside of the eye caused by an abnormal immune response. It is generally triggered by local or systemic infection. One of the common causes is leptospirosis, a bacterial disease spread by rodents and other animals that contaminate feed and water with their urine. Other causes include trauma to the eye, bacterial infection with brucellosis, streptococcus, *Rhodococcus equi*, and *Borrelia bergdorferi* (which causes Lyme disease). Some viral infections can also trigger an attack, as can certain parasites and systemic infections. The recurring episodes of inflammation within the eye may develop weeks or months after the initial uveitis (inflammation in the uveal tract — iris and tissues that line the part of the eye behind the iris) subsides.

Often, the initial infection or eye injury goes unnoticed until the reaction occurs in the eye. Attacks of inflammation often recur, and unless it is treated properly in early stages, this eye problem may lead to blindness in the affected eye. It's not the infection itself that causes the inflammation, but the eye's immune response. Leptospirosis commonly causes uveitis because the molecular makeup of these pathogens is similar to some of the structures in the eye. When the body reacts against the lepto bacteria, the immune response also attacks the eye itself. If the immune response recognizes those same proteins again, this triggers another episode.

Symptoms

Early symptoms are a single watery eye and a constricting pupil; the horse holds his eye shut because it is sensitive to light. The membranes are red and swollen, and the cornea becomes opaque and cloudy. Pus may form in the lower part of the eyeball. Attacks last for a few days or weeks, then the eye seems to get better. The problem recurs, however, a few weeks or up to a year later, and each attack further damages the eye. Close observation of the eye between attacks shows it is not normal; there is a bluish ring around the cornea, and the pupil may stay somewhat constricted except in advanced cases — when it is widely dilated and cloudy. Occasionally a horse eventually develops a similar problem in the other eye.

Causes

One of the most common causes of moonblindness is leptospirosis (see chapter 8). Other causes include streptococcus infections such as strangles, and a filarid worm (a parasitic larva) whose immature stage is spent in the bloodstream of the host horse; the worms are spread to others by biting flies. The eye condition is not contagious. It may not occur for a year or more after the horse has leptospirosis or other infections; a second or third exposure to the bacteria may bring it on with a more serious reaction in the eye.

Treatment

A horse with an attack of moonblindness should receive veterinary attention. Administering medication and keeping the horse in a dark barn during treatment are important. The goal is to try to preserve vision and relieve pain. A preparation of sulfacetamide (an antibiotic) and prednisolone acetate (steroidal human eye medication) can be effective. Put two or three drops in the eye twice daily for 10 days to 2 weeks. A drop of atropine in the eye at the beginning of treatment, to keep the pupil dilated, is also helpful.

Various steroids and nonsteroidal anti-inflammatory drugs will work in about 90 percent of cases. A more recent innovation involves a drug implant that delivers cyclosporine (to keep the self-attacking immune system at bay) over a 3- to 4-year period. The implant is put beneath a flap of tissue on the outside of the eyeball, where it produces a constant level of the drug directly into the eye, eliminating the need for the horse owner to continually medicate the animal. Treatment is most successful if begun early, before the eye has been severely damaged by several attacks. Untreated horses eventually become blind.

Prevention

Moonblindness is somewhat preventable by protecting horses against leptospirosis, streptococcal diseases such as strangles, and parasitic larvae. Because leptospirosis is often a mild disease in horses, many people do not realize their horse has had it unless it produces serious side effects such as abortion or moonblindness.

Night Blindness

Moonblindness should not be confused with night blindness, which is an inherited abnormality of the retina in some horses, predominantly Appaloosas. It is inherited as a recessive trait, which means that both parents must be carriers of the abnormal gene for it to show up in the offspring. Most horses affected with night blindness begin to show nighttime visual impairment as foals. In some individuals, vision is affected in daylight as well. The problem does not worsen (as with moonblindness) and is unlikely to result in total blindness. Affected eyes look normal; electronic tests are necessary to make a diagnosis. There is no known treatment for night blindness. You usually can prevent it by not breeding any mare or stallion known to produce a foal with this disorder, although an affected foal may crop up when unknown carriers are bred.

APPLYING EYE MEDICATIONS

When giving eye drops or ointment, have someone hold the horse's halter to keep his head still. Then, gently insert the ointment or the eye drops into the front corner of the eye nearest his nose. The horse will likely close his eye, but you can usually hold the eyelid apart enough at the corner to insert the medication. If the horse is uncooperative, use a Stableizer or twitch to keep him calm and relaxed while you apply the medication (see chapter 5).

Ear Problems

The horse's ears are very sensitive and delicate. The ear canal at the base drops straight down about 2 inches (5.1 cm), then turns sharply before reaching the eardrum. This sharp turn keeps insects, dust, and dirt from getting down against the eardrum. The fine inner hairs also provide protection against foreign material. If a horse becomes sensitive about having his ears handled because they hurt, he may have an infection or irritation within the ear. Flies or crawling insects may have gotten into the ear canal, or dirt and trimmed hairs may have fallen down inside. You should never trim the fine inner hairs.

Internal Infections and Trauma

Horses rarely suffer hearing impairment but occasionally lose their hearing from damage to the eardrum. Penetration by a sharp object, a hard blow to the head, or infection in the middle ear can cause a punctured eardrum.

Inflammation of the guttural pouch sometimes causes swelling beneath the ear. This pouch is a blind sac that opens into the Eustachian tube between ear and throat. The guttural pouch may fill with pus if it becomes infected, causing chronic nasal discharge.

Occasionally, inflammation of the salivary gland below the ear causes pain and swelling. This makes the horse carry his head in an abnormal position and shy away from having his ear handled.

Middle-Ear Infection

Infection is the most common ailment of the middle ear. It may be caused by bacteria or viruses that enter the ear through the bloodstream or from the throat through the Eustachian tube. Pus may develop in the ear, and the horse may have a fever, carry his head at an odd angle, or experience difficulty chewing. Middle ear infection makes him ear shy. At first, the infection may not produce any symptoms except ear rubbing and head tossing. As it gets

EFFECTS OF PARTIAL PARALYSIS

Infection in the middle ear irritates surrounding tissues and can cause abnormal bone growth or damage facial nerves, paralyzing the muscles on that side of the horse's face. The ear and eyelid droop, the muscles of the face and lip sag, and the horse may have trouble swallowing.

EAR INFECTIONS
DIFFICULT TO DIAGNOSE

It is hard to accurately diagnose an ear infection. The head-tossing, ear-rubbing, and other expressions of pain are often mistaken for other problems or bad manners. After bony enlargements develop, they show on X-rays; otherwise the head-tossing, pain at the base of the ear, and — in some cases — eating problems, are the only outward signs. Middle-ear infections, if discovered early enough, should be treated with antibiotics. Surgery will relieve stress and pressure on the ear bones and prevent eventual skull fracture.

worse, the infection builds up pressure inside the middle ear, affecting the horse's balance and behavior. An untreated infection can eventually lead to skull fracture, convulsions, or death.

The eardrum is tough and unlikely to break; therefore, infection cannot drain to the outside. Instead, it may spread to bones surrounding the ear at the base of the skull. Inflammation causes new bone growth, which fuses the joint between the base of the skull and the bones that extend to the tongue. Eventually, the movement of the tongue breaks the new bony fusion, causing a skull fracture and subsequent contamination of tissues around the brain. This kills the horse.

External Ear Problems

Most ear problems are minor ailments involving the external ear; they cause discomfort rather than hearing loss or threat to life. External ear problems include frostbite, lacerations, bites from another horse, skin infections, warts, tumors, and insect irritations. Skin infections and insect bites may cause the horse to shake his head in annoyance.

Flies

Flies and other external parasites sometimes get into the ears and cause discomfort. Horn flies, which are very small, crawl in past the protective ear hairs to suck blood. Black flies leave bloody scabs and crusts inside the ear where they have bitten. Some small flies leave crusts that itch and become sore. You can relieve this soreness by applying medications that contain corticosteroids to the insides of the ears.

To protect the delicate insides of the ears from fly bites, gently smear a small amount of petroleum jelly on the inner surfaces of each ear as far down as your finger will reach but without contacting the ear canal structure at the bottom. The flies can't bite through it and will leave the ears alone. The jelly is also soothing if the horse already has itchy fly bites. Ear nets drenched in fly repellent also keep flies from getting into the horse's ears.

Ear Ticks

Several ticks, including the spinose ear tick, occasionally get into the ears. The larvae of the spinose ear tick burrow into the ear canal and grow there for several months, causing discomfort. A horse with ear ticks shakes his head a lot, rubs his ears, and has a droop-eared, unhappy look. Some types of mites also cause problems in ears.

To get rid of ear ticks or mites, use an insecticidal ointment from your veterinarian. It goes directly into the ear canal with a special syringe or an oil can that has been

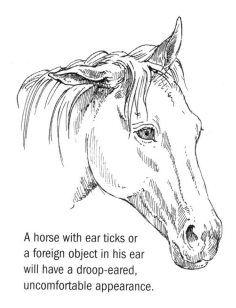

A horse with ear ticks or a foreign object in his ear will have a droop-eared, uncomfortable appearance.

thoroughly cleaned with soap and water. The long neck on the oil can gets the medication deep into the ear. If the horse is sensitive about having his ears handled, use a twitch or Stableizer to keep him mellow so he will be less apt to move his head during treatment (see chapter 5). The ointment kills ticks and mites and keeps the horse from becoming reinfested for about 2 to 3 weeks. Worming a horse with ivermectin also eliminates blood-sucking parasites and controls ear ticks, lice, and some types of mites.

Ear Mites

Excessive infestations of ear mites eventually destroy a horse's hearing. The inflammation and build-up of debris in the ear canal interfere with the animal's perception of high-pitched sounds. Mites and ticks also make the ear secrete too much earwax. Your veterinarian may need to administer a special solvent to dissolve the excessive wax so the horse can shake it out of his ear.

Warts

Warts are a common problem in ears, especially in young horses. Horses usually don't mind ear warts, so they are best left alone unless they become large and begin to bother the horse. Most warts eventually go away on their own; however, they cause discomfort when irritated. If your horse suddenly becomes difficult to bridle for no apparent reason, look in his ears for warts or other irritations. Your veterinarian can remove the warts.

Foreign Objects

Water, small twigs, leaves, or even pebbles can fall into an ear under certain circumstances and cause instant and constant pain. It can be hard for a horse to dislodge a foreign object from his ear because he cannot turn the ear completely upside down. All he can do is shake his head from side to side, which may drive the object in even farther. The hairs inside the ear usually keep foreign objects from penetrating too deeply or quickly so the horse can shake them out. If a horse gets water in his ears, he is usually able to shake most of it out and the rest evaporates.

PAY ATTENTION TO EARS

If a horse starts doing a lot of ear-rubbing and head-shaking or becomes reluctant to have his ears handled, try to discover the cause of discomfort. Have your veterinarian examine any swelling, lumps, or discharges. Don't ignore a horse's attempts to tell you he is uncomfortable. A thorough examination of his ears, and proper diagnosis and treatment of the problem, may be crucial to his future well-being — or even his life.

15

Digestive Tract Problems and Colic

THE TERM *COLIC* SIMPLY MEANS "abdominal pain"; it refers to symptoms rather than to a specific disease or cause of discomfort. However, colic due to digestive tract problems is one of the most common causes of death in horses. Many things can cause abdominal pain in horses; therefore, there are many "kinds" of colic. The term is often used to describe a problem originating within the digestive tract, but the same painful symptoms may also occur with uterine contractions in a foaling mare or with malfunction in the kidneys, bladder, or other internal organs. This chapter examines sources of pain within the digestive tract.

Horses are particularly susceptible to colic arising from digestive tract malfunction. In fact, they are more likely to have colic than many other animals, partly because they cannot vomit to unload the stomach of harmful contents or burp as easily as a cow to relieve pressure from gas. The horse's digestion is not as efficient as that of a cow, so he is more likely to have problems with poor-quality feeds or feed changes. Colic sometimes occurs because of the way the digestive tract is constructed — some portions can become displaced and adversely affect or twist other parts.

Digestive-tract colic in horses is sometimes fatal because of blockage, twisted intestine, or shock from digestive tract shutdown and complications. Not all cases of colic are fatal, but a horse with signs of abdominal pain, even mild pain, needs constant watching. Call your veterinarian to determine the cause.

Signs of Colic

The horse may be merely uncomfortable or in severe pain, depending on the seriousness of the problem. Mild colic may just make him restless; he may switch his tail, look at his belly, paw, stamp his feet, or lie down. Severe colic will make him noticeably uncomfortable — getting up and down, staggering, or throwing himself on the ground to roll violently or thrash around.

Types and Causes of Colic

Pain accompanies many types of digestive problems — inflammation in the intestines, pressure from gas, spasms, or

A horse with gut pain may stagger, with legs buckling.

obstructions. The pain is caused by distention of the stomach or intestines. This may be persistent when the tract is filled with too much feed, gas, or fluid, or intermittent when there is localized, periodic pressure from spasms and increased activity in certain parts of the digestive tract.

Spasmodic Colic

The intermittent cramps of spasmodic colic are sometimes caused by over-excitement and nervousness but more commonly result from eating spoiled or moldy feed, making a sudden change in diet (going from hay to green grass, from grass hay to alfalfa, or adding a lot of grain or other feed the horse is not used to eating), taking a big drink of cold water when the horse is overheated from exertion, or doing anything that suddenly upsets the digestive tract. The affected horse suffers from short attacks of abdominal pain. He may roll, paw, and kick at his belly for a few minutes, then seem normal for a while until the next wave of pain hits. Sometimes, noisy intestinal sounds are audible when the gut is hyperactive.

Symptoms of spasmodic colic often disappear without treatment; the horse may be fine in a few hours. If attacks of pain do not subside, he should have medication to relax the digestive tract and ease the pain. Your veterinarian can administer or prescribe pain medication. Flunixin meglumine (Banamine) is often given for colic. If this does not ease the pain, the veterinarian may administer stronger painkillers, along with a tranquilizer. If the horse rolls too violently during an attack of pain, he might twist an intestine, which is much more serious than colic — and fatal unless surgically corrected.

Twisted Intestine

Whenever an intestine becomes twisted or kinked, pressure on blood vessels at the site of the twist block the blood supply to that part of the gut and food can't pass through. If the problem is not corrected quickly, that portion of gut will die from inadequate blood supply. The resulting toxins, shock, gangrene of the bowel, or rupture of the intestine will kill the horse.

Causes

A primary cause of twisted intestine is violent rolling in response to the pain of colic from some other problem in the gut such as impaction, excessive gas caused by indigestion, or parasite damage that hinders proper working of the gut. Regardless of the original cause of colic, twisted intestine becomes the fatal sequel if the horse rolls too violently in his distress and flips part of the gut over on itself or displaces it. Even after-pains from foaling (uterine cramps) may make a mare so uncomfortable that she rolls violently. Any abrupt action by the horse that makes his intestines shift can cause a twist if conditions are right. If part of the gut is quite full and the horse rolls frantically or plays hard (bucking and running, making fast stops and spinning around), the heavy gut may not turn as fast as the horse and, therefore, become twisted.

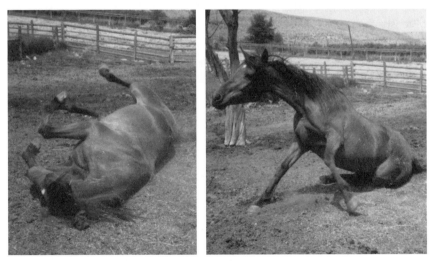

A horse with abdominal pain will often roll *(left)*. A horse with colic may keep getting up and down *(right)*.

Gas can also cause a twist even if the horse is standing quietly. A normal by-product of digestion, gas usually passes. Indigestion, and excessive gas may distend part of the gut, however. The distention itself may create a twist or displacement, or lead to pain and rolling, which can result in a twist.

Before the advent of ivermectin, bloodworms played a major role in causing colic and twisted intestine in horses who were not regularly dewormed, but this type of colic is not so common today. When bloodworms are the culprits, the adult worms attach to and feed on the intestinal lining, causing ulcers and irritation and, in turn, abnormal contractions and indigestion, which occasionally leads to colic. Even more damaging are the actions of worm larvae within the arteries that supply the intestines. These larvae often create blood clots, which can block an artery and cause death of the portion of gut that is supplied by the artery. Bloodworm damage also decreases the flow of blood to the intestines, resulting in abnormal intestinal movements.

Endotoxic Shock

If blockage from an intestinal twist is complete (food can no longer pass through) and blood supply to that part of the gut is crimped off, the horse suffers severe pain and goes into shock very rapidly. The damaged gut wall (where the intestine is twisted) and the bacteria within the tract will create a fatal endotoxemia unless the condition is promptly treated. If blood supply to a section of intestine is impaired, cells in the gut lining die and slough off, causing the protective barrier to be lost. Toxins then gain access to the bloodstream and circulate through the body: the horse goes into shock. This can be difficult to reverse.

Diagnosis

Spasmodic colic from indigestion causes waves of pain. The horse shows symptoms of colic and increased pulse rate when the pain hits, but is relatively comfortable between spasms. Twisted intestine creates constant pain and consis-

---

WHEN ROLLING IS DANGEROUS

A colicky horse with gut partly distended from gas or food overload is more likely to suffer a twist from rolling than is a healthy horse who rolls just to scratch his back or get rid of flies. The healthy horse doesn't have a distended gut. In contrast, the overly full gut has more weight and swing to it and may turn over on itself.

tently elevated pulse. Because blood supply to the gut is shut off, the horse soon shows signs of shock; his mucous membranes (gums) become pale and dull or dark muddy red instead of normal healthy pink. He becomes dehydrated as he loses circulating blood volume. Immediate veterinary attention is essential to combat shock and keep his condition stable. Actual diagnosis of

Signs of Endotoxic Shock

- Increased heart rate
- Weak pulse pressure
- Dehydration
- Cold legs, feet, and ears from circulatory failure
- Paleness or reddening of the mucous membranes

twisted or displaced intestine can't be made until the horse has been surgically opened, but sometimes this condition can be felt during rectal palpation. If a twist is suspected, immediate surgical correction is the only hope for the horse.

Treatment

Most intestinal twists cannot correct themselves, making surgery the only hope for the horse. Whether he survives the surgery depends on how much intestine is involved, how much it is twisted and damaged, and how long it has been that way. The earlier the surgery is performed, the better chance the horse has; once he goes into endotoxic shock, his condition swiftly deteriorates. And the longer the wait before surgery, the greater the chance of the blocked gut rupturing. The horse's condition is also critical. If he is exhausted from pain, he may not have a chance, especially if he must be transported a long way to a veterinary hospital.

Every colic episode should be treated promptly to prevent rolling and the risk of twisted intestine. If a twisted intestine is suspected, the treatment given by a veterinarian may prolong the horse's life and keep his condition stable until surgery can be done.

Prevention

Horses are less apt to suffer twisted intestine if the underlying causative factors can be prevented. This includes a proactive deworming schedule and instituting feeding practices that prevent colic. Feed only high-quality feeds; avoid spoiled, moldy, or fermentable feeds that create a lot of gas. Follow a regular feeding schedule to keep the horse from overloading his stomach and digestive tract.

Avoid overloading on grain, especially just before exercise. During exercise, the blood supply shifts to the working muscles and away from the gut, which may result in indigestion. Inadequate blood supply to the full gut leaves the undigested grain sitting there, and toxins may build up because grain sits too long in one place in the tract. Feeding a lot of grain right after a hard ride can also cause indigestion and colic because the gut is nearly shut down at that point. It's safer to feed grain at least 4 to 8 hours before a ride, depending on how strenuously the horse will be used, or 2 or 3 hours after the horse has been cooled out and is back to normal.

Enteritis (Inflammation of the Intestines)

Gut pain can be caused by inflammation of the intestines. Because of the inflammation, the horse does not absorb fluids traveling through the gut and suffers from pain and diarrhea. This happens when there are too many parasites in the digestive tract or when sand, poison, or infection irritate it. Indeed, bacteria such as *Escherichia coli* or *salmonella* produce toxins that are injurious to the intestines. If this is the case, the horse will need immediate veterinary attention and IV fluids.

Sometimes, the intestinal irritation is caused by a protozoan or a chemical agent such as lead, arsenic, phosphorus, mercury, copper, salt, oxalates, nitrates, or poisonous plants or if the horse eats too much soil or sand.

A horse with acute enteritis with abdominal pain kicks at his belly, rolls, and shows other signs of colic. Enteritis is always a serious matter. If a horse develops diarrhea, have him checked by your veterinarian immediately.

Sand Colic

Sand colic is caused by an accumulation of sand or small gravel the horse has swallowed, usually over a long period. This can happen when feed gets blown full of sand during a windstorm or when the horse eats hay on sandy ground (he picks up sand while eating the last wisps of hay). It can also happen on sandy pastures where plants pull up easily and sand clings to the roots. The sand irritates the gut and accumulates within the digestive tract. Being heavier than the horse's feed, the sand lodges in the cecum or large intestine instead of passing on through, a condition that can be fatal.

Diagnosis involves checking manure for bits of sand — diluting the manure with water and seeing whether any sand settles out. Pick up a handful of manure with a plastic obstetric glove or sleeve, turn it inside out so the sample is inside it, add water and shake it until the manure particles float. Any sand that was in the manure will settle in the finger tips of the plastic glove. Remove accumulations of sand in the digestive tract by treating the horse with mineral oil (the vet may give 1 or 2 quarts [1–2 L] by stomach tube) and feeding a product containing psyllium powder. The psyllium swells when wet and traps the sand, carrying it out through the gut. Bran has been used for this purpose for many years, but psyllium works even better. Serious sand accumulations that block the tract, however, must be surgically removed.

Flatulent Colic

This type of colic is caused by excessive gas in the digestive tract. The buildup of gas may be from intestinal obstruction or highly fermentable feeds such as lush, green grass or too much clover and alfalfa. Excessive gas in the small intestine is generally the result of some kind of obstruction, whereas gas in the large intestine is usually caused by fermentable feeds or obstruction.

Intestinal Obstruction

Occasionally, the intestine becomes completely blocked, causing dehydration, colic, and death unless the block can be removed or dislodged. Many things cause intestinal obstruction. Sometimes, a horse eats indigestible material such as hay twine, rubber fencing material, fibers from old rubber tires or rubber tubs, and hay nets. If the indigestible object or fragment does not pass through, it may block the intestine. If it doesn't cause a blockage immediately, it may sit there for a while and become the start of an enterolith (intestinal stone), a concretion that may grow until it blocks the tract.

Enteroliths

These intestinal stones develop when foreign material builds up layers of minerals around itself, like a pearl in an oyster, creating a mass of minerals, salts, and enzymes. Some horses carry enteroliths in their guts for years without any problems. The stones, some larger than a grapefruit, are discovered only after the horse dies from another cause. Other horses suffer blockage of the gut, which must be relieved by surgical removal of the obstruction.

Impaction

Obstruction can also be caused by extremely dry feed or improperly chewed food becoming impacted in the intestine. Some feed, such as dry beet pulp, makes the manure too firm and leads to impaction. A horse who bolts his feed without chewing may become impacted, as can a horse with bad teeth who cannot chew properly. Eating dry feed without adequate drinking water is a common cause of impaction, especially in winter, when horses don't drink enough because of the cold weather or frozen water. In these cases, the horses become constipated; what little manure they do pass is small and hard, covered with mucus.

If food in the cecum (the large fermentation pouch located between small and large intestine) is too dry for proper digestion, it ferments and swells and does not move on through. The fermentation creates gas and pressure, producing mild, persistent pain. Excessive gas causes acute pain and colic.

When impaction occurs in both the cecum and large colon, it is more difficult to alleviate. Several days of treatment and medication are necessary to lubricate the impacted material and get it moving. Sometimes a serious impaction can be corrected only by surgery.

Constipation

A simple fecal impaction may be caused by a combination of dry feed, lack of exercise, insufficient water, or eating of bedding. This type of impaction may build up over several days. The horse becomes increasingly sluggish and eventually stops eating (because he is "full") or just picks at his feed. He may lie down a lot. He stops passing manure or passes only small amounts. As the overly full intestine puts pressure on his bladder, he may frequently try to urinate, but he passes only a small amount of urine.

Often, the blockage is moved with lubricants — mineral oil administered by stomach tube and water to soften the impacted mass. The earlier the problem is discovered, the better the chances for simple correction without resorting to surgery. IV fluids may be necessary to rehydrate the horse.

WHILE WAITING FOR THE VET

If a horse shows signs of colic, call your veterinarian and describe the symptoms, pulse, and any gut sounds. The veterinarian may advise you on emergency treatment measures and tell you to keep the horse on his feet while you wait for help. Tend to the horse in an open area with soft ground so that if he does crash down to roll, he cannot injure himself. If the horse is not trying to go down, you don't need to force him to walk; let him rest whenever you can. If he wants to roll or collapse to the ground, keep him moving so he'll be less likely to actually go down. Sometimes, the exercise of gentle walking helps the horse pass a bowel movement, which relieves some of the pressure and pain.

16

First Aid and Medical Treatments

HORSE INJURIES AND ILLNESSES require prompt and proper care to ensure recovery. Always try to assess the severity of a problem to determine whether you can treat it yourself. If it's a serious problem or has the potential to become quite serious, you'll need veterinary help. What you do for the horse as you wait for the veterinarian may make all the difference, especially in wounds that are bleeding profusely, fractures, or some types of colic.

Emergency Situations

Be prepared to give a vet accurate information over the phone about the horse's vital signs and symptoms. This will help him or her determine whether the situation is something you can handle and monitor with advice or the horse needs professional diagnosis or immediate emergency treatment.

Some situations, such as bleeding, fractured leg, or colic, carry a large responsibility for the horse owner; your first-aid actions may well mean life or death for the horse. In most instances, you have time to call the veterinarian

---

TETANUS SHOTS ARE IMPORTANT

Minor injuries can be taken care of by the horse owner, so you must know how to treat them. In addition, after any wound that breaks the skin, the horse should have a tetanus shot if his vaccinations are not up to date. To be effective, it should be given within 24 hours of the horse sustaining the wound.

---

and then give first aid until he or she arrives. However, with serious bleeding or possible leg fracture, you should give emergency treatment first and then call the veterinarian (if you are alone) because time may be crucial. If possible, send someone else to call while you work on the horse.

The first things to do are to clean, then assess the wound. You might not discover an injury while it is still fresh and clean; it may be hours old, with caked blood and dirt in it. If it's swollen and painful, the horse may resent the handling it takes to clean it. You may have to resort to simply holding a running hose on the wound, if the horse will stand still. Use a Stableizer or some other restraint if necessary. The cold water will dull the pain as well as soften and loosen the dirt and dried blood. Then you can scrub the area gently with a cloth and examine the wound as you remove the debris. Hair and dirt can disguise a deep wound.

Don't worry about new bleeding as you scrub. Unless the bleeding becomes extensive, finish cleaning the wound before applying a bandage. Once cleaned, you can decide whether the wound is something you can treat by yourself or whether you should call the veterinarian. Any cut over a joint or into a tendon should receive veterinary attention, as should any deep cut below the pastern.

If bacteria get into a joint wound, infection can damage the joint. Tendon injuries are slow to heal, as are cuts in the heel. Damage to the coronary band may cause permanent changes in hoof growth. Injuries to other body parts may need stitches. If you have any questions about a wound or how to treat it, call your vet.

Dealing with Wounds

There are basically five types of wounds: bruises, abrasions, incised wounds, lacerations, and punctures. Fortunately, most wounds are not life threatening, but most will need some attention.

Bruises

A bruise is a surface injury that doesn't break the skin, but the severity of the blow may damage the muscle underneath. A horse may get a bad bruise from a fall, a kick, or running into something solid. Bleeding under the skin, as well as lymph seeping from injured tissues, causes swelling. Small bruises are best left alone; they reabsorb in a few days and the swelling disappears. Large bruises may need to be drained because there is too much fluid to be reabsorbed. After shaving and disinfecting the area, your veterinarian will insert a large-diameter sterile needle into the lower part of the swelling to drain out the fluid. If the fluid accumulation is extensive, this area may need to be opened up, drained, and flushed, and a temporary drain tube installed.

A bruised leg benefits from soaking in cold water or ice. Reducing the swelling immediately after the injury can keep a knot from forming on the bruised bone. A bony lump can be an unsightly blemish at best or, more seriously, may interfere with movement of tendons.

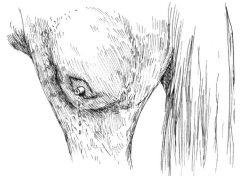

Draining a fluid-filled bruise

Abrasions

An abrasion is scraped skin that may bleed or ooze and be slow to heal. Usually, the best treatment is to clean the scraped area and apply antibiotic ointment to combat infection, keeping the area soft instead of crusty as it heals.

Incised Wounds

An incised wound is a clean cut. The usual problem, if the cut is deep, is excessive bleeding. Horses have a lot of blood and losing even a pint or two won't be fatal. Small vessels readily clot and stop bleeding. If the bleeding does not cease after a few minutes, or if the wound is spurting profusely, you should intervene. When a 1,000-pound (454 kg) horse loses more than 2 gallons (7.5 L) of blood, the situation could become fatal.

Pressure Bandage to Control Bleeding

Don't panic. Apply the cleanest pressure bandage you can create quickly; put it directly over the spurting artery or flowing vein. Towels, shirts, quilted

---

ARTERY OR VEIN?

A cut artery squirts bright red blood with each beat of the heart. A cut vein oozes dark blood continually (the blood hasn't been through the lungs yet for color-changing oxygen). There can be significant blood loss from a cut major vein, but the bleeding is slower than from a major artery; you have more time to control blood loss from a vein.

bandages, and the like will work. Your barn first-aid kit should always contain clean cloths, gauze sponges, or old sheets torn into strips and rolled as bandages.

You can make a pressure bandage by placing a roll of bandage or a hard object such as a piece of wood or even a rock (wrapped and padded with clean cloth) against the cut, then wrapping firmly with bandage or strips of old sheet. Continue wrapping until all bleeding stops. It may take many layers wrapped fairly tightly. Applying a firm bandage will allow you plenty of time to contact the veterinarian. Leave the bandage in place until he or she arrives.

Don't use a tourniquet! It can cause permanent damage because it completely blocks blood flow to the area below it. Always call your veterinarian if you can't get the bleeding to stop within several minutes.

Lacerations

A laceration is a cut with torn, irregular, edges. Healing is slow because of extensive tissue damage, and there is usually danger of infection. If there's no emergency bleeding, clean the wound and look to determine whether stitches are needed. Don't medicate the wound until it is clean. Plain, warm water is best for cleaning a wound; many antiseptics are irritating to tissues. When in doubt, don't use them, especially if the horse might need stitches. A disinfectant that burns the tissues will make it impossible for a vet to stitch the wound; it won't be able to heal properly.

Running water from a garden hose is often effective for cleaning a wound. Cold water can help numb the pain. If necessary, you can use a washcloth, gauze sponges, or even paper towels to gently scrub the wound and remove hair, dirt, dried blood, and dead tissue. Continue scrubbing until you are down to clean, pink flesh with all dirt and debris removed.

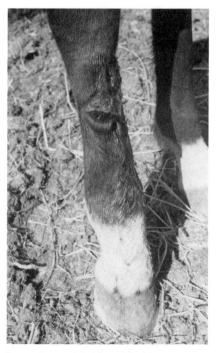

Laceration on hind leg with skin torn away.

Proud Flesh

In deep cuts below the knee or hock, proud flesh may develop unless steps are taken to prevent it. Proud flesh is an unsightly rubbery growth of excessive granulation tissue — the tissue that grows to fill in and close up a wound. As the wound fills in, this tissue keeps growing, extending above the surface and preventing the skin from growing over the wound — essentially making too much scar tissue. Traditionally, horse people used a healing medication such as scarlet oil the first few days until the wound gap began to fill in. Then, until the wound was healed and the skin had grown in from the edges, they treated it with a caustic medication to retard and prevent growth of proud flesh.

Scarlet-oil wound dressing and the like still work, but avoid follow up with the harsh anti-growth medications; apply alum instead. Alum is an inexpensive white powder used in making pickles, is available at grocery stores, and works well to prevent proud flesh. Its astringent action retards growth of proud flesh and is much less painful than traditional caustic remedies. Just sprinkle the powder over the wound, enough to lightly cover the surface. You can use your fingers — a pinch of alum is usually enough.

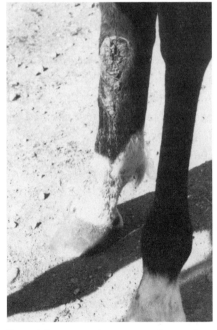

Same injury as previous page, a few weeks later; we treated it with alum to avoid the growth of proud flesh (excess scar tissue).

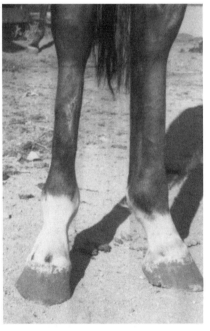

Four months after the injury, it is nearly healed.

The best treatment for retarding the growth of excess proud flesh is application of a cortisone ointment (that contains triamcinalone) early in the healing process. Cortisone slows the healing process because it reduces the inflammatory response, and hinders growth of proud flesh. If the wound is large, however, or a big hole that needs to knit together as quickly as possible, don't use the cortisone ointment until it has healed to the point where vital structures (such as tendons or joint capsules) are covered. Also, do not use cortisone in an infected wound until the infection is cleared up because it will also hinder the antibody response to the infection.

STITCHES AND STAPLES

Some wounds and cuts heal more readily, and with less scarring, if the edges can be sutured or stapled together. Others are best left to heal by letting new skin grow in from the edges. For successful stitching and stapling, the wound has to be relatively fresh and an adequate blood supply must be flowing to the damaged tissues. If the wound may require stitches or staples, do not medicate it in any way until the veterinarian has seen it. Stitching and stapling should be done within 2 hours of injury for a head wound, within 6 hours for a leg wound, and within 8 hours for and upper body wound.

The wound must be clean. If dirt and foreign material or dead flesh are still in the wound, the chances of successful suturing and healing are greatly reduced because of the likelihood of infection. The skin pieces must also still be intact, not torn so badly that the edges can't be put back together. Stitching and stapling work best if the wound involves only skin or the top layer of muscle. A deep muscle cut usually can't be sutured; it must heal on its own from the inside out. Likewise, if a joint or tendon sheath is also damaged, suturing may not be possible. Deep lacerations usually need to drain and there may be some sloughing away of damaged tissue; therefore, it is better to leave them open.

Even if a wound is stitched or stapled, the injured area may need additional bandaging to support the stitches or staples and to keep the area from opening up again.

Punctures

A puncture wound doesn't bleed much and may heal first on the outside, leaving a pocket of infection that will abscess or infect the horse's body if it gets into the bloodstream. To treat a puncture, first open the hole a little more and clean it with hydrogen peroxide or a disinfectant recommended by your veterinarian (such as dilute povidone-iodine [Betadine]). Then put sterile cotton or gauze soaked in antiseptic into the hole. Change this dressing daily until the wound has healed.

Never use tincture of iodine or other strong medication on open wounds or punctures in the muscle. These strong disinfectants burn the tissues. Tincture of iodine should be used only for treating thrush, ringworm, punctures in the foot, and disinfecting a newborn foal's navel stump.

A. Treat a puncture wound in the foot with iodine.

B. Wrap and bandage the treated foot to keep the puncture wound clean between treatments.

Kick Wounds

Because horses who live together often squabble, kick wounds are common. A flying hoof has enough force to shatter bone, but most of the time the damage is not that severe. It's important to evaluate the injury in terms of the horse's disability rather than the size of the wound. For example, a blow from the edge of a hoof may seem only to graze the skin but, in fact, may crack or chip the bone underneath. A kick wound may be an open laceration or cut, swelling, or bruise. It can also cause lameness. Sometimes, there is irreversible damage under the skin to nerves, blood vessels, muscles, tendons, or ligaments. Call your veterinarian if there is excessive swelling, lameness, a large area of torn flesh, or exposed muscle tissue, bone, or tendons.

You can help prevent kick wounds by separating horses at mealtime, introducing newcomers to the herd gradually, taking hind shoes off horses who are going to live together at pasture, and separating quarrelsome individuals.

Fractures

Most fractures occur below the knee or hock when a horse missteps or falls, gets cast on his back against the wall in his stall, or is kicked. Whenever a horse won't put weight on a leg, suspect a possible fracture. Assume a fracture until proven otherwise, or you may jeopardize the horse's future. Don't move the horse. Have the most experienced horse person available take charge of the horse and control him until the veterinarian arrives; a horse in pain may unintentionally hurt a young or inexperienced handler. Don't give any medication until a vet has examined the horse; painkillers may enable the horse to put weight on the leg, causing a more serious injury. If the horse will stand quietly, do nothing until the vet arrives.

Supporting a Fractured Leg

If the leg must be supported, do it properly and carefully. Improper splinting or bandaging can make the problem worse, and trying to bandage a frantic horse is dangerous. A fracture below the knee or hock requires a heavy support wrap, such as 10 or more quilted leg wraps under a firm bandage, or terrycloth towels or pillows wrapped tightly with bandages to hold them in place. No matter how high the break, start low on the leg and work up it

---

HELP YOUR HORSE RELAX

While waiting for the veterinarian, keep calm. Your horse will do best if he is as relaxed as possible, and he'll stay more secure and relaxed if you are. He can sense your nervous worry and will become more upset and difficult to manage if you are uneasy, which may make his problem worse. Part of your first-aid job is to comfort and console him, being mindful of his mental state as well as his physical condition. If he senses confidence, he'll stay more relaxed and easier to handle as you treat him or wait for the veterinarian. Talk soothingly, and stroke him if that is something he likes.

with the wrapping. The finished bandage should be at least 10 to 15 inches (25.5–38 cm) in diameter.

You can add additional stabilization with 3-inch-wide (7.5 cm) boards of appropriate length (making sure they are well padded) placed on the outside of the bandage, not next to the horse. Once bandaging is complete, don't move the horse unless absolutely necessary; wait for the veterinarian to examine him. The horse may or may not have a chance for recovery, but often, the difference between success or failure depends on the first aid given.

Head Injuries

The horse is prone to head injuries because of his emotional and reactive nature. An upset horse will fling up his head, increasing the risk of striking something solid. Simple injuries include skin lacerations that must be sutured. A blow to the face can fracture the facial bone over the sinuses, even though the skin may not be broken. The horse will need reconstructive surgery to realign the bones and antibiotics to prevent sinus infection. If facial nerves are injured, his face may become paralyzed on that side, causing the lip and nostril to droop — and also the eyelid and ear if the damage occurs high on the face.

A common head injury occurs when the horse rears over backward, as when breaking his halter or rope while tied. If he lands on his poll (the top of his head) a fracture may occur at the base of his brain, causing permanent damage or death. After hitting his head, the horse may stagger or be unable to stand or get up. Other signs of serious head injury are bleeding from the nostrils and involuntary rapid eye movements.

A horse suffering from severe head injury must receive immediate veterinary attention. If the skull has been injured or bruised, prompt treatment with DMSO given by IV may keep the brain tissue from swelling.

Choking

Choking occurs when a mass of food or a foreign object becomes stuck in the esophagus. A horse can choke on an apple or piece of carrot if he swallows it without chewing it first. He may also choke on hay or pellets; a wad of food may be too dry and bulky to slide down easily.

The choking horse drools saliva, and feed may come out of his nose. When he tries to drink, water flows out of both nostrils. When he coughs, he sprays food from nose and mouth. If the blockage is not relieved, he becomes dull and depressed, standing with head down. Sometimes a choking horse shows signs of discomfort similar to colic.

When a horse suffers complete blockage — being unable to swallow or cough it up — he is in danger of getting material into his windpipe and lungs because saliva and feed fill the esophagus and overflow into the windpipe. This can lead to aspiration pneumonia, which is almost always fatal; because foreign particles can't be removed from the lungs, they serve as a constant source of irritation and infection.

First Aid for Choking

Some horses panic, slinging the head and throwing themselves around in attempts to avoid the discomfort. Call your veterinarian as soon as you discover the problem. Try to keep the horse calm until the veterinarian gets there. Don't let the horse drink. Let him put his head down so material can drain out of his nose and mouth. The horse is in no immediate danger because he can still breathe. Just keep him quiet (so he won't injure himself), either by slowly walking him or holding his halter and talking calmly, and gently massage the throat area to help move the blockage on down.

Shock

Sometimes when a serious injury occurs, the body reacts by going into shock. Measures should be taken to prevent or treat shock as soon as it is suspected. Horses often die from shock rather than from the trauma that caused it.

Shock is a condition in which circulation in the smaller blood vessels and capillaries shuts down. The horse in shock has pale mucous membranes, subnormal temperature, falling blood pressure, feeble but rapid pulse, and shallow respiration that may be fast at first and then slow. He may be restless and anxious or very dull. He sweats but has cold, clammy skin.

Hemorrhagic Shock

If shock is from excessive blood loss, the horse should be kept warm with blankets and leg wraps while you try to stop the bleeding and get veterinary help. His circulating blood volume must be restored with IV fluids or blood transfusions, and medication should be given to control the shock.

Secondary Shock

If shock is from colic, severe burns, or some other serious body tissue injury or toxic infection, there is no actual blood loss. Circulatory failure in the small

HAVE VET TREAT SHOCK

It is very important to carefully monitor all vital signs when a horse is injured or colicky. If a horse ever appears to be going into shock, call your veterinarian immediately. If you have steroids such as dexamethasone on hand, ask whether you should give the horse an injection. The sooner the shock can be treated, the better the chance of saving the horse. While you wait for help, try to keep the horse warm and quiet. Blanket him, if possible, to retain whatever body heat he has. Do not move him. Conscientious first aid may make the difference between life and death during that critical short time when the effects of shock are still reversible.

vessels is triggered by pain and tissue damage. Fluid from the bloodstream seeps into other tissues and mucous membranes become pale because no blood reaches them. There is only a short time in which this condition can be reversed — by giving the horse IV fluids, epinephrine, and steroids to constrict the small blood vessels (stopping any further loss of fluids through the capillaries), thereby preventing irreversible damage to tissues from lack of oxygen.

Burns

A burned horse needs immediate first aid, especially if the burns are deep. A severely burned horse may go into shock, collapse, and die. Shock may occur soon after the burn accident or several hours afterwards, brought on by circulatory failure from loss from the many small blood vessels injured.

First-aid treatment for a serious burn includes applying ice packs or cold packs, even cold water from a garden hose, as soon as possible to reduce the depth of the burn and minimize tissue damage. A serious burn from high heat keeps penetrating the tissues for a while, and the sooner you halt this the better. The horse needs painkillers, and antibiotics to prevent infection in the damaged tissues. The veterinarian will treat the burned areas and show you how to care for the horse in the following days. Although the procedure is very painful, raw areas must be cleaned and treated daily to prevent infection.

Toxemia and infection are likely to follow an extensive burn because the damaged skin cells die and produce toxic substances. Germs may also invade the raw and damaged tissues and start multiplying, producing toxins. This is why a burn should be treated carefully and kept very clean. A burn that covers

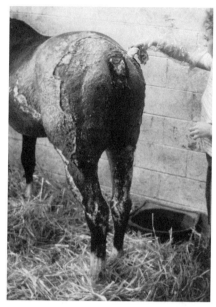 

Treating a horse who was severely burned in a barn fire, losing the skin off much of his body

The same horse a year later after skin grafts and healing, with some hair growing back

more than 5 percent of the body surface requires immediate attention: treat the burn, give the horse painkillers and steroids to prevent shock, and administer IV fluids to replace what is leaking out of damaged tissues.

Rope and Friction Burns

If a lead rope or longe line gets caught under the horse's leg or tangled around his body, it can cause a rope burn, whether the line is soft cotton, nylon, or a leather lead shank. The same kind of burn can also occur when a horse skids along the ground after falling or slides on the backs of his legs.

Rope burn is often harder to heal than a cut or wound, simply because it is a burn. The friction from the rope grating along the flesh (or the flesh sliding along the ground) produces heat as well as scraping and laceration. The scraping damages more skin tissue than does a simple cut, and the heat from the friction tends to kill the skin cells. This kind of damage takes a long time to heal and without proper treatment, may leave a thick scar, which may hinder normal movement of the leg.

Immediate treatment is crucial. Any hair remaining around the burn should be clipped away and the burn well scrubbed to get out any embedded foreign material and dead skin. Scrub until you get down to clean, healthy, pink tissue, or have your veterinarian shave out the dead tissue with a scalpel. The horse may need to be restrained or tranquilized for this painful treatment.

Clean and bandage the burn daily to prevent infection, and use a soothing, anti-inflammatory medicated ointment between cleanings. Diligent treatment can reduce a serious rope burn to a simple, easy-to-heal wound within a week; then you won't have to change the bandage so often.

Embedded Particles

Another problem involving rope burns is irritation from particles of rope; the tiny fibers rubbing into the skin are like little needles breaking off and embedding in the damaged, swollen flesh. These can delay healing. For these reasons, a serious rope burn may require more medical attention than a bad cut to heal properly and without extensive scar tissue or persistent inflammation.

Snakebite

Horses are occasionally bitten by snakes. The seriousness of a bite depends on the type of snake, amount of venom injected, and location of the bite. For example, a rattlesnake bite is generally not fatal if the horse is bitten on the leg because a large animal such as a horse is not as adversely affected as a small animal. If the bite is on the nose or face, however, swelling caused by toxins may interfere with the horse's ability to breathe and he may die of suffocation. If you suspect a horse may be snakebitten, consult your veterinarian immediately. The horse may need antivenom (depending on the type of snake) or treatment to reduce swelling and tissue damage and to control infection.

Porcupine Quills

Quills in the nose are a problem when curious horses get too close to a porcupine. The best way to remove quills is to have one person hold the horse by the halter and distract him while another person pulls out the quills one at a time with needle-nosed pliers. A quick, straight jerk works best. If you pull sideways they'll break off and be harder to get out. Once you get them all out, feel the skin to see whether there are any broken-off quills. Also

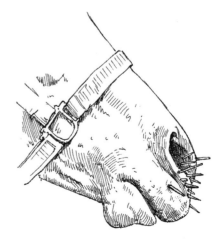

A horse with porcupine quills in his nose

feel inside the lips in case there are some in the mouth. Do not clip off the ends of the quills; this pushes them deeper into the tissue and makes them more difficult to remove. If the horse is too upset to stand still while the quills are being pulled, use a Stableizer (see chapter 5) or have your vet give him a tranquilizer.

Urinary Stones

Occasionally, a horse suffers from urinary stones, which are formed by crystallized mineral salts in the urine. A stone may block the outlet from the bladder or lodge in the urethra, causing discomfort, colic, and painful urination. This is more common in male horses because their urethras are long, narrow, and curved — more easily obstructed. The affected horse may take

---

FIRST-AID KIT

Keep all first-aid supplies together in a handy place, except for injectable antibiotics and epinephrine that must be kept cool. Reserve a place in your refrigerator for those. Also keep cold packs handy in your freezer for use as cold compresses. In a pinch, you can use a package of frozen vegetables. Other items should be at room temperature (not in an unheated barn or tack room in subzero weather). Also, be sure to keep on hand clean buckets (for carrying wash water) and towels. If you don't have your veterinarian's phone number memorized, keep it posted in the barn or by your phone. A first-aid kit should contain:

- Wound dressings such as nitrofurazone ointment (Fura, Fura-Zone), povidone-iodine, scarlet oil, aloe vera cream, chlorine dioxide ointment, or wound spray (a mixture of various oils and alcohol)
- Chlorhexidine (Nolvasan) ointment or zinc oxide (Desitin) for fungal skin problems
- Povidone-iodine solution in a spray bottle

- Tincture of iodine for thrush, quicked hoof, or hoof puncture
- Dexamethasone
- Chlorhexidine solution, hydrogen peroxide, and furacin solution (the latter must be mixed for you by a veterinarian); these three ingredients may be mixed to make a flushing solution for a puncture wound or for cleaning a deep, dirty wound

a stretching stance to urinate but hold the stance a long time, passing only small amounts of urine. If a stone is caught in the urethra, the veterinarian may be able to insert a catheter to remove the obstruction. If the stone is in the bladder, surgery may be necessary.

Sheath Swellings

Sometimes, sheath swells or becomes sore, usually from a buildup of debris (dirt and urine) inside the sheath. This may lead to infection and urinary problems. Glands in the sheath lining also produce a dark secretion called *smegma*. Sometimes these secretions build up into a soft, waxy deposit or dry hard flakes. If the lining of the sheath becomes irritated, pain and swelling may make it hard for the gelding to let down his penis to urinate.

- Phenylbutazone (an anti-inflammatory to relieve pain and swelling) in tablets, powder, or paste form
- Disposable syringes and needles (various sizes for injectables; large syringes for flushing and other purposes)
- Gauze pads
- Strips of clean old sheets rolled into bandages
- Sterile nonstick wound dressings
- Cotton batting or cotton leg quilt
- Conforming stretchy crepe bandaging that sticks to itself (Vetrap)
- Stretchy adhesive tape
- Duct tape
- Bandage scissors
- Animal thermometer
- Stethoscope
- Flashlight and fresh batteries
- Pocketknife
- Twitch or Stableizer
- Petroleum jelly
- Bottle of mineral oil
- Protective boot for a hoof
- Bottle of alcohol
- A watch with a second hand if you don't wear one
- Disposable razor
- Well-washed squeeze bottles (ketchup and dish detergent bottles will do) for applying warm soapy water or rinsing a wound
- Well-washed deworming syringe for administering dissolved tablets

Signs of Sheath Problems

A horse with a sheath problem may try to urinate without letting down his penis. His sheath may become swollen, or there may be flakes or deposits of smegma clinging to the sides of the penis or hind legs. If the horse has constant discomfort, he may be cranky just because he is hurting. Pain may make him colicky after a ride because he needs to urinate but has trouble doing it. If your gelding becomes grouchy, it may mean he's hurting. The older the gelding, the more likely he has a painful buildup that needs to be washed out.

Any difficulty in urinating calls for examination. Swelling of the sheath also can be caused by recent castration, local injury, local erupting of a skin problem, and bladder stones; therefore, it is important to determine the cause.

Cleaning the Sheath

Some geldings manage fine without ever having the sheath cleaned; others need periodic care. A buildup that is not washed out may form a claylike ball of debris at the end of the penis in a small pocket just inside the urethra. This "bean" can cause infection or interfere with urination if it becomes larger than the end of your thumb. You can feel into this pocket with the end of your finger to tell whether there's a bean there.

A horse with a bean may urinate in any of several ways. He may spray urine in an obstructed stream, or just dribble, or start and stop suddenly because of the discomfort; he may try several times to urinate before he finishes the job. The bean may become as large as a walnut and cause infection. You can work out a small bean with your finger, but a large one may be so painful that the horse resists. In fact, a large bean may have to be broken into several pieces for removal. If this is the case, you might need help from your veterinarian, especially if the horse must be tranquilized. The tranquilizer may be needed to make the horse relax enough to let down his penis.

Washing the Sheath

The few horses who develop beans regularly need routine cleaning of the sheath to prevent buildup. Some need cleaning every few weeks; others once or twice a year. It's a good idea to get the horse used to being handled in this area whenever you groom him, so he won't be ticklish and likely to kick. When you are ready to wash him, squirt warm water into the sheath with a soft-tipped rubber syringe to soften and loosen debris. Lubricate your hand with mineral oil, put your fingertips together, and gently enter the sheath, using your hand to scrub away deposits and debris. You can check for beans at the same time.

If your horse is used to being hosed with water, an easy way to clean the sheath is to irrigate the sheath with a low-pressure flow from a garden hose, as long as the water is not too cold. You can get him used to the hose by putting a slow trickle on his legs on a warm day, working up the legs gradually, and eventually directing the water into his sheath. A simple irrigation with the

ESTIMATING THE HORSE'S WEIGHT FOR PROPER DOSAGE

Most medications and dewormers are given according to body weight, so you'll need to know roughly how much each horse weighs. Although most medications have some leeway for safety, overdosing can be wasteful or dangerous (depending on the drug) and underdosing may not provide adequate treatment. Weight tapes are helpful but give only a rough estimate.

The most accurate way to determine a horse's weight is by using livestock or truck scales. At a truck weighing station, you should first weigh the towing vehicle and horse trailer empty, then weigh it again with only your horse in it. Subtract the difference to determine the horse's weight.

If you can't weigh the horse on a scale, the next best alternative is to use a formula based on two body measurements: heart girth and body length. Measure heart girth just behind the elbow, taking the reading right after the horse exhales. Measure body length from point of shoulder to point of buttocks in a straight line. Don't use a cloth measuring tape because it may stretch. A metal carpenter's tape measure is accurate but noisy and might spook the horse. A plastic coated tape is best, but if you don't have one, you can use a piece of cord (or string with no stretch) and mark the spot on the cord with a pen. Then measure the cord with a carpenter's tape or yardstick.

Multiply heart girth (in inches) by itself (heart girth x heart girth), then multiply that figure by body length (in inches). Divide the total by 330) to determine the approximate weight of the horse in pounds. For example, if a horse measures 75 inches around heart girth and 64 inches in body length, you would multiply 75 x 75 x 64 to come up with 360,000. Dividing that by 330, you arrive at an approximate weight of 1,090.9 pounds.

hose can wash away a lot of debris; for some horses this is adequate cleaning if it is done periodically. If your horse won't tolerate water from a hose, use a large syringe (without needle) to gently squirt warm water into the sheath.

An even easier way to wash him is when he relaxes his penis after a ride. He may leave it down long enough for you to clean it gently and quickly with a soft, wet cloth when you unsaddle him. If he lets down or urinates, take gentle hold of his penis as he finishes and quickly clean it. A few quick cleanings like this after a ride may be enough to prevent serious build-up. You can check for beans at the same time because the end of the penis is visible. A little care in this area now and then can prevent problems later.

Alternative Treatments

Many horse owners are interested in alternative therapies for horses — treatments other than traditional medications, drugs, and surgery. These treatments, which are outside the realm of traditional veterinary practice, include herbal therapy and homeopathy, chiropractic and physical therapy, massage therapy, and acupuncture. In 1992, the American Association of Equine Practitioners (AAEP) acknowledged these unconventional therapies as valid alternatives when carried out or overseen by a licensed veterinarian to make sure they are used ethically and not misused.

Herbal Therapy and Homeopathy

Both herbal therapy and homeopathy make use of natural substances to treat specific ailments. Herbal treatments are used like traditional medicines, aimed at destroying disease agents or easing a specific ailment. A number of products are available, such as Cough Free to alleviate coughing, and yucca pellets for relief of pain and inflammation in joints. Homeopathy (which literally means "like cures like") involves treating illnesses with substances that, in larger doses, would produce the same symptoms of the illness. The highly diluted substances are given to promote the body's self-healing efforts by stimulating the immune system.

Chiropractic

In chiropractic therapies, spinal adjustments are performed to relieve pressure on nerves. The theory behind this type of treatment is that the body's ability to maintain good health relies on the nervous system, and that misalignments of vertebrae cause disease by putting pressure on the spinal cord. Chiropractic treatment is aimed at returning proper balance to the total body.

BACK PAIN

The horse's spine includes roughly 170 joints, each of which has only slight capability of movement — creating a rigid backbone. Back pain occurs if a vertebra is out of line, suffers inflammation, or is pulled by muscle tension and spasms.

The same is true of physical therapy in which the horse's limbs and body are manipulated to rebalance the whole animal and do away with stiffness, soreness, and lameness.

Equine chiropractors can often help horses with chronic back and neck soreness. The chiropractic evaluation and diagnosis include a history of the horse's lameness or stiffness, gait abnormalities, saddling objections, or poor attitude about work. They also take into account unnatural head and neck position, back outline (humps, sways, or curves in the spine), and deviations from normal gait when moving. As trouble spots are identified, the chiropractor adjusts them with rapid thrusts of the hands along the backbone.

Massage Therapy

More than 60 percent of the horse's body weight is muscle tissue. Muscles can become tight, less flexible, and vulnerable to damage when fatigued or stressed beyond their capability. This can lead to muscle spasms, stiffness, and pain. After locating the area of soreness, a massage therapist can use deep palpations and body manipulation to break up muscle spasms and restore proper circulation of blood and oxygen to the affected areas. This reduces pain and restores normal motion and mobility. Once the tension in his muscles is reduced, the horse's comfort and general attitude greatly improve. Massage can help prevent as well as treat muscle injuries.

Acupuncture

Acupuncture is a healing science that holds as a central principle the concept of the individual animal as living energy rather than as a catalogue of various signs of illness. This ancient Asian healing art is now used to treat many equine health problems, ranging from arthritic joints to heaves and colic. It has proven beneficial in correcting reproductive problems and in relieving pain and chronic conditions (arthritis, respiratory ailments, allergies, and intestinal disorders) that have not responded well to standard medications. Effective veterinary acupuncture is based on both natural and scientific aspects of healing. No one is sure why it works, but acupuncture has been effectively used in medicine for more than 3,500 years.

In Chinese cosmology, everything maintains a balance. The two sides of this balance are yin and yang (dark and light, front and back, cool and warm, and so on). When an animal is healthy, its yin and yang aspects are balanced. Part of this balance involves the circulation of vital energy, which follows certain pathways through the body. The Chinese believe that all illnesses and physical problems are caused by a disruption of the yin/yang balance and this flow of vital energy, with a resulting pattern of disharmony. Stimulating or unblocking the disharmony allows the normal energy flow to be reestablished. The energy flow is connected to all vital organs, muscles, bones, joints, and so on; therefore, acupuncture treatments in specific locations along the energy pathways can influence all areas of the body.

Acupuncture Points

The Chinese developed a systematic method by which to stimulate relationships between external body points and internal organs as they relate to normal bodily function. Acupuncture treatment consists of stimulating certain points by means of needle insertion, massage, pressure, or heat. Modern veterinary acupuncture also includes other methods of stimulation such as weak electrical currents, injection, cold laser, and ultrasound. The stimulation of certain acupuncture points can produce changes in blood vessels, nerve

---

TRADITIONAL AND NEWER THERAPIES ARE COMPLEMENTARY

Alternative therapies are neither magic bullets, nor better than conventional medicine in controlling disease. One type of therapy is not a cure-all. There is a place for antibiotics, surgical repair, and alternative treatments, using the complementary skills of a number of practitioners. Today, the horse owner can take advantage of specialists who are well trained in both alternative therapies and conventional medicine. For readers who want to learn more about alternative therapies, a good overview is *Healing Your Horse: Alternative Therapies* by Meredith L. Snader et al., published by IDG Books Worldwide, 1993.

impulses, and hormone levels. Its action can be more specific than that of most drugs — and without side effects.

Most veterinary acupuncturists combine the practice with traditional medicine for more complete benefit to the animal being treated. The two methods work well together, being very different but complementary concepts of medicine. For example, a horse with respiratory infection can be treated not only with antibiotics but also with acupuncture — to stimulate the immune system, ease the pain and discomfort so the horse will feel better and eat better, and improve the function of the lungs.

Rehabilitation Facilities and Techniques

Many horses suffer injuries, partly because they are large, strong animals best suited to roaming free in large areas and programmed for flight as their main defense against predators. When we confine them, they adapt quite well but may be injured when coming into contact with fences, trailers, or some other hazards in their artificial environment. Because we also use them in many athletic events and competitions, lameness and leg or foot injuries may occur if a horse overworks his limb structure. Modern technology has produced various ways to treat injuries and help them heal faster, and with less scar tissue or other permanent damage. Following is a sampling of some of the methods now being used in equine rehabilitation centers and some veterinary clinics.

Cold Saltwater Spa

Cold water has been used for many decades as a simple way to reduce heat, swelling, pain, and inflammation. For example, cold hosing or ice packs can help minimize the swelling, pain, and damage in a leg injury. In 1998, an Australian veterinarian invented a cold saltwater "spa" bath in which the horse is put into a fiberglass tank, which is then filled with very cold water that is saturated with sea salt and Epsom salts. The saltwater is from a built-in holding tank that keeps it at 35°F (1.7°C). The spa is filled to a point above the level of the leg injury (often as deep as possible — to provide pressure from the water — but never up to the level of the belly) and kept circulating. The salty, cold water acts as a poultice, drawing infection out of wounds, and pulls all the heat out of the limb.

The horse is kept in this bath for no more than 10 minutes. By then, the legs are as cold as they will get. Afterward, the increase in circulation (the body's reaction to try to warm up the limb) brings more blood, oxygen, and nutrients to the injured area. The dual effect — reduction in swelling and inflammation, and the increase in blood flow — speeds healing.

Hyperbaric Oxygen Chambers

Oxygen chambers—small enclosed rooms filled with pure pressurized oxygen—in which human patients spend time, have been used for several decades to treat many types of injuries and infections. Under these conditions, oxygen becomes physically dissolved in the blood and is more readily utilized by the body, even in tissues that have been deprived of oxygen because of injury. Pressure chambers are being used to treat human burns, stroke, coma, lung abscesses, difficult wounds and infections (such as bone infections), diabetes, traumatic brain injury, and so on. Beginning in the 1990s, larger chambers were created for horses and now a number of these are available for treating horses at rehabilitation centers, university hospitals, and private veterinary practices. They are proven to be very beneficial in helping heal difficult equine wounds and injuries.

Other Healing Aids

There are numerous new techniques being used to treat horse injuries, including pulsed electromagnetic therapy (such as the P3 machine, which stimulates increased blood flow to the area being treated), and cold laser therapy (also called photo-therapy), in which a small, focused beam of light is directed at the injury to stimulate cellular activity.

Several effective methods for physical rehabilitation after an injury include swimming the horse, underwater treadmills, and the like. These exercises help a horse regain fitness and leg strength, increase range of motion, and more quickly bring him back to athletic soundness after injury or surgery.

Euthanasia

Anyone who owns horses must eventually deal with the hard reality of losing one. After all, horses do become old and unable to function, seriously ill with no chance of recovery (as in an acute and terminal case of colic), or fatally injured in an accident. You may someday face the difficult and unpleasant task of putting your horse down and ending his pain.

Making the Decision

With an old horse, there is time to think about it and plan for it in advance, realizing that, at some point, you must determine when life becomes more pain than pleasure — and mercifully end it for your equine companion. You'll also have to decide where and how that life should end, who will be responsible for it (usually the veterinarian), and what to do with the remains afterward.

It's hardest when you must deal suddenly with an unexpected euthanasia decision, as when a young horse is critically and irreparably injured, suffers incurable disease, or is moving into the realm of hopeless and painful colic; at this point it becomes a matter of mercifully ending his life rather than prolonging the suffering. Another difficult situation is when the horse's condition is not life threatening but his future athletic ability or general health is seriously compromised and he faces a long retirement. You must ask yourself if the prospects for his future are continuous discomfort or pain. Can you justify keeping him alive just because of your own personal feelings and reluctance to part with him?

The circumstances that force these decisions will vary greatly in each situation, but generally, they involve assessing the following: (1) the amount of suffering the horse is going through; (2) the horse's chances of recovery, if any; (3) the possibilities of the horse being useful to someone else, even if his age or impairment have made him useless to you; and last, but not least, (4) the cost of treating him or maintaining him in his present debilitated condition. Sometimes a medical "cure" is available for what ails your horse, but the cost is completely out of your financial reach, no matter how much you love him.

Ending His Life

Once the decision has been made, the next step is to determine how and where. With an old horse, you may have the option of taking him to a prearranged location to put him down, but with an injury victim or colic case, you'll likely have no choice but to end his suffering wherever he happens to be, dealing with removal of the body later. In some areas, a rendering plant will take the body. Some landfill dumps will accept large animals, but others won't. Ask your veterinarian or county health department what the local regulations are for animal disposal and who to contact.

The most instantaneous, and therefore the most humane, method of ending his life is with a bullet or "captive bolt" into the brain. The latter device is used for humane killing of cattle in slaughterhouses and is often the easiest way to shoot a horse; however, it is generally quicker and easier to locate a gun. Most horse owners have a hard time emotionally dealing with the task and prefer to have a veterinarian euthanize the horse. A veterinarian can administer an overdose of anesthetic (barbituate) by IV to induce a deep sleep and stop the heart and other vital functions. The horse becomes unconscious and his life ebbs away painlessly.

Although this method is probably easiest on the horse owner, there are times when it's not feasible to wait for a veterinarian. Sometimes a horse is mortally injured in a situation far from a veterinarian (as on a back-country ride or pack trip), or you live a long distance from the veterinarian and a horse is going through terrible suffering in the last throes of serious illness or colic, which can happen even under the influence of painkilling drugs. In these cases, you have no choice but to end his suffering yourself and must know how to mercifully end a life. Even more agony can occur (for both you and the horse) if you do not know how to properly shoot a horse and he does not die instantly.

By Gunshot

If you must take on the difficult task of killing your horse by gunshot, it's necessary to aim so the bullet goes into the brain and enters the spinal cord, or at least passes through the center of the brain. The mistake most people make when shooting a horse is aiming too low, such as between the eyes. In these locations, the bullet misses the brain and merely injures the horse and makes him frantic. Keep in mind that the horse's brain is small and high, located at the top of his head. The best way to shoot a horse is to aim for the middle of his forehead about 4 or 5 inches (10 or 12 cm) above the level of his eyes in the spot where two lines drawn in an X from eyeball to ear would cross in the center of the forehead. To make sure you aim correctly, it's wise to actually make those lines on his forehead with chalk, tape, or marking pen if you can.

A handgun is better than a rifle if there is no one available to hold the horse for you; a rifle takes two hands to shoot. On a small horse, a .22-caliber gun may be adequate, but with a large horse, it's better to use a larger caliber bullet; you need a 9 mm or a .38-caliber gun. Hold the weapon close to the forehead, almost touching it, and aim slightly upward rather than horizontally, so the bullet will enter not only the brain but the top of the spinal cord. When shot in this way, the horse will drop down immediately, without a struggle, without pain. Sometimes the greatest gift you can give your horse is a merciful end, a release from the pain that binds him.

PART III

Breeding Horses

17

Selecting Breeding Stock

IF YOU'VE DECIDED TO BREED A MARE or two for your own purposes, the stallion and mares you select should have desirable qualities that will enable their offspring to excel in whatever activity you hope to pursue. Or perhaps you want to raise horses to sell, breeding the type of horses you really like—for which you think you'll have a good market—and selling them to folks who have the same goals as you. No matter what your objective, it is important to make informed decisions when you select breeding stock.

The goal of any horse breeder is to raise quality animals who will be outstanding athletes in the careers they are trained to pursue, with the conformation and durability to stay sound while doing it. Animals chosen for breeding stock should be above average in the characteristics you are looking to perpetuate. The responsible breeder strives to improve his or her horses by reducing or eliminating faults. You don't want to perpetuate weaknesses or problems that will mar a horse's chances of being a useful animal. When raising horses to sell, you have an ethical responsibility to your potential customers; you should not produce horses who may cause hardship and heartache because of genetic defects or conformational faults that lead to unsoundness.

This chapter reviews the basic aspects of conformation and the key issues in selecting broodmares and stallions; an overview of genetics appears in chapter 18.

What Breed?

There are good horses and poor ones in every breed; there's no "best" kind of horse. Each breed has outstanding and versatile individuals. If you have a favorite breed, you will probably choose horses from within its ranks. If partiality for a certain breed is not a factor, look for horses of the breed or type best suited for the activities you want them to do. When raising horses for

BE A CRITICAL JUDGE OF YOUR HORSES

There are some basic rules to follow when selecting breeding stock or choosing a stallion to be bred to your mares. Avoid the following genetic defects in the breeds known to have produced individuals with these heritable diseases.

Genetic Defect	Breed
night blindness	Appaloosas
severe combined immune deficiency (SCID)	Arabians or part-Arabians
hyperkalemic periodic paralysis (HYPP, a metabolic muscle problem)	Quarter Horses and related breeds
hereditary equine regional dermal asthenia (HERDA)	Quarter Horses, Appaloosas, and Paints
glycogen branching enzyme deficiency (GBED)	Quarter Horses, Appaloosas, and Paints
polysaccharide storage myopathy (PSSM)	Quarter Horses, draft horses, Paints, Appaloosas, warmbloods, and several other breeds
lethal white	Paints and Pintos
degenerative suspensory ligament desmitis (DSLD)	Peruvian Paso and several other breeds

Note: Several of these breeds are working to eliminate or minimize instances of these genetic problems by testing potential breeding stock, utilizing DNA tests.

Also avoid conformational faults such as crooked legs that will lead to lameness and unsoundness with hard use. Never let emotion — fondness for a certain animal, bloodline, or color — overcome good judgment. A favorite animal should be judged as critically as any other when it comes to faults or virtues that could be passed on to offspring.

yourself, you may be interested in crossbreeding to combine the best traits of two or more breeds.

Visualize Perfection

Knowing the areas of weakness in your own horses is the first step to becoming a good breeder, along with having a picture in mind of what "perfection" would be. No horse is perfect, but a breeder should always strive toward

┌───┐

DON'T BE "BARN BLIND"

A good breeder is ruthless when judging a horse, seeing faults as clearly as recognizing the good attributes. Too many breeders are "barn blind," unable to see weaknesses in their own horses. Learning to look objectively at horses (your own as well as someone else's) and analyze their good and bad traits is crucial to breeding.

└───┘

a goal, applying a working knowledge of how to balance faults and strong points in choosing a stallion who will improve on the attributes of a given mare. Visualize what could result from a prospective mating, choosing animals who complement each other to produce offspring who are even better than the parents.

Look Farther Than the Pedigree

Many breeders are obsessed with pedigree, seeking popular bloodlines or horses who carry the blood of a favorite stallion or stallions. Yet, no matter how perfect the match seems on paper, good offspring are not produced by pedigree alone. A foal is no better than his sire and dam's physical or temperamental compatibility and the conformation and disposition that could result. The study of pedigrees, however, can be a useful tool if a breeder knows how to determine the ways in which different lines cross with each other and what they are likely to produce. A study of pedigrees is a science and an art. It should never be merely a collection of names. You want to make sure there are excellent individuals in the horse's background, but a desirable pedigree without good conformation and disposition in the individual animal is worthless. (For more information on genetics and heritability, see chapter 18.)

Conformation

Conformation relates to body proportions and angles — how the horse is built. The way his bones, tendons, muscles, and other parts are put together plays an important role in his ability to perform well and remain sound with steady work. A sound horse has no weakness or injury to impair his abilities or make him lame. It is very important to critically evaluate the conformation of breeding stock. For a more complete discussion, see *The Horse Conformation Handbook* (Storey Publishing, 2005).

JUDGING CONFORMATION

Stand directly in front of (and then behind) the horse as he's being led toward you and away from you, at both a walk and a trot. Lameness, certain types of unsoundness, and bad conformation are more evident when he's moving. If his front feet paddle outward or wing inward, or if any of his feet come too close together when in motion, the horse does not have good conformation.

A horse should be well balanced, with a smooth and pleasing overall appearance. He's an athlete; his main purpose is motion. So judge him while he's moving as well as when standing still to see whether he is clumsy or agile. He should travel "straight and clean," with straight feet and legs that move forward in straight, not crooked, lines.

The Feet

A good place to begin your assessment of a horse's conformation is at the ground because a horse is only as good as his feet and legs. The feet should be well shaped and of appropriate size. A large or heavy horse needs feet proportionally large to support his weight. A small horse needs smaller feet. Some horses have feet too small for their bodies. Small feet may look neat and dainty, but a heavy horse must have sufficient foot structure to support his weight or he'll become lame; the small foot is subjected to too much concussion. Indeed, small feet on a large horse can lead to navicular disease (see chapters 4 and 9). Alternatively, feet too large for a horse will make him clumsy.

Hooves should be wide at the heel, not contracted. Wide heels have some give, springing apart when the horse's weight is placed on the foot and absorbing some of the concussion when the foot hits the ground. Otherwise, all the shock and jarring would be transmitted directly to the bones of feet and legs and to the joints.

If a horse has good conformation, his feet wear evenly. He doesn't wear one side lower than the other or "dub off" one side of the toe more than the other. Conformation of legs often determines the shape of the horse's feet, the way they wear, and their flight — how they are picked up, swing through the air, and are put down. When the horse moves, the foot should break over (leave the ground) squarely — directly over the center of the toe, not off to one side.

A base-wide horse (feet wide apart) or with feet pointing outward (front or hind) will wing inward as he travels.

A base-narrow horse (feet too close together) or one who is pigeon-toed (front or hind) will generally paddle outward as he travels.

Breaking squarely. When a horse moves, his feet should break over (leave the ground) squarely at the center of the toes and travel forward in a straight line.

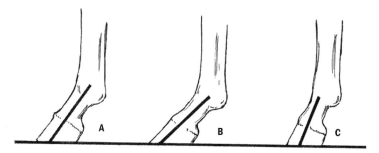

Pastern angles: **A.** Normal **B.** Too much angle (weak construction) **C.** Too steep, resulting in choppy gate and excessive concussion

Pastern Angle

Ideally, the angle of foot and pastern should be approximately 50 degrees. Hind-leg angles can be a little steeper. Foot and pastern angle should not be too sloping or too upright. If it slopes too much, the pastern will be weak, with the fetlock joint sometimes hitting the ground when weight is placed on the leg. If it does not slope enough, the foot won't have enough give; the horse will have a choppy, jarring gait, which is uncomfortable to the rider and likely to damage the horse's feet and legs with excessive concussion.

The Body

When you judge the body of a horse, look for a deep heart girth and well-sprung ribs (a wide barrel). A horse needs plenty of room for heart and lungs. Horses who are wide through the ribs and deep in the chest usually have more endurance than narrow, shallow-bodied horses. When viewed from the front, the horse's ribcage should be wider than his shoulders. There should be sufficient width between his front legs (about 6 to 12 inches [15–30.5 cm], depending on the size of the horse), yet the chest and front end should never be too wide in a riding horse because that will reduce his speed and agility. Wide shoulders are more desirable in a draft horse, for power in pulling and fitting a collar.

Muscling

Good muscling between the front legs looks like an inverted V. All muscles in a riding horse should be long and lean. Extremely heavy muscling on thighs, arms, and forearms hinder free action. A heavily muscled horse may be able to give a quick burst of speed but will tire much sooner than the trim horse and may have problems with muscles tying up if he is worked hard. Speed and stamina come from long, lean muscles, not bulky ones.

Neck conformation: **A.** Normal **B.** Too short and thick (cresty neck) **C.** Thin and weak (ewe neck)

Head and Neck

The horse uses his head and neck for balance and to collect and extend. A long neck is more helpful to athletic ability than a short, thick one. The horse must be able to swing his head up and down like a pendulum to shift his weight and balance at each stride. The head should be set onto the neck at such an angle that the horse can flex at the poll (the top of the head).

BODY PROPORTIONS

To be in the best form for athletic ability, a horse should be the same height at the withers as at the highest point of his hips (croup), and his height should be proportionate to his length. A long-bodied, short-legged horse or a short-bodied, long-legged horse is not desirable. If a short-backed horse has legs too long, he will overreach when he moves, striking his front feet with his hinds, unless he has a long hip and shoulder, which gives him a long underline. The shorter the back, the better, provided it's because of a long hip and shoulder. A horse who is too long-backed may go lame because he can't "round" his back to carry weight; a long back is generally a weak back.

In young horses up to three years of age, the croup may be higher than the withers; the latter do not attain full height as quickly. Yearlings and 2-year-olds often have no withers but, by the time they are 3 or 4, the withers catch up with the croup. By then, the horse looks balanced and holds a saddle better. In some horses, the croup remains higher than the withers.

Also, the upper and lower jaws should match; the horse will have dental problems if the upper and lower teeth do not meet properly. You may have to open his mouth to see how his teeth line up.

Shoulders and Withers

A horse with good withers and a long, sloped shoulder holds a saddle well, without needing a tight cinch and breast collar to keep it in place. A shoulder too short and straight (up and down instead of sloping back) limits free action and gives a choppy stride. A horse with short, upright shoulders usually has a short neck and short pasterns, creating more jar and concussion.

A good horse is square. This means he is roughly the same height at the withers as the distance from the point of his shoulder to the point of his buttocks; therefore, length of his body is the same as his height at the withers. Except for his head and neck, his body fits into a perfect square because the withers and croup are the same height.

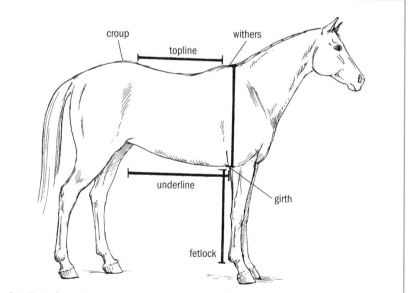

A well-conformed horse has a short topline and longer underline and is the same distance (depth) from fetlock joint to girth as he is from girth to withers. If he is older than 3 years, he should also be the same height at croup and withers.

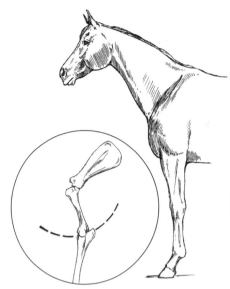

A. A well-sloped shoulder is usually associated with proper pastern angle and good withers. The sloping shoulder allows greater freedom of leg movement and longer swing of stride.

B. An upright shoulder is associated with upright pasterns, which result in more jar and concussion. The upright shoulder produces a short, choppy stride and less swing of leg.

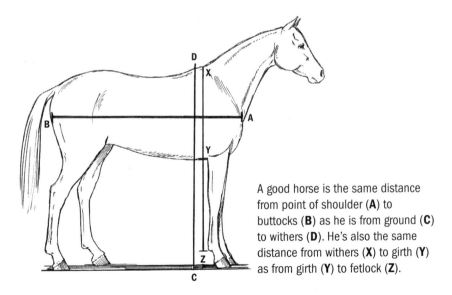

A good horse is the same distance from point of shoulder (**A**) to buttocks (**B**) as he is from ground (**C**) to withers (**D**). He's also the same distance from withers (**X**) to girth (**Y**) as from girth (**Y**) to fetlock (**Z**).

Front Legs

Front legs should be perfectly straight when viewed from the front. A line dropped from the point of the shoulder should go straight down the center of the leg — through forearm, knee, cannon, fetlock joint, pastern, and hoof. Toes should point directly to the front; feet should be exactly the same distance apart as the distance between forearms where they come out of the chest. Knees should be large, flat, and well proportioned — as large as possible without looking clumsy. A small, pinched-in knee crowds bone and tendons and hinders smooth action. The front of the knee should be flat with no roundness, and the outer edges should look square.

FRONT VIEW

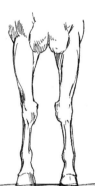

| Bowlegged | Knock-kneed | Bench-kneed (cannon bones set too far apart) |

SIDE VIEW

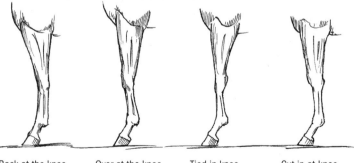

| Back at the knee (calf knees) | Over at the knee (buck knee) | Tied in knee | Cut in at knee |

Examples of front leg conformation

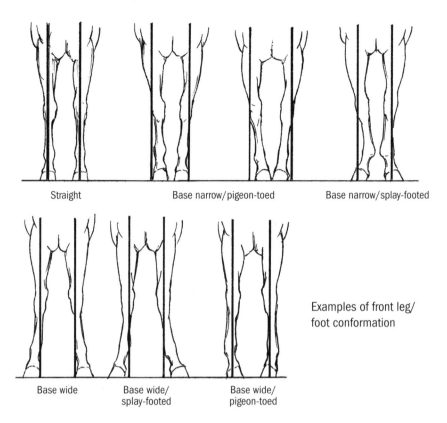

Straight   Base narrow/pigeon-toed   Base narrow/splay-footed

Base wide   Base wide/   Base wide/
                   splay-footed   pigeon-toed

Examples of front leg/
foot conformation

Cannon Bone

The cannon bone should be centered under the knee. If it is offset, it will put more strain on the splint bones (the small bones that sit on each side of the cannon bone) and make the horse likely to develop splints (bony lumps on the cannon). If the cannon bone is too far back, the horse is **calf-kneed**, a very weak conformation that leads to knee fractures. The opposite problem is called **over at the knee** or **buck knees**.

Flat Bone

Cannon bones should be wide when viewed from the side. This is called **flat bone**; it describes the combination of bone and tendon that gives the lower leg more depth from front to back. The tendon should be set well back of the cannon bone for best action and strength. A common fault is being "tied in" close to the knee or hock; this means the tendon is too close to the bone, causing friction, wear, and eventual unsoundness. Undesirable **round bone** occurs when cannon and tendon are too close together all the way down. This type of leg does not hold up well.

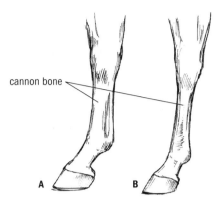

cannon bone

"Flat bone" (desirable: **A**) versus "round bone" (**B**) in which the cannon bone and tendon are too close together.

A   B

Base-Narrow or Base-Wide

A base-narrow horse has feet too close together — a condition found most often in horses with large pectoral muscles and a wide front. Base-narrow conformation is usually accompanied by pigeon toes, which point toward one another instead of straight to the front. This puts strain on pastern and fetlock joints, often leading to windpuffs in the tendon sheath or joint, or ringbone or sidebone on the outside part of the foot. **Windpuffs** are soft enlargements in the fetlock joint area, toward the rear; **ringbone** is a bony enlargement at the pastern joint or just above the hoof at the front; and **sidebone** is bony enlargement at the cartilage just above the hoof, on the sides. The hoof wears too much on the outside edge because it tends to land on the outside wall instead of squarely flat. Base-narrow, pigeon-toed horses usually paddle with their front feet, swinging them outward as they travel, which cuts down on speed and agility. The pigeon-toed horse usually paddles whether he is base-wide or base-narrow.

GOOD BONE IS PROPORTIONATE TO THE HORSE'S SIZE

Coarse bone means a heavy, clumsy cannon. Fine bone means insufficient support under the horse — cannon bones too small in diameter. A horse should have about an 8-inch (20 cm) circumference below the knee per 1,000 pounds (454 kg) of body weight; therefore, a 750-pound (340 kg) Arabian has sufficient bone at a 6-inch (15 cm) circumference, whereas a 1,250-pound (567 kg) Quarter Horse needs a 10-inch (25.5 cm) circumference to be equivalent in strength.

The base-wide horse has legs farther apart at the feet than at the forearm. This is usually found in horses with narrow chests. Most base-wide horses are also splay footed, which causes winging inward as the legs move. The inside of the limb is under great stress; windpuffs are common, and ringbone or sidebone can occur on the inside of the foot. Splay-footed horses wing to the inside, whether they are base wide or base narrow, and wear the feet excessively on the inside wall.

Hind Legs

Hind legs play a major role in how the horse travels — whether he is fast or slow, clumsy or agile. Stifle and hock joints work together in unison. The angle of the stifle and hock will be the same in any given individual; a horse with a too-straight stifle will also have a too-straight hock.

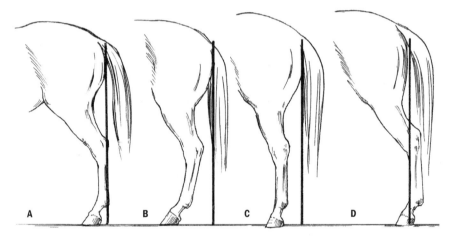

Hind-leg conformation viewed from the side: **A.** Proper hock and stifle angle **B.** Sickle hocked (too angled) **C.** Post legged (too straight) **D.** Out in the country or camped out (too far back)

The Hock

Because this is the hardest worked joint in the body, it must be large and sturdy. It should be flat on the outside and only slightly rounded on the inside edge. Proper angle is very important for speed, agility, and soundness. Viewed from the rear when the horse is standing squarely, the hind leg should be perfectly straight. Viewed from the side, the back of the hock and back of the cannon should be perfectly straight and perpendicular to the ground. If the hock joint is too straight (set too far underneath the buttock) or too angled (hock and cannon not perpendicular to the ground — hind feet too far under the body and cannon bone angled), there is too much strain on the joints.

A horse with too straight a leg (at stifle and hock) is likely to suffer dislocation and locking of the stifle joint, that is, upward fixation of the patella. Too much angle in the hock, called **sickle hock**, puts strain on tendons and joint and makes a horse likely to develop curb or spavin. (Curb is strain and enlargement of the tendon at the back of the hock; spavin is enlargement of certain bones in the hock joint from excessive strain and arthritis, causing hind-leg lameness.)

UNSOUNDNESS LOCATIONS

An unsoundness is any physical defect that interferes with a horse's usefulness, making him lame, unsafe to ride, or unable to perform as he should.

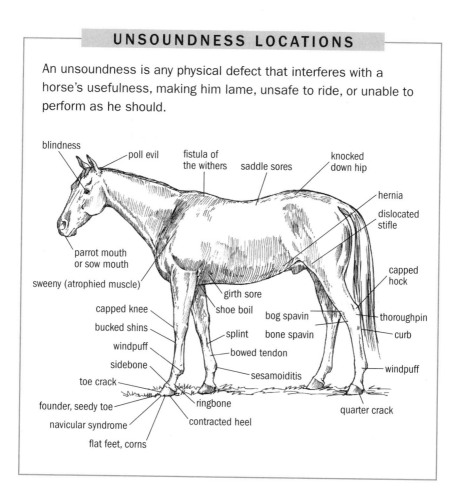

DISPOSITION AND TEMPERAMENT

Do not overlook disposition when selecting a horse. Even if a horse has excellent conformation and athletic ability, he will be difficult to work with if he has a bad disposition — and so might the foals. Temperament (boldness or timidity, friendliness or independence, calmness or nervousness, and so on) is inherited, so choose animals with a pleasant basic nature. This applies to both stallions and mares.

Base-Narrow or Base-Wide Hinds

As in the front legs, hinds can be base wide or base narrow. These deviations from straight (when viewed from behind) put more strain on hock and fetlock joints, leg bones, tendons, and ligaments, leading to possible breakdown and unsoundness as well as loss of agility and coordination.

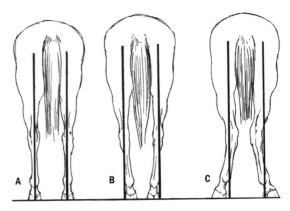

Hind leg conformation viewed from the rear:
A. Straight (normal)
B. Base narrow
C. Base wide (cow hocked)

Selecting Broodmares

Find mares who embody attributes you want in the offspring — good conformation and disposition, with the athletic ability to perform as you want. If you are raising horses to sell, also pay attention to pedigree and bloodlines; popular bloodlines are more sought after than little-known individuals, even if the latter are excellent animals; however, never sacrifice conformation, disposition, or athletic ability for name alone. A few "top" horses got to the top mainly because their owners had a lot of money to publicize them; such horses don't always produce top offspring. When buying breeding stock, look at a lot of horses before making any final decisions. Buy the best you can afford and then try to improve your animals by selective breeding. Your mares will determine the success of your breeding program. Only good mares produce good foals, no matter how much you spend on stud fees.

Age and Price

The highest priced females are fillies with good parents and young to middle-aged (6- to 13-year-old) broodmares who have produced good foals. The most economical age groups are the very young (unproven) and older matrons. Beware of buying an older mare who hasn't had a foal for several years. Older mares can become infertile; one who has consistently produced foals may be more likely to keep producing than an old mare who hasn't had one lately.

Stallions

Many breeders like to have a stallion of their own, but this is not necessary. You can send a mare to be bred to someone else's stallion, or have semen shipped and breed your mare at home by artificial insemination. More and more breeders are now using the latter option.

Buying a Stallion

If you do buy a stallion, buy the best you can afford; you are counting on him to improve your stock. He will produce many foals, whereas a mare will only produce a dozen or so in her lifetime, unless you use embryo transfer to have surrogate mothers raise extra babies from her. If your stallion is very good, you may be able to raise foals who are better than their mothers. You may decide to gamble on a weanling colt. This is probably the cheapest way to buy a stallion, but also very uncertain — you can't always tell how he will look as a mature horse. Judge him closely. If you like what you see, look at his parents and siblings to get a glimpse of traits he might display when he grows up. Whenever you buy breeding stock (male or female), try to see as many of the horse's relatives as possible. If there are any characteristics you don't like, don't gamble.

Age and Price

A stallion who is 2 to 4 years old may cost more than a weanling or yearling because he's starting to mature, you can see what he'll look like, and he has had more training. The mature stallion age 5 to about 15 will probably be the most expensive because he is at his peak and easily judged, and has already produced offspring. It's always an advantage to see his foals. Sometimes, the best-looking stallions don't sire the best foals, and plain animals may sire exceptional ones.

Before you buy any animal for breeding purposes, mare or stallion, have it checked for breeding soundness and fertility by a veterinarian.

Choosing a Stallion for Your Mare

If you have only a few mares, it's easier to breed them to an outside stallion than to own one yourself. This eliminates the expense and work of keeping a stallion, and it allows you to choose a different stallion for each mare to more perfectly complement her individual attributes. You can select a stallion who is strong where she is weak, with correctness where she has faults. This increases the likelihood that the foal will have better conformation or disposition than she does. You can probably produce better foals by choosing a different stallion for each mare than by breeding all of them to the same one, unless your mares are a uniform group with similar genetics (closely related).

Shipped Semen

If you don't want to send a mare to be bred, shipped semen is an alternative. Many breeders are now offering this option. It greatly increases your choice of stallions and eliminates the expense of transportation and care of the mare as well as a lot of risk — especially if the mare has a foal at her side. The mare must be palpated daily by a veterinarian to determine her time of ovulation, after which the stallion owner will collect the semen and ship it by overnight air to be deposited in the mare the next day. The veterinary expense is cheap compared with the money saved in other ways. Check to see whether your breed registry allows this type of breeding.

Frozen Semen

Another option today is use of frozen semen, which is helpful if you live in a remote area where timely delivery of cooled, shipped semen is not feasible, or when you want to have it on hand until needed. If kept properly frozen (stored in liquid nitrogen at -321°F [-196°C]) it will last forever. The disadvantage is that it requires more expensive shipping and storage tanks, proper procedure and technology to freeze and ship it, and great care in getting it out of the tank and thawing it. Also, the term of viability of frozen semen once thawed is shorter than that of cooled, shipped semen; proper timing is crucial. It must be put into the mare as close to ovulation as possible, which requires more intensive management and monitoring via palpation and/or ultrasound.

CROSSBREEDING

If you are breeding for excellence in a certain sport, consider crossbreeding (unless it's a sport that requires a purebred, such as Thoroughbred racing). For performance activities or any sport that demands a combination of agility, speed, and endurance, a good crossbred can do well. For versatility in English or Western pleasure riding, competitive trail riding, or speed events at a horse show, a good crossbred can be outstanding.

When crossbreeding, choose outstanding individuals who will enhance your goals with the attributes you want. For example, breeders of sport horses for jumping, dressage, and eventing often enhance their goals through careful crossbreeding to retain the athletic ability of the Thoroughbred while adding more substance and the mellow nature of a heavier framed breed, such as a draft breed or warmblood. A warmblood is any European breed of large horse that includes some Arabian blood in its ancestry. The term is also used to describe the result of crossing a heavy (draft) horse with an Arabian or Thoroughbred.

Fertility Record

Be sure to check the fertility record of a stallion you are considering for purchase or to impregnate your mare. Some stallions have less ability to settle (impregnate) mares than others. When using shipped semen or sending a mare to be bred, even if you get a live-foal guarantee, don't close your eyes to a possible fertility problem. If the mare comes home empty, you can always send her back, but that costs extra in transportation, board, and veterinary bills. You may also have the expense of keeping her another year with no foal if she doesn't settle by the end of the breeding season.

When selecting a stallion for your mare, weigh as many factors as possible and use your best judgment. Breeding is a gamble, but you can increase the odds of success with careful homework.

18

Genetics

WHEN BREEDING HORSES, the mating matchups determine the quality of the offspring. You need to have knowledge of genetics to predict what traits the foal will inherit from his parents. You are shooting in the dark if you look only at the individuals when selecting the mare and stallion to be mated; you must also know something about their genetic inheritance. Some recessive traits are not evident in the parents but may appear in offspring if the foal inherits a recessive gene from each parent. Some perfectly normal and outstanding horses are carriers of undesirable genes and may produce an abnormal foal; therefore, it is important to know as much as possible about the genetics of the horses you wish to mate.

Inheritance of Traits

An individual's characteristics are determined by genes and chromosomes. Genes are chemical codes that transmit various traits. They are located on chromosomes, which are strands of genetic material carried in every cell of the body. Chromosomes occur in pairs. As cells divide, half the genetic material goes with the new cell; it is a perfect replica of the old one (except when chromosomes become twisted, damaged, or misplaced, resulting in mutations).

Each cell contains chromosome pairs that carry the code of inheritance. Egg and sperm cells have only one chromosome from each pair, so when they unite, the newly formed pairs are a joining of one from the male and one from the female — the offspring gets half his genetic material from each parent. Because there's such a variety of genetic material in the many genes and chromosomes, the possibilities for different matchups are great. No two foals (even full brothers or sisters) are exactly alike unless they are identical twins.

LOOK AT ANCESTORS AND RELATIVES

The ancestors of a horse contribute to his makeup. Try to see as many as you can, in photos if not in the flesh. They'll give a clue as to what the foal might look like. The first fourteen ancestors (sire and dam; grandparents and great-grandparents on each side) are the most important to consider; any ancestor further back won't have as much influence on the characteristics the foal might inherit, unless the ancestor is present in the pedigree several times (which would be an example of linebreeding). Study the entire pedigree as far back as you can. If most of the individuals in a horse's pedigree have desirable traits, that horse will most likely produce a good foal when mated with an individual of equally good pedigree — provided both are typical representatives of the horses in their pedigrees. The sample pedigree chart below goes back four generations.

Pedigree Chart of Arabian Stallion "Surrabu"

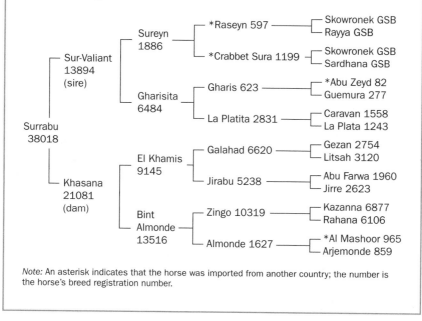

Note: An asterisk indicates that the horse was imported from another country; the number is the horse's breed registration number.

RECESSIVE TRAITS MIGHT APPEAR

A recessive trait might appear in the next generation even if it was not apparent in the sire or dam, if it is carried as a hidden recessive by both parents. Their mating might produce a foal whose genes line up to combine the two recessives.

Dominant and Recessive Traits

Most characteristics are determined by several sets of genes. It's impossible to predict what the foal will be like, except in traits affected by only one pair of genes. Some genes are dominant and others are recessive. When chromosomes join to form the new individual, two genes (one from each chromosome — one from each parent) control the trait. If one is dominant, the trait it expresses will appear in the foal. The trait represented by the recessive gene won't be expressed with the dominant gene present but might be expressed in future offspring if the gene is not masked by a dominant gene in that next union.

In some instances, genes work in tandem to express combined traits. There are also "dilution factors" involving some of the color expressions, which make things more complicated than simple dominants or recessives. These genes are usually incompletely dominant, so you can often tell the differences between two copies of one, two copies of the other, or one copy of each gene.

Dominant and recessive genes can be combined in three different ways:

1. The two dominants can come together, producing an animal that is "homozygous" dominant for that trait (homo means "same"). In this case, the only genes it carries for that trait are dominant; therefore, it not only expresses that trait but can pass on no other trait to its future offspring.
2. The two recessives can come together, producing a homozygous recessive individual that expresses the recessive trait and can only pass on this recessive trait to its future offspring.
3. The foal can inherit a mixed pair of genes — dominant and recessive — and be heterozygous. In this case, the foal itself shows the dominant trait because any dominant gene always masks the presence of a recessive one, but this animal can pass on either gene (dominant or recessive) to its future offspring.

Genetics of Color

Traits controlled by dominant and recessive genes, such as those for color, are readily seen. An example of color inheritance can help explain how traits are passed from parent to offspring. For instance, let's say "G" is the gene for gray and "g" is the one for non-gray. The only time a horse can be a color other than gray is when a "gg" combination (two recessives) exists in the gene matchup because gray is the dominant trait. The foal may inherit from one parent a recessive chestnut gene or a bay gene (which is dominant over chestnut) but will always turn gray if he inherits the "G" gene for gray.

Gray, Bay, and Chestnut

Gray, bay, and chestnut are easy to predict because they are controlled by fairly simple dominants and recessives. Of these three colors, gray is always dominant. A gray horse is born bay or chestnut and becomes gray later because of the dominance of the graying gene. The gray horse must have at least one gray parent; gray can never skip a generation as can chestnut.

Bay is dominant over chestnut but still recessive to gray. Chestnut is always recessive; a chestnut bred to a chestnut can produce no other color but chestnut. Chestnut may appear from a mating of two bays or two grays, or a bay and a gray, if each parent carries a recessive chestnut gene.

Sample Crosses

In the first cross (below), the "pure" dominant parent will decide the color of the foal when mated to a recessive. The foal will then carry a recessive color gene that can be passed on to its offspring.

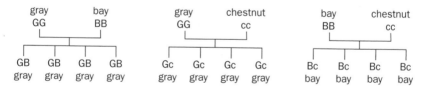

A gray horse who carries a bay gene will produce mostly gray foals but can still produce a bay foal. When mated with another gray horse who carries a chestnut gene, the foal has these four possibilities (below): pure gray, gray but carrying a chestnut gene, gray but carrying a bay gene, and bay carrying a chestnut gene.

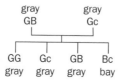

If a gray horse who carries a recessive chestnut gene is mated to a bay horse who carries one also, the foal has these four color possibilities (below): gray carrying a bay gene, gray carrying a chestnut gene, bay carrying a chestnut gene, and chestnut — if he inherits both recessive genes.

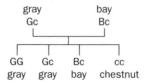

Note: Gray = dominant over all colors; bay = dominant over chestnut; chestnut = recessive; capital letters denote dominant gene; lowercase letters denote recessive gene.

Color Inheritance: Gray. A pure gray mated to an impure gray (carrying a bay gene) will produce nothing but gray foals, but some of them may carry a bay gene and pass the bay color to their offspring if mated with another horse who carries a bay gene. A pure gray mated to a bay carrying a chestnut gene will produce nothing but gray foals, but some will carry bay genes and some will carry chestnut genes. Two gray horses carrying bay genes can produce a bay foal, but the chances are one in four.

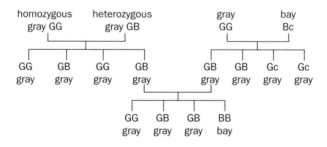

Color Inheritance: Bay and Chestnut. A pure bay mated with a bay who carries a chestnut gene will produce bay foals, but some of the foals may carry the recessive chestnut gene. Two bay horses carrying a chestnut gene can produce a chestnut foal.

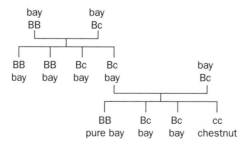

Color Inheritance: Gray, Chestnut, and Bay. Gray is dominant and chestnut is recessive, but two gray parents can produce a chestnut foal if each carries a recessive chestnut gene.

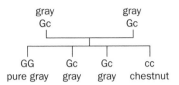

Chestnut to chestnut *always* produces chestnut.

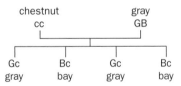

Chestnut to any other color (provided the other color doesn't carry a chestnut gene) will always produce the other color or its genetic possibilities. In this case, a chestnut was bred to a gray who carried a bay gene. The foal will be either gray or bay and will carry a chestnut gene.

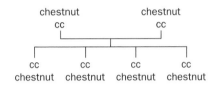

Note: These general rules are relatively consistent, although somewhat oversimplified. Because these genes appear at different loci (locations on the chromosome), the results may occasionally be skewed.

COLORS AND BREEDS

Some colors are common to all breeds, whereas others are found only in certain breeds. For example, there are no true palomino, buckskin, or dun Arabians, but these colors can be found in Quarter Horses. A chestnut Arabian with a very light mane and tail may be mistaken for a palomino but is genetically chestnut.

Other Colors

Breeding for black, palomino, dun, Paint, or Appaloosa colors is more complicated. Roan is dominant over all colors except gray and white. Some colors appear because of recessive factors that "dilute" another color gene. There are several dilutions, each genetically different. When one of these factors (sometimes called "cream") is present, bay dilutes to buckskin, brown becomes red dun, and chestnut becomes palomino. When both recessive dilution factors are present, each color is diluted to some shade of cream with blue eyes. Bay becomes perlino (nearly white, with rust-colored mane and tail) and chestnut becomes cremello (creamy white with light mane and tail and blue eyes). Another dilution factor involves dun, which dilutes in one dose or two. This can produce grulla from black or brown, change bay to dun, or chestnut to red dun.

To get a desired color in a foal, you have to know whether that color is dominant, recessive, or heterozygous (mixed) or dependent on dilution factors. You also must know the mare's and stallion's genotypes. This is easy if they have already produced offspring because you can see the color of their foals. Checking the colors of ancestors (in their pedigrees) also gives clues as to the presence of recessive genes.

Dominant colors are bay, gray, tobiano paint, blanket Appaloosa, roan, and line-backed dun. Black must have multiple traits present; it can occur without black parents. Recessive colors are chestnut, perlino, and cremello. Mixed (heterozygous) colors are white, buckskin, and palomino. Breeding for these is complicated because heterozygous colors cannot breed true; there is a 50–50 chance of producing the same color as the parent. Pure white at birth exists only in heterozygous form because these genes are lethal to the embryo when doubled up (homozygous) in dominant form (see Lethal Genes, page 339).

THE GENETICS OF COLOR

Color is a less important trait than conformation, athletic ability, soundness, disposition, or intelligence; however, understanding the genetics of color makes it easier to understand the inheritance of other traits because the expressions of color genes are so visible. If you're raising horses in a "color" breed such as Appaloosa or Paint, or for customers who want a certain color in their favorite breed, then having a working knowledge of color genetics is extremely important.

Types of Breeding

There are several genetic "tools" that a breeder can use to reach certain goals and produce outstanding individuals. A good knowledge of genetics helps a breeder use these techniques properly.

Inbreeding

Inbreeding involves the mating of closely related individuals and it should be avoided, except in special cases, because it increases the chance of doubling up undesirable recessive traits or genetic defects that are hidden in the parents. Most undesirable traits are recessive homozygous; therefore, breeding closely related animals (ones who may carry the same recessive trait) brings them out. Inbred animals are also generally less hardy; inbreeding decreases the genetic variations within a bloodline. If undesirable or lethal genes are present, inbreeding presents a greater chance of birth defects and problems. Inbreeding concentrates the genes available, both good and bad.

Many breeders use inbreeding as a tool to intensify and "fix" certain good traits in a family line. If a horse is outstanding and has no known faulty genes, a breeder may double up the good genes by mating sons and daughters (half-siblings) of the outstanding horse. This offers the chance that the offspring will be exceptional, especially in genetics that can be passed on. Inbreeding is always risky, but with luck, a horse who is a product of inbreeding from outstanding parents will carry only doubled-up, good attributes in his genetic makeup. He'll be genetically pure for these characteristics, serving as an outstanding sire (or dam) and passing on those traits consistently to the foals.

Linebreeding

Linebreeding, a form of inbreeding, involves the mating of relatives (the descendants of one particular outstanding horse) in an effort to double up and "fix" the traits of a superior ancestor so the animals will breed true and pass on the traits of that ancestor. Because the total number of gene combinations in a linebred animal are limited (owing to less outcrossing to bring in different genes), the linebred (or inbred) animal is fairly predictable in traits it will pass to offspring. Linebreeding, like inbreeding, must be done cautiously, with careful examination of all known traits of the bloodline to make sure there are no undesirable traits that might be doubled up as well. Some of the devastating genetic defects described later in the chapter, such as SCID, HYPP, HERDA, and GBED, that have cropped up in various breeds during the past several decades are the result of linebreeding to a popular ancestor who

---

NICKING

Nicking is a term that refers to the production of superior offspring from mating two individuals or family lines. If the two nick well, they produce offspring who are better than would be expected from those parents. The mating results in more favorable gene combinations than if they were mated to other individuals. Nicking is a lucky phenomenon — an unexplainable good combination of genes.

---

carried a defective recessive gene. The defect appeared in later descendants when these genes were doubled up by linebreeding.

Outcrossing (Outbreeding)

Outbreeding refers to the mating of unrelated individuals within a breed. This often produces superior offspring, but their genetic makeup is more varied than that of the inbred or linebred horse, and there may be more variance among their offspring.

Crossbreeding

Crossbreeding involves breeding individuals of different breeds to obtain offspring with characteristics of both breeds. Crossbred animals also may have certain traits superior to those of both parents through hybrid vigor (heterosis). When the best characteristics of each parent are combined, the foals have more versatility and hardiness. The key to successful crossbreeding is to choose animals with traits you want; outstanding parents usually produce a crossbred foal that is exceptional — partly because of hybrid vigor and partly from combining the traits you desired from each breed.

Most of the breeds in existence today were originally created by crossbreeding. For example, many of the non-draft-horse breeds have some Arabian blood in their background. In earlier times, the "breeds" of certain locales were improved by crossing with Arabians. Selective crossbreeding can produce versatile athletes. For example, crossing a Thoroughbred with a Quarter Horse or warmblood may produce a superior jumper. Crossing an Arabian with another breed will often produce a good all-around athlete with more endurance.

If your goal is a specialized sport like racing, however, you are usually better off sticking with breeds that have been selectively bred for speed — such as the Thoroughbred for fastest speed at a mile, or a Quarter Horse for the fastest burst of speed in a short sprint. If your goal is exceptional endurance, the Arabian (or an Arabian derivative such as the Spanish Mustang) is best. For certain gaits, choose the gaited breed that has been bred for that specialty. Crossbreeding, however, often works well to create a more all-around athletic horse.

Genetic Defects

Some individuals and family lines carry genetic defects; be wary about using these horses for breeding. The most common inheritance pattern involves a simple recessive trait; the defective foal inherits a recessive gene from both sire and dam — individuals who are otherwise completely normal. A few defects are caused by genes with incomplete dominance, and some are caused by two or more sets of genes, but the most common problems are caused by simple recessives.

Defects run in families; therefore, inbreeding or mating closely related animals is risky unless you know there are no bad genes in that family to cause a defect when doubled up. For example, the parents of a defective foal often have at least one ancestor in common. Today, with the availability of DNA (genetic material) testing, there are tests for some of the most common devastating genetic defects, and you can make sure that individuals are free of these defects before you breed them.

Lethal Genes

A few genes, when doubled up, cause death of the fetus before or shortly after birth. Some of these genes, when doubled up as an expressed trait, result in a fetus whose organs are improperly formed (such as lack of an anus, absence of

---

SELECTIVITY IN BREEDING

Work toward genetic improvement by selectively keeping as breeding stock the animals who inherit traits you want and selling animals who don't. You can do this by outbreeding or linebreeding. Linebreeding concentrates the desired traits more quickly and a little more safely than inbreeding; it still leaves enough genetic variation to minimize (but not eliminate) the chances of genetic defects becoming evident.

eye sockets, or water on the brain). The most common lethal gene is the one that creates a white coat. It is inherited as a dominant trait and is only lethal to the embryo when appearing on both chromosomes.

Lethal White Genes

White hair or coat, which results from lack of hair pigment, can lead to foal mortalities. Lethal whites die soon after conception, and lethal overo whites (a white foal produced by breeding two ovaros, a type of pinto) die soon after birth because of faulty intestinal tracts; the foal looks normal at birth but cannot survive without a functioning gut. Not every white horse carries the lethal genes for early fetal death, and not every overo-patterned Paint or Pinto produces white foals who die young. The lethal situation occurs when two horses carrying the gene are mated.

True White. These horses are usually entirely white with pink skin and brown eyes. Albinos (which do not exist in equines) would have pink eyes, but none are ever born. Lethal white is a dominant gene; but because all true whites are heterozygous, they produce 50 percent white and 50 percent colored off-spring when mated with colored horses. When true whites are bred to other true whites, the result is 25 percent colored foals who inherit the recessive color gene, 50 percent heterozygous true whites, and 25 percent homozygous white who do not survive.

Lethal Overo. Paint breeders have learned to beware of mating frame-type overos to each other. A frame overo is basically another color (like chestnut) with jagged-edged splashes of white over the body. A rare few white horses from overo breedings survive, but these are usually sabinos and not frame ove-ros. Foals with two copies of the frame overo gene are white and die soon after birth because of nonfunctioning guts as a result of failure of nerve develop-ment to the digestive tract. These foals may seem normal at birth, and may get up and nurse, but nothing can move on through the digestive tract and they soon die

Severe Combined Immunodeficiency (SCID)

SCID is a fatal disorder of Arabian foals (and sometimes part Arabians if both parents have the defective gene); it is a defect of the immune system inher-ited from two carriers who otherwise appear normal. SCID foals are normal at birth, but a portion of their immune system fails to develop. As soon as

ORIGIN OF SCID

SCID may be the result of a genetic mutation that occurred in a single horse in the early 1900s. Because of its rapid increase in incidence, it is thought that the mutation appeared in a relatively popular horse. Today, SCID may be present in almost all Arabian bloodlines and may occur to a lesser extent in other breeds created during this century that were crossbred with Arabians during their formation. During the past two decades, breeders have been working to reduce or eliminate this defect, and this effort became much easier after the advent of the DNA test.

their temporary protection from maternal antibodies in colostrum begins to wane (a few months after birth), they die from infections.

When two SCID carrier horses are mated, there is a 25 percent chance that the foal will inherit the recessive gene from each parent and be affected with SCID, a 50 percent chance the foal will be normal but carry the recessive gene, and a 25 percent chance of not inheriting the gene at all. If a carrier horse is bred to a noncarrier, none of the offspring will develop SCID but half will be carriers.

There is no treatment for SCID; the only way to avoid this problem is not to breed two horses who are known carriers. You can test for this defect in a horse's DNA through an oral swab plus a blood or hair sample. Hair samples consist of 20 to 30 mane or tail hairs, pulled out rather than cut; the hair root is needed for the DNA test. This sample can be sent to a laboratory to be analyzed.

The test kit can be ordered through your own veterinarian or from a genetics lab that does this type of test, such as VetGen. A carrier horse can still be used for breeding if the potential mate is free of SCID because they will never produce a SCID foal. Half of all offspring from such a mating would be carriers but the other half would not have the recessive gene; therefore, you would not have to sacrifice the bloodline from an outstanding individual who happens to be a carrier. If you test all offspring, you would know which ones could be safely bred. The others could be used in performance careers.

Glycogen Branching Enzyme Deficiency (GBED)
GBED is a mutation that causes inability of tissues to store sugar properly, and foals with this defect are usually aborted, stillborn, or weak, dying within

the first days or weeks of life. The glycogen in their skeletal muscles, heart, and liver can't be mobilized very well, so they have no energy reserves. This inherited defect was first recognized in 1997 and is now present in about 10 percent of all Quarter Horses and related bloodlines (Paints, Appaloosas, and so on, that have incorporated Quarter Horse breeding). This mutation traces back to King, a popular foundation Quarter Horse born in 1932.

Some of the affected foals may be alive at birth but need help to stand and nurse. They may seem healthy for the first hours or days and then die suddenly, develop seizures, or become weak and unable to get up. With intensive care, some have lived to be two or three months old, but all have eventually died or been euthanized.

Researchers discovered that GBED appears only in offspring who inherit the recessive, mutated gene from both parents. If a carrier is mated to a normal horse, the offspring have a 50 percent chance of being a carrier, but none would be affected themselves. But when a carrier is mated with another carrier, resultant foals have a 50 percent chance of being carriers, 25 percent chance of being affected by GBED (and dying), and 25 percent chance of not receiving the defective gene at all. There is a genetic test for GBED, and if any descendent of King produces a foal lost to abortion, stillbirth, or early death for any unknown reason, that animal should be tested. If breeders test their horses and choose not to mate carriers with carriers, no more GBED foals will be produced. Testing can be done at VetGen laboratory (see above) or at UC Davis.

Other Serious Genetic Defects

There are other defects that are not as lethal, but some of them can eventually lead to death of the horse. These include abnormalities in muscle metabolism that may result in collapse of the horse, muscles chronically cramping, defects in the skin, and defects in the ligaments that support the fetlock joints.

Hyperkalemic Periodic Paralysis (HYPP)

HYPP is a muscle disorder that occurs in certain family lines of Quarter Horses, Appaloosas, and Paints. Affected horses usually have heavy muscling, as is popular in halter classes. The problem is characterized by sporadic attacks of muscle tremors (shaking), weakness, and/or collapse and inability to get up. Attacks may be accompanied by noisy breathing from paralysis of muscles in the upper air passages. Sometimes, sudden death follows an attack, with the horse dying from heart failure or paralysis of the respiratory muscles. The problem is sometimes mistaken for colic, choking, muscles tying up, or a respiratory ailment.

HYPP is inherited as a dominant trait. Homozygous individuals who inherit the defective gene from each parent (H/H) are severely affected (many do not survive to adulthood); heterozygous individuals with one normal gene and one defective gene (N/H) are more moderately affected and may survive if they have proper temperament and metabolism. With careful management (feeding a low-potassium diet and maintaining a steady, careful exercise program), many of them survive and function, but this can be a headache for the horse owner.

Breeding a heterozygous animal (N/H) to a normal animal (N/N) will result in offspring with a 50 percent chance of being normal, and a 50 percent chance of having the defective gene (N/H). Breeding a homozygous (H/H) animal, sometimes called a double positive, will result in all offspring inheriting the HYPP gene, regardless of the status of the other parent. Breeding an H/H horse to an N/H horse will produce offspring with a 50 percent chance of being H/H and a 50 percent chance of being N/H. Breeding N/H to N/H gives a 50 percent chance of being N/H, a 25 percent chance of being completely normal (N/N), and a 25 percent chance of being H/H.

The HYPP-affected horse must be kept on a low-potassium diet (little if any alfalfa hay). Some owners give the horse frequent doses of a diuretic (acetazolamide), which helps the body eliminate potassium. Treatment of an attack will vary with its severity. Giving the horse glucose (such as Karo Syrup) may resolve a mild attack, but immediate veterinary treatment and IV fluids will be needed for a more severe attack in which the horse collapses.

ORIGIN OF HYPP

First identified in 1985, HYPP appears to have originated as a mutation in the popular Quarter Horse stallion, Impressive. During the 1970s, 1980s, and early 1990s, this famous halter champion produced many offspring, being bred to Appaloosas and Paints as well as Quarter Horses — so this genetic defect is widespread. By 1992, Impressive had more than 100,000 descendants, and that number has mushroomed since then. Breeders who wish to avoid HYPP can use a DNA test to make sure they breed only animals who do not carry the defective gene.

Polysaccharide Storage Myopathy (PSSM)

The primary symptoms of *PSSM* are muscle cramping, trembling, unwilling-ness to move forward, and tense and swollen muscles of the hindquarters (see chapter 9). In some affected animals, however, the owners merely notice poor performance, lack of energy, difficulty in backing up, stiff gait or gait abnormalities, unwillingness to lift the feet for hoof care, or sensitivity to pressure over certain muscles of the back and hindquarters. PSSM occurs in many breeds — mostly heavily muscled horses — and is a dominant trait; therefore, it only takes one copy of the mutant gene to express it. This is dif-ferent from GBED (see above) or HERDA (see below), which are recessive (the horse must inherit two copies, one from each parent). The risk of having affected offspring from a horse with PSSM is much higher because there is always at least a 50 percent chance of the foal inheriting the dominant gene. And if a foal gets two copies of the PSSM gene, that foal will often be more severely affected.

Over many decades, horse breeders have inadvertently selected for PSSM, especially in draft horses and Quarter Horses, in an effort to get more muscling — often doubling up these defective genes. There is now a DNA test that can detect one type of PSSM (which involves about 80 percent of affected animals), and researchers are working on a test that will detect another type that has recently been recognized.

Malignant Hypothermia (MH)

This genetic defect has been recognized in certain family lines of Quarter Horses. It is similar to MH in humans, in that it generally isn't noticed until

---

TESTS FOR GENETIC DEFECTS

There are several university labs and a few private labs around the country that offer tests for genetic defects, coat color, parentage verification, and so on. For example, the lab at Cornell University currently tests for HERDA; UC Davis tests for coat color, lethal white overo, GBED, HERDA, HYPP; the University of Minnesota tests for PSSM type 1 and MH; and the University of Kentucky tests for lethal white. Test availability changes from time to time because, in some cases, this will be dependent on patents and licensing, so you may need to check.

the individual is exposed to gas anesthesia (as for surgery). An affected horse may have a severe reaction and die. Some horses have *MH* along with PSSM, which makes their muscle tying-up episodes even more severe. The horses who die during a muscle cramping problem are likely to also be affected with MH. The genetics of this problem are still being studied, but there is now a DNA test for this defect.

Hereditary Equine Regional Dermal Asthenia (HERDA)
Also called *hyperelastosis cutis*, this skin defect was first documented in the 1960s and traced back to the famous Quarter Horse sire, Poco Bueno. It is an inherited connective-tissue disorder characterized by abnormal skin that tears easily and separates readily from the underlying tissue. Any pressure or trauma can pull the skin apart. The problem is often not discovered until the young horse goes into training and the simple act of wearing a saddle creates massive injury to the skin.

All affected horses are related but the defective gene did not cause problems in the earliest offspring of Poco Bueno because they only carried half the equation; the defect did not show up until some of his descendents were bred to each other. During early research, skin biopsies were used to diagnose this problem, but now there is a DNA test and breeders can readily determine whether a horse is a carrier.

Degenerative Suspensory Ligament Desmitis (DSLD)
DSLD is an untreatable genetic defect that causes abnormalities in certain body tissue, particularly the supportive structures of the legs. The most noticeable problem is weakness of the suspensory ligaments and dropping of the fetlock joints, especially in the hind legs. Originally identified in Peruvian Paso horses, it was first thought to be a problem only in that breed, but this condition has since been seen in several other breeds, including Quarter Horses, Thoroughbreds, warmbloods, and draft horses. Research on this defect is still ongoing; the genetic basis for this problem is not yet known.

Know Your Horses' Genetics
To be a successful breeder, you must not only be a good judge of horses when assessing conformation and ability but also know their genetics. You will be mating not only two individuals but also two separate ancestries, combining their genes. You must be able to predict with reasonable accuracy what the product of each mating will be, complementing the sire's and dam's

---

A LESSON IN GENETICS

Genetic defects are common but because most of them are recessive, they may not show up in offspring unless carried by both parents. Defects show up more frequently in domestic animals than in humans because these animals have a much smaller gene pool; they usually have arranged matings, within a certain breed, often within certain popular family lines.

Mutations in humans and animals occur frequently — because of chemical insults to cells, physical insults (such as too much sun exposure), errors in DNA replication, and so on. If the error happens in the earliest stage of an embryo's development, it ends up in all the tissues, including the gonads, and can be passed along to that individual's offspring. Under natural conditions, you might never see the defect unless an animal with that hidden DNA error mated with another who has a similar hidden error — which is most likely to happen only in small, isolated populations where inbreeding occurs. An animal born with a severe defect might not survive to have offspring. But with domestic animals, we often perpetuate a hidden defect, especially if the carriers have traits we desire; therefore, some of the popular bloodlines in cutting horses, halter horses, racing, and other disciplines have passed along some inherited problems.

All too often, a horse who can't perform athletically because of a problem is bred instead because he comes from a popular and expensive bloodline. Sometimes, the horse industry is its own worst enemy; a valuable horse who can't perform is often retired to breeding!

---

characteristics as well as creating foals who inherit the best of both ancestries. There are no perfect horses; therefore, you must try to mate sound individuals who will produce even better offspring.

If a horse has faults, mate it with one who has strengths in that area. If you mate two horses with the same faults, you will most likely produce a foal who is worse than his parents. Never double up bad points of conformation or any known defects. Genetic tests for certain defects are now available and can help horse breeders make informed decisions.

Breeding good horses is a time-consuming job that involves much work and expense. It costs no more to keep and feed a good broodmare than a poor one, so make every effort to have the best you can afford. If you have a stallion, a good one will help sell your foals. If you consistently have good foals, people will come back for more; but if you sell a few poor ones, or some who carry a serious defect, the word will get around. If you are conscientious about breeding good horses, you won't let fads popular in the show ring overrule your best judgment regarding what is best for the breed. If you strive to attain consistently high goals, there will always be a market for your horses even if you aren't chasing the latest fad.

FURTHER READING

Genetics is a complex subject. These books are helpful:

- Ann T. Bowling, *Horse Genetics* (New York: CAB International, 1996). A look at genetics and heritable characteristics
- Philip Sponenberg, *Equine Color Genetics* (Blackwell; Third edition, 2009). A simplified review that is easy to understand

19

Keeping a Stallion

CONSIDER THIS: You may never need a stallion. Many breeders today send their mares to be bred or use shipped semen. But if you do decide that owning a stallion best suits your purposes, spend some time at someone else's breeding farm observing and learning how to handle a stallion before you try to buy or raise your own. This will help you immeasurably.

If breeding horses is a business venture you want to try, keep in mind that the stallion is a crucial element of the operation. He is the horse who comes most before the public eye, giving your enterprise its reputation — especially if he excelled in a performance career before being used at stud. When people come to see your breeding program, they are usually most interested in the stallion — his looks, ability, condition, and the ability and conformation of the foals he sires. Good mares are equally important; but, because it's the stallion who many people come to see, he carries more than his share of responsibility for making a good impression.

If you have a stallion, extra attention must be given to his training, care, handling, and living quarters. The stallion is more aggressive and exuberant than other horses, requiring increased responsibility in care and handling.

Facilities for a Stallion

The stallion's enclosure should be high and well built with strong, smooth materials. If he spends part of his time in a box stall, it should be large: at least 16 × 16 feet (5 × 5 m). The stall should have an adjoining paddock or pasture where he can be turned out to exercise himself. If you want him to have room to graze, the pasture should be large enough that he won't tramp out most of the grass by constantly walking the fence. Most stallions expend a lot of energy worrying about what goes on around them and spend hours pacing or cavorting along the borders of their enclosure.

The fence around a stallion pen should be high and strong.

The stallion's fence should be 5 to 6 feet (1.5–2 m) high or higher, depending on the individual — high enough that he won't try to jump out. It can be made of poles, boards, or pipes, or anything else that is safe, smooth, and durable enough to withstand kicking and leaning. Space the poles close enough to keep him from getting his head between them (3-inch [7.5 cm] space), but wide enough for him to get a foot back out again easily if he puts it between the poles. If the enclosure is made of wood, he might chew it, but an electric wire will keep him away from the fence if he likes to gnaw. It will also prevent him from kicking it, pawing it, and trying to get over or through it.

King of His Own Turf

A stallion is very sensitive about his surroundings. He considers all the mares on the place to be his and is very concerned about them. He notices more things than most mares and geldings do; all the goings-on around the place interest him, for he feels they are his personal concern. If you bring a mare in

SHELTER

On cold or hot days, the stallion should have access to shelter or shade. Standing all day in the heat will sap his energy. In a hot climate (temperatures above 90°F [32°C]), you should run a fan in his stall or shed during the heat of the day or give him a bath (with cool, not ice-cold water) after his exercise to refresh him.

from the pasture, he will notice; if you take one to ride, he'll be watching and probably will be the first to become aware of your return. He knows where each horse on the place belongs. Changes upset him, such as when strange horses are brought in or when his daily routine is changed. If the stallion is in a paddock next to another enclosure, make sure the other place is always occupied by the same horse — one he gets along with.

If you have more than one stallion, it's best to have them widely separated, secure in their own "turf." Horses are creatures of habit; they like the security of sameness in their surroundings, daily routine, and neighbors. If you can, keep the mares a certain stallion will be breeding near him. Some stallions feel possessive about every mare on the place, but many come to develop a sense of "herd" and territoriality about the mares who live near them. A stallion may fuss and call to each of "his" mares when they are worked or turned out to pasture, only half-heartedly showing interest in another stallion's mares.

Don't Isolate a Stallion

A stallion can be stabled or penned near other horses without problems, but if his stall is next to another horse, there should be a solid wall between them. Keep in mind, however, that it's not wise to socially isolate a stallion. He will be emotionally healthier and easier to manage if he has company, even just horses in the next field or some he can watch from his stall door. Young stallions should be raised with other horses as much as possible to keep them well socialized. They are less of a problem to handle later and can usually be housed in fairly conventional facilities. However, if you buy a mature stallion who already has certain habits, pay close attention to how he was housed and make his new home similar to the old one if that arrangement was working and he was happy there.

Handling a Stallion

Stallions must have special care in handling because they have an attitude very different from that of mares or geldings. The stallion is proud, strong, and bold. Because of his highly exuberant and sensitive nature, he should always be treated with firmness and respect but also with sympathy and intelligence. To abuse a stallion is to invite trouble — potentially, very dangerous trouble!

Proper Training Is Essential

Proper handling and training of a young stallion are essential for developing and keeping a good disposition; from the time he is a foal, you must teach and insist on good manners. If a stallion has a good temperament to start with and is handled properly—with tact, understanding, and respect as well as firmness —he won't be a problem to handle, especially if he is well trained while young and gets plenty of exercise as an outlet for all his energy. The confined stallion with no way to burn off energy may take out his frustrations on his surroundings or the people who handle him.

Understand Him and Gain His Respect

The stallion handler must understand him and maintain a good rapport. He must be able to demand and get the horse's obedience with a mere word or touch. You should rarely have to discipline a well-trained stallion, but if you do, you and he both know he deserves the reprimand and he does not forget the punishment or the reason for it. Any disobedience such as charging, rearing, striking, or biting must be halted immediately while the young horse is first experimenting with the limits of his privileges. In fact, most young stallions will try to see who is really the boss; it's their nature to establish pecking order and try to gain dominance. Early on, if the handler establishes who is in control and lays out the limits of behavior for the horse, a good working relationship can be developed.

The person handling a stallion must never lose his or her temper and punish the horse in anger. When the horse responds and behaves, the reprimand should stop immediately and the handler should resume what he or she was doing, with no continued anger over the misbehavior. The sooner the relationship again becomes friendly and relaxed, the less resentment the horse will carry.

A good stallion handler is tactful and kind but also constantly mindful that the horse is a stallion. The handler must be able to command respect and instant obedience whenever necessary.

Demand Obedience without Roughness

The stallion handler must demand obedience but should never beat on the animal to demonstrate superiority. This happens all too often when a person is afraid of the stallion. The reason for punishment should always be clear to the horse; he should never be goaded unnecessarily or he'll fight back. Anticipate confrontations so you can avoid them if at all possible. Try to keep the upper hand psychologically. If you establish a good foundation of respect during the young horse's training, you can maintain good rapport and "mind control" because he knows the limits of what he can and cannot do. If you are consistent in your handling, leaving no room for doubt or trying to get the best of you, you'll rarely have to discipline him harshly.

Rough handling is a bad experience for any horse, but especially for the stallion because he is so proud and bold. When treated with roughness, he will fight back. Abusive handling makes him irritable, bad-tempered, and more dangerous. It is better to be gentle but firm, communicating to the stallion that you have respect for and confidence in him. Most stallions enjoy human companionship and want to please you, but if you're nervous or afraid, a stallion will sense this lack of confidence and become nervous also, reacting adversely if you resort to force.

No one who is afraid of a stallion should ever handle one, for fear is usually accompanied by forceful, rough treatment and a loud voice in an attempt to cover it up. A self-respecting stallion will resent that nonsense and retaliate. Use of force just leads to more force. If there is a certain stallion you are afraid of, he should be handled by someone else.

Build Trust

The relationship between stallion and handler should be one of mutual admiration, trust, and respect; a spirit of teamwork and good communication is extremely important. The more you treat your stallion just as you would any other horse — riding him, giving him regular exercise, using good judgment (and always remembering his male characteristics of aggressiveness, desire to dominate, intense interest in females, and so on) — the easier it will be to keep him gentle, good-tempered, trustworthy, and easy to communicate with.

WHO SHOULD HANDLE A STALLION?

Some people have the ability to handle a stallion and others don't. It's a matter of personal aptitude and the individual horse. Some horses need a strong and steady hand, whereas others require tact, patience, and sympathetic understanding. Not every horse person or trainer gets along well with every horse. Individual personalities of horse and handler sometimes clash. If you find you can't get along with a certain stallion or become afraid of him, someone else should handle that horse.

<table>
<tr><td>

VERBAL COMMANDS

Proper response to verbal commands should be ingrained early, from the time of halter and leading lessons as a foal and during early groundwork. The stallion must know the meaning of *whoa*. You can also teach him *back up*, *easy* (slow down), and so on. Proper response to commands should be taught long before he is ever used for breeding.

</td></tr>
</table>

Extra Training

Because he'll be used for breeding, you should prepare the stallion ahead of time for this role, training him as a young horse so he can be handled later with good control and no problems. Being a creature of habit, the stallion can be taught to respond to certain patterns with expected behavior. If patterns are changed, he becomes insecure, reacting unpredictably. It's important to establish certain ways of doing things early on and then be consistent. The young stallion must learn to respond to you instead of his environment. Even if he'd rather charge after a mare in heat, he knows he must obey you.

Have a Different Halter for Breeding

Do not use the same halter for breeding and regular work. The stallion should know that when his working halter or bridle is put on, he is to keep his mind on being exercised or ridden and forget about the mares. It's usually not wise to use the same headgear for regular handling and for breeding. If you can keep the tack and tasks separate in his mind, he'll be more apt to behave himself. If you use a halter with a nose chain for breeding, it should be different from the halter and lead you use for ordinary purposes. If you use a breeding bridle for more control during breeding, it should be different from the bridle you use for riding. The most important factor is consistency of routine. Remember that the stallion will not automatically associate the different headgear with the proper job unless you keep the process consistent.

Use Strong Equipment

Equipment used on the stallion should be well made and strong. There should always be enough lead shank or rope to allow extra length to work with (as when breeding mares and giving the stallion room to do the job), but not so

---

GROOMING TIP

One aspect of training that you shouldn't neglect is getting the young stallion accustomed to having his scrotal and sheath area washed. Some individuals resent this at first, but you can avoid problems by making it part of the normal grooming routine from the time he is a foal.

---

much that there is danger of you becoming tangled in coils of rope. You may want a section of chain in the lead — to go over the bridge of his nose — to keep him under control while going to the mare, teasing, and so on. In working with the young, excitable stallion during early training, you may need a chain over the nose to make him mind his manners. Many stallions like to rear and strike when they are being led near other horses, and the young stallion must learn this is unacceptable behavior.

Discourage Rearing

If a stallion wants to rear, you should nip this bad habit in the bud, being prepared for it while leading him. A sharp jerk with nose chain before he actually gets off the ground is best. If he rears before you have a chance to halt him, allow him to complete the rear — a sharp jerk while he is going up or at the peak of his rear may make him throw himself over backward or sideways in his attempt to avoid more pain. It could also cause him to strike at you. It's better to wait until he's starting back down and then punish him with a jerk on the nose chain, with you well off to the side for safety. Split-second timing is best for good control and punishment.

Let an Older Horse Educate the Young Stallion

Some young stallions are so aggressive and difficult to handle that they need more "education" than a human being can physically accomplish. The best solution for this problem is not to use more force — which merely creates a confrontational situation and precludes a good working relationship — but to turn the young stallion out for a while to live with an older, domineering gelding or a few pregnant mares. They'll quickly put him in his place and teach him humility. Once he learns about being at the bottom of the pecking order instead of the top, he'll be a lot more manageable and responsive to his human handlers as well. He'll know that someone else can boss him.

Provide Daily Exercise

Daily exercise is important. Because the stallion is, by nature, active and aggressive, if he is cooped up, you will have trouble. He needs room to exercise himself or have regular supervised exercise. Once he is old enough to be ridden, the best exercise is just to ride him regularly. This will do him good both mentally and physically, keeping him fit and also constantly handled and disciplined; he'll be a lot happier, less bored, and more apt to behave like a regular horse than a monster stallion. Work out some method of daily exercise, especially if your stallion does not have room to exercise himself.

Lead a stallion for exercise on a regular basis.

BOREDOM IS YOUR BIGGEST ENEMY

Boredom and lack of exercise, especially while confined in a stall or small paddock, can make a stallion sluggish or ready to explode — and bad tempered. Any high-strung horse who is well fed and confined needs an outlet to burn up extra energy. If that horse is an underexercised, undisciplined stallion, there will be serious trouble when you try to handle him. Keep in mind that regular exercise and handling will do more to mellow a horse than anything else — and will keep the stallion more physically fit for the breeding season. He'll also be less bored, doing less cribbing and masturbating.

Pasture Breeding

Most valuable stallions are never allowed loose with mares; breeding occurs in a closely supervised situation. In some breeds, many mares are now bred through artificial insemination (AI), often using shipped or frozen semen (see chapter 17 for more information). Thoroughbreds, however, can only be bred by natural cover; AI is not allowed.

Pasture breeding has many advantages over hand breeding (supervised breeding with both mare and stallion under the control of handlers), or AI, especially if you have a small group of mares and aren't using a stallion to breed visiting mares or to frequently collect and provide semen for shipping. Pasture breeding often results in higher conception rates than hand breeding because the stallion courts and breeds the mares at the proper time. There are no "shy" mares who get missed because of silent heats in which they ovulate but give little external sign of being in heat (see chapter 20); the stallion knows when each mare is in heat. He can smell the changes in her sweat and urine even when she gives no behavioral indications of being in heat. Pasture breeding also gives the stallion plenty of exercise, eliminating that daily chore from your management tasks.

A stallion who grows up interacting with other horses and learning about mares at an early age (and how to stay away from kicking feet) can do very well in this situation. He never rushes into anything and is wise enough to never approach a mare from the rear, taking his time and teasing her properly until she is ready to breed. It is difficult, however, for an inexperienced older stallion who has only been used for hand breeding to suddenly be put out with a group of mares. He will invariably get hurt because he has not learned the patience of courting.

Some Stallions Will Not Pasture Breed and Hand Breed

Not every stallion will pasture breed and hand breed. You are fortunate if you have an easygoing individual who lives with his mares but also lets you catch him and take him away from them to hand breed a visiting mare now and then, or to collect semen for AI breeding. Some stallions who have been used only for pasture breeding do not cooperate for hand breeding or semen collection; they are accustomed to courting and breeding a mare at their own leisure, as nature intended, and may be hesitant to breed a stranger (or a "phantom" dummy for collection) that is suddenly presented for breeding. Being handled and confined at breeding time rather than being able to do it their own way is distasteful to some stallions.

RECORDS

Complete and accurate records are an essential part of any breeding program. Keep individual records on every mare: results of teasing (how the mare reacted to the stallion), visits by the veterinarian, breeding dates, vaccinations, dewormings, special considerations in handling the mare, and so on. A good set of records is one of the best management tools you'll have when breeding horses.

The Pasture

If you pasture breed, a large, clean, grassy pasture with safe fencing and no hazards for horses to run into is best. (See chapter 1 for fencing information.) There should be no horses through the fence in adjoining pastures unless they are young ones (weanlings or yearlings) or pregnant mares. Never place another stallion or any mares who might come into heat in an adjoining pasture or paddock. Geldings are also taboo, for a stallion may hate a gelding as much as he would another stallion. If you have an adequate and well-fenced pasture with no adjacent troublemakers, and if you select a group of mares for your stallion to be with during the breeding season, pasture breeding is an ideal way to settle the mares. It saves time and labor and usually ensures getting the mares in foal.

Be careful when pasture breeding to avoid problems caused by personality clashes. You may not be able to add mares to the stallion's group without upsetting the pecking order. It's best to give him a small group at the start of the season and make no changes to disrupt the social order. Ways to determine the number of mares to give to a stallion is discussed in the box on the next page.

Age and Fertility

The stallion is not mature until age 4. Some people start using a stallion for breeding at age 2 or 3, but it's better to wait until he's 4 because overuse of a young stallion can be physically and mentally damaging; he may lose his desire for the job. Even at age 4, he should have a limited number of mares — usually less than 20.

Be aware that if a stallion is used too often, you can't count on him settling every mare; he's more fertile if used sparingly. Also, an older stallion should not be used hard. As with a young one, his number of mares should be limited.

At the start of the breeding season, the stallion shouldn't be expected to be perfectly ready for his job. It takes a little time for him to become fit and fertile; moreover, sperm production is lower in early spring than later in the season. Have his semen checked by your veterinarian before breeding season starts. If there are any problems, it's better to find out beforehand rather than after several mares have been bred fruitlessly.

Sperm Production

Many factors affect sperm production, including age of the stallion, size and condition of his testes, disease, fever, trauma to his reproductive organs, and time of year. For example, daily sperm production is directly related to testicle size. As the stallion matures and his testicles become larger, he produces more sperm. The testicles "shrink" during winter (when stallions as well as mares are not as fertile), and also may suffer degeneration with advancing age.

Sperm production and sex drive peak in May and June. If you're trying to breed mares early to bear early foals, you may find fertility problems not only in the mares — because they are still in winter anestrus — but also in the stallions. Most stallions produce only 50 to 75 percent of possible sperm early in the year compared with their output in June.

Sperm Life

Each stallion's sperm has a unique life span; the amount of time the sperm will remain alive within the mare's reproductive tract is different for each stallion. Average sperm life is at least 40 to 60 hours for most stallions, but a few have sperm that are viable up to only 24 hours. By contrast, other stallions can breed a mare as many as 7 or 8 days before her ovulation and still get her in foal. A stallion must have long-lived sperm if you plan to ship semen. Your vet can do a semen test to determine the longevity of the sperm.

Number of Mares

A mature stallion can breed many mares, the number depending on the individual, his physical fitness, semen quality, and situation. Forty mares usually is considered an adequate number for any one stallion to breed by natural service during a breeding season, but with careful hand breeding (trying to breed each mare at the proper time and not having to breed her more than once or twice at most), a stallion can successfully breed many more than that. For best conception rates, one a day is plenty. If he must service two, the breedings should be spaced out, one in the morning and another in the evening. If your stallion has good semen quality, he can service even more mares through semen collection and AI. A single ejaculate can often be divided into several batches for shipping, for example, if that particular stallion has a high number of viable, good-quality sperm.

Illness and Injury

Infertility in stallions can be caused by fever from illness or inflammation from injury. Anything that raises the temperature of the testicles for any length of time interferes with sperm production. The higher and longer the fever, the more severe the effect — but it may not be evident for several weeks

CONTRACTS

If you breed mares owned by other people, be sure to sign off on a contract specifying the responsibilities of both parties. A breeding contract protects both mare owner and stallion owner. It ensures that the mare owner knows the breed fee, terms of payment, daily charges for care and feeding while the mare is at your farm, and whether the mare will have a return privilege to be rebred if she doesn't settle or produce a live foal.

A live foal guarantee means the mare must produce a live foal before all or part of payment is due. *Return-in-season* means the breed fee is due at end of the breeding season regardless of results. Other subjects to be clarified include liability for veterinary services and any other possible points of contention between mare owner and stallion owner.

A veterinary exam for breeding soundness of the mare is beneficial to both owners because, if any problems are found, they can be cleared up before the mare is bred. The exam also ensures that only healthy mares are bred to your stallion.

Use of shipped semen will require a slightly different type of contract and there are some different expenses. It is not necessarily cheaper than sending the mare to a stallion. There are usually charges for the collection and preparation of the semen. The mare owner must also pay the costs for transporting the semen — usually by air courier, so that it can be received the same day or the day after it's collected.

because of the large amount of sperm already produced and maturing. Those sperm may still be viable, enabling the stallion to settle mares for a while until he runs out of mature sperm and has no more coming on because of the gap in production while he was sick or injured. Formation and maturation of sperm cells take 60 to 70 days; therefore, it is important to do a periodic semen check following any illness or injury to determine whether the stallion will have an infertile period and how long it will last.

Breeding Fee

A stallion's breeding fee should be reasonable, especially if he is young, unproven, or not well known. You can start a stallion's fee low and raise it later if he becomes desirable as a good sire. If you start too high and later want to lower the fee, this can be embarrassing, and challenge your reputation as a good horse breeder. If the fee is reasonable, more people will be able to afford and use your stallion. When trying to promote your stallion and make him well-known, start him at a reasonable fee, then, as his foals begin to make a good name for him, you can go up from there.

20

Breeding the Mare

WHETHER YOU WANT TO BREED A FAVORITE MARE to raise a foal who might be like her or raise horses as a hobby or a business, to be successful, you must understand equine reproduction. It is also vital that you give the broodmare proper care for conceiving, carrying, and raising a foal.

A mare should not be bred until she is mature and fully grown. Breeding when she is too young is hard on her; if she's still growing, she can't maintain herself and the foal she's carrying. Some mares mature earlier than others and can be bred at age 3 to foal at age 4, but for others, it's better to wait until age 4 for foaling at age 5. In most cases, 2-year-olds should not be bred.

If a mare is old, she may be hard to settle, but it is not impossible to get older mares in foal unless they have a fertility problem. If a mare is over 10 years old and has never had a foal, she may be difficult to settle, but even some 20-year-olds have been known to raise foals successfully when bred for the first time. Regardless of age, have every mare checked by a veterinarian first to make sure there are no problems that might interfere with her ability to conceive or carry a foal.

The Reproductive Tract

The mare's genital tract begins at the vulva, the external opening just below the rectum. Inside the lips of the vulva is the clitoris, a walnut-size protrusion that is visible when the mare urinates or "winks" the folds of the vulva when showing signs of heat. Beyond the vulva is the vagina, extending to the brim of the pelvis and leading to the cervix — the opening into the uterus. The cervix remains tightly closed except when the mare is in heat or ready to foal. When she is in heat and ready to breed, the cervix opens up to enable sperm from the stallion to enter the uterus. There are two horns to the uterus that extend to either side from the main body of the organ; at the tip of each horn, the oviducts extend to the ovaries, which are bean-shaped and roughly 1 to 2 inches (2.5–5 cm) wide by 2 to 3 inches (5–7.5 cm) long.

MARE'S REPRODUCTIVE TRACT

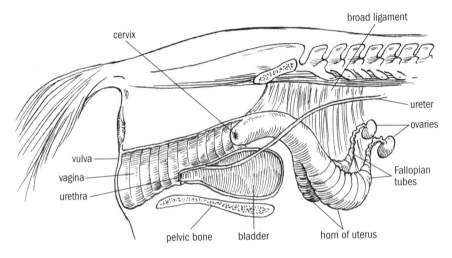

A HEALTHY, SOUND MARE

The mare should be in good physical condition before she is bred — not too thin or too fat — otherwise, she may be hard to settle. She should be on a regular deworming schedule and have up-to-date vaccinations. After all, several diseases cause abortion in mares (rhinopneumonitis, leptospirosis, equine viral arteritis, and so on), some of which can be prevented by vaccination, and you do not want her losing the fetus just because you neglected to vaccinate her.

Before a mare is bred, have her teeth checked and make sure she has no lameness or chronic problem that might cause her discomfort. Any serious leg problems could become worse with increased body weight during pregnancy. The mare should be in proper body condition as well — if she is too fat she may have unpredictable heat cycles or none at all, and if she is too thin, she may not cycle. A mare with a hormone problem also may not have normal cycles. Suspect a thyroid problem if a mare stays fat on reduced rations.

Breeding-Soundness Exam

The best mare in the world can't be a broodmare if she has a physical problem that interferes with conception or carrying a foal to term. Have your veterinarian examine the mare before she is bred. Then, if there's a problem, you'll know about it. If it's something that can be corrected, you can address it before taking the time and expense to have her bred.

Rectal Palpation

When examining a mare for breeding soundness, the veterinarian checks her uterus and ovaries by means of rectal palpation. He or she reaches gently into the rectum with a well-lubricated, long-sleeved plastic glove and can tell a great deal about the health of the uterus and ovaries from this examination. If the ovaries are hard, fibrous, or excessively large, the mare may have problems with reproduction. Sometimes, a cyst or tumor in the ovary interferes with fertility. If one ovary is greatly enlarged, the other may be small and atrophied. A single hard ovary is usually not a problem in terms of fertility if the other is normal. While palpating the ovaries, the veterinarian also checks the oviducts for adhesions that might interfere with the passage of eggs.

The Uterus

The veterinarian checks the cervix and uterus by rectal palpation — feeling them through the rectal wall — and may also use ultrasound. Enlargement of the uterus or thickening of its walls may mean the mare has had an infection that left scar tissue. A uterus that lacks proper tone, feeling flabby to the touch, is abnormal and may reduce the chances of a successful pregnancy. Any adhesions of the uterus to the surrounding structures, or hematomas on the wall of the uterus or near the oviducts, also limit the mare's breeding ability.

SAFETY TIP

Most veterinarians prefer to have the mare confined in stocks (a restraining "chute" or framework) for rectal palpation, but if this is not possible, be assured that many mares will stand quietly with no restraint. Some may need a twitch or Stableizer, or a front leg held up (see chapter 5). If a mare is quite nervous and upset, she may need a tranquilizer. A mare who is prone to kick may still do so even when tranquilized.

These blood-filled swellings may be from an injury or rupture of a blood vessel during a previous foaling. An old mare may have degenerative changes in her uterine lining, even if she has never had any foals. A thickened cervix from injury at foaling or from infection also makes a mare unsound for reproduction. An ultrasound examination along with the rectal palpation can help clarify the presence and risk of any uterine or ovarian abnormalities.

The Vagina

The veterinarian may take a closer look at the vagina and cervix through a speculum. This examination should be done while the mare is in heat. A sterile, well-lubricated speculum is inserted through the vulva, expanding its folds so the inner tissues of the vagina can be viewed with a flashlight. This enables the veterinarian to check for adhesions that might indicate previous tears and injuries. Color and health of the vaginal tissues can indicate the presence of infection. Any abnormal discharge or drainage should be noted.

Infection

To determine if there's an infection, a cervical culture (sample of fluid and tissue) should be taken during the heat period. At that time, the cervix is open and the vet has access to uterine fluids. These should be cultured for 48 hours. If bacteria grow in the culture, they should be tested with various antibiotics to determine which will most effective treat the infection. It also helps to examine the uterine cells from a swab under a microscope because the presence of white blood cells may indicate infection. Any genital tract infections should be treated as soon as they are discovered. If they are allowed to persist, they may cause permanent infertility from uterine damage. Infections are most common in older mares who have had numerous foals.

Abnormalities

Occasionally during a breeding soundness exam, it is discovered that a mare does not have a properly developed reproductive tract or has an inherited or congenital defect that will prevent reproduction. These cases are rare, but if your mare has such a problem, it's best to find out before you try to breed her. The veterinarian should also look at the mare's udder to see whether there are abnormalities or scar tissue from previous mastitis.

Imperforate Hymen

More common than a congenital defect is a maiden mare — one that hasn't been bred before — with membranes still intact at the opening of the vagina.

She may have an imperforate hymen or an obstructing membrane partition. These problems can be corrected with any sharp, sterile instrument during the breeding soundness examination. Otherwise, the membranes would be torn at breeding, which could lead to infection. It is better to correct them with sterile techniques, allowing the mare at least 2 or 3 weeks to heal before being bred so there's no soreness.

Caslick Repair

Proper conformation of the mare's vulva is important. A mare with a tipped vulva (sloped inward rather than up and down) is likely to develop vaginal infection by constant contamination from feces falling through the opening. Correcting this problem involves suturing the lips (all but the very bottom inch or two) of the vulva. The veterinarian uses a local anesthetic so the mare won't feel pain while he or she trims the edges of the vulva and sutures them so they will grow together. This simple Caslick repair can prevent the most common type of infection that causes infertility in older mares. The grown-together vulva lips must be reopened for breeding (and then restitched) and again for foaling so the vulva does not get torn. In some cases, artificial insemination can be accomplished with a high Caslick in place; this may be the preferred breeding method for such mares because there's less chance of introduced infection.

Sending a Mare to Be Bred

Some breed registries (such as Jockey Club) do not allow use of AI or other assisted reproduction techniques. All Thoroughbred mares, for example, must be bred by naturally in order to register the foal. Thus, you must send the mare to the stallion, unless she will be bred to your own stallion at home.

For best success, sending a mare to be bred should take place well ahead of when she's due to come into heat. She'll be more likely to come into heat

---

CHECK FOR A CASLICK REPAIR

Before a mare is bred, check to see whether she's been sutured and if there's enough room to permit easy access by the stallion. If there's not enough room, her vulva must be opened in time to allow the area to heal so breeding won't be painful. If she's sutured and there is room to breed her, have the veterinarian put a breeding stitch (cross stitch) at the end of the suture seam to avoid tearing during breeding.

properly and conceive if she has a chance to get used to the unfamiliar place and the stallion beforehand. If she is pregnant and you are sending her to be foaled and rebred, send her at least 2 months before she is due to foal.

Have her well groomed and clean with feet properly trimmed and no hind shoes (for the stallion's safety). Also send her with a good halter. Remember to include records of worming, vaccinations, and heat cycles.

Breeding to Your Own Stallion

Mares can be bred at pasture, hand bred (with mare and stallion both under supervision), or artificially inseminated. The method you choose will depend on the mare and your circumstances. Pasture breeding is discussed briefly in chapter 19. Most breeders prefer to hand breed because there is less risk of injury to the animals and it is possible to take precautions against infection.

Conscientious breeding practices are crucial to success. If hand breeding is overseen in a careless manner by inexperienced people, it can result in distress and injury to the horses or injury to people. It may also distort the animals' attitudes to such an extent that they don't behave normally. For example, a stallion may become impotent or aggressive, and a mare may become overly fearful and unwilling to be bred. She may also become so aggressive that she injures her foal. A mare deprived of normal social interaction before breeding can become uncontrollably violent when forced to accept the stallion's unwanted advances, even if she is biologically ready. Proper teasing and determination of heat cycles is vital for breeding success.

Understanding Heat Cycles

When breeding by hand or AI, you must determine when the mare is in heat and most likely to accept the stallion or conceive with an AI process. A mare may be in heat from 1 to 10 days, but there is a certain time in that period when she's ready to conceive. Breeding too early will not be successful. Ovulation usually occurs about 24 to 48 hours before the end of heat. It is traditional to breed on the third day of heat and every other day thereafter as long as the heat lasts, even though one good breeding is enough if it is done at the proper time. The mare's ova live only about 12 hours, but sperm can survive in her reproductive tract for 60 hours or more; therefore, breeding every other day keeps viable sperm present for fertilization of the egg.

Knowing how long a certain mare tends to be in heat helps you determine when to breed her. On farms where there are many mares to be bred and the owners don't want to overuse their stallions — or irritate a mare's tract with repeated breedings, which could lead to infection — the mares are palpated (or checked with ultrasound equipment) by a veterinarian to determine the time of ovulation so they need only be bred once. If you are using shipped or frozen semen, you must determine the proper time to inseminate the mare because you'll be doing it only once.

The Cycle

To breed a mare successfully, you must be familiar with her heat cycles. No two mares are alike in their reproductive cycles, and any given mare may show different characteristics from one year to the next. For example, she may have winter anestrus (lack of cycles) one year and cycle all through winter the next. Mares are unpredictable.

The cycle is divided into two parts, estrus (the period of heat), and diestrus (the time when she is not in heat). The average length of the combined periods for most mares is 21 to 23 days. During estrus, a few follicles (containing immature eggs) in the ovaries grow rapidly and secrete estrogen, causing the mare to come into heat and display signs of sexual responsiveness. Near the end of heat, one of the follicles ruptures and releases a mature egg into the oviduct, where it can be fertilized by sperm. Half of all mares go out of heat within 24 hours of ovulation.

Diestrus

The diestrus period begins immediately after ovulation and varies from 5 to 33 days, but for most mares, it lasts 14 to 19 days. The length of diestrus is actually more predictable than that of estrus; most mares will come back into heat 15 to 17 days after they go out. Estrus periods tend to be longer in early spring, however, and get progressively shorter throughout the summer. Most mares are fairly consistent in their cycles through the breeding season, and the pattern and timing can be determined with a good teasing program.

Silent Heat

Some mares are obvious about their heat cycles, but those who are not may have

---

Signs of Heat

- Relaxed vulva
- Spasmodic winking (opening and closing of the vulva)
- Mucous discharge
- Slightly raised tail
- Frequent urination
- Gregarious attitude (wanting to be near other horses)

to be teased regularly by a stallion to show any changes. A mare who never sees a stallion may not show obvious heats. If she has a diestrus longer than 33 days, she may be having regular heats but not showing. This is called "silent heat." Sometimes, an observant person who knows the mare well can detect the subtle changes.

A mare suspected of having silent heats should be regularly palpated by a veterinarian to determine her time of ovulation — then, she can be bred artificially if she will not accept the stallion. At times, silent heats are a sign that something else is wrong in the reproductive cycle or that the mare has a uterine infection. Silent heats may also be intermingled with normal heats.

Some mares with foals at side do not cycle at all until the foal is weaned, and others fail to show heat because they are preoccupied with their foals. Estrus in these mares can often be detected by means of a careful and diligent teasing program. Shy mares may become more obvious about their cycles if a stallion is nearby and some breeding activity is going on. Horses are very social animals, and often a little more "social stimulation" causes a mare to become more obvious about her cycles.

Erratic First Cycles

As spring days lengthen, mares who were in anestrus start cycling again. Breeding season for most mares begins soon after the vernal equinox (March 21, when there are equal hours of daylight and dark) and reaches its peak by summer solstice (June 21, the longest day of the year).

The first cycles in spring are usually erratic and the mare is not particularly fertile, because follicles may not ovulate. Heat periods during this transitional period are often long however, because the follicles persist. They may wax and

---

MARES ARE SEASONAL BREEDERS

The most important aspect of equine reproduction is the fact that mares are seasonal breeders; unlike cattle and certain other mammals, they usually cannot be bred throughout the year. Fertility varies greatly during different seasons. About 75 percent of mares stop showing signs of heat during winter, when days are short. This period of winter rest, called anestrus, occurs in most mares in the northern hemisphere from November through February, although, in some mares, it may begin as early as September.

wane but never quite reach maturity to ovulate. Mares may come into heat for a day or two, go back out for a week, then come in again for 2 weeks — or some other erratic pattern. Try not to breed mares early in their heat cycles. If a mare is going to be in heat for 20 days, it's a waste of time. Have her palpated first, then breed only if a follicle is close to ovulation. Transitional heats are usually not fertile but sometimes produce a viable follicle at the end.

Breeding in Late Spring and Summer

As heat cycles become shorter and more normal toward summer, conception rates improve. By late April, nearly 80 percent of mares are cycling, and by June, almost all mares are coming into heat regularly with properly functioning cycles. Heat periods in June are short (3 to 5 days); getting mares pregnant during May, June, and July is quite easy compared with that of early spring. Some breeders initiate teasing in early spring, or use artificial lighting to extend winter "daylight" hours and help stimulate the mares' reproduction.

Teasing Mares

If a mare is not obvious about her heats, she's more likely to show if you walk her close to the stallion every day (or bring him to her) to discover her reaction. This is called "teasing." You can put the stallion in a box stall with top door open so he can see and talk to the mare as she's led up to him, but he is protected if she suddenly decides to strike or kick. Introduce the mare to the stallion several times before she is actually ready to breed, so she'll be acquainted with him and accept his services when the time comes. Many mares do not submit to breeding unless they undergo a lengthy teasing and courtship first.

Most mares show stronger estrus the more they are teased, and some won't accept a stallion without this teasing process. The act of teasing stimulates the release of oxytocin in the mare. This hormone has many purposes, such as stimulating uterine contractions and milk let-down in the broodmare. It also initiates several processes that enhance ovulation and the transport of egg and sperm to their destinations. If you tease problem mares often, and long enough to stimulate oxytocin release, they will ultimately show signs of heat.

A good teasing stallion is one who simulates the actions of a stallion under natural conditions, being patiently aggressive. A stallion living with mares is interested but controlled; the mares are not haltered or hobbled, so he has a healthy respect for their view of the situation. He's not about to barge right up and get kicked. He has learned how to be patient, knowing that when a mare does come into heat, he will be the one to breed her.

KEEP RECORDS OF COURTSHIP

Keeping a teasing record throughout the breeding season is very valuable; it shows when the mare's cycles become normal and indicates the most favorable time to breed her. It helps you determine the number of days per heat so you can keep the number of services on each mare to a minimum, with maximum likelihood of conception. If the mare is to be bred at foal heat (her first heat after foaling), teasing should begin on the fourth day after foaling.

Teasing Routines

Large farms usually identify one stallion to do the teasing so their breeding stallions do not have to spend so much time and energy at this job. In contrast, the breeder who has only one stallion generally uses him for teasing as well as breeding. There are advantages to this, because some mares object to being bred by a stallion different from the one that did the courting.

When using your stallion for teasing, establish certain teasing routines that are distinctly separate from breeding routines. Use different headgear on the stallion for the different jobs. He should have a specific piece of tack for each job he does, so he knows what to expect — whether he is to be ridden or taken to a show, or is going to tease mares, or is going to breed. For instance, some people tease with a chain over the nose and breed with a headstall and rubber-covered bit (a bridle and bit never used for riding). Whatever you choose, use it consistently with that particular stallion. With his teasing headgear on, he knows he is only going to make "small talk," which is a large part of his springtime work. Having one site for teasing and another for breeding also helps the stallion keep his jobs straight.

Teasing Methods

Some breeders tease with one of the horses in a stall. By bringing the stallion to the mares, a manageable stallion and one handler can tease a barn full of mares in a short time, making this an efficient method for reliable mares. However, some mares feel threatened when another horse invades their space; they show aggression when approached in their stall by an inquisitive stallion at the door. This is often true of mares with foals. Be aware that a foal might be injured inadvertently by his overprotective mother charging at a stallion. In this case, it's better to take the mare to the stallion, with the foal left behind in the stall out of harm's way. Then, the mare can focus on the stallion.

Some mares need to meet the stallion on neutral turf (neither her stall nor his) and require more stimulation than just nose-to-nose at a stall door. Use trial and error to determine whether a certain mare does best being teased with her foal beside her, within sight but out of reach of the stallion, or completely out of sight. The site and method of teasing should be tailored to the individual mare and stallion, not the convenience of the handlers. What's easiest for you may be inadequate for a particular mare.

Teasing Mares at Pasture

If mares are kept outside, the stallion can be walked past their paddocks or pastures. This reveals a great deal about the status of the mares' heat cycles. A mare who is not in heat will usually go on about her business after glancing up or coming over to say hello, whereas a mare who is in heat and ready to be bred will come over and show interest. A shy mare or one who shows no obvious signs of heat may have to be brought to the stallion every day for more intensive teasing at closer quarters.

Teasing Cage

An effective way to tease mares at pasture is to build a small pen or teasing cage (like a portable stall) in the field or pasture where the mares are kept.

Always use the same headgear on the stallion when teasing a mare; this helps establish a teasing routine. Use a different piece of tack for each of the stallion's jobs.

The stallion is put into the pen for a while each day and the mares approach him at their leisure. This is very similar to pasturing a stallion with the mares — but without risk of him being hurt. Many mares are more likely to show signs of heat while in a natural and relaxed herd situation. Even the timid ones can approach the stallion if they wish, when the more dominant mares have grown tired of socializing and wandered off. An alert observer can tell a lot from the mares' actions. If there is a stallion nearby, even the shyest mare may give a clue that she is in heat while calmly grazing. The constant presence of the stallion is helpful in getting timid mares to cycle and show heat.

Teasing Chutes

Some breeders use teasing bars or a teasing chute to check a mare at close quarters. The chute has a solid partition about chest high between mare and stallion. Once the mare is in the chute, the stallion is led up to her. She can reach over the top to sniff noses with him and show her inclination or objec-

BE OBSERVANT

Experience will enhance your ability to interpret the actions of the stallion and mares. If a good teasing stallion loses interest in courting a certain mare, trust his judgment. The veterinarian will probably confirm that she isn't ready to breed. If the stallion continues to insist that she ought to find him attractive despite her vigorous protests, the veterinarian will probably confirm that her reproductive tract has all signals "go."

If the stallion doesn't give any clues, you will have to read the signals from the mare. A mare who is in heat may demonstrate obvious willingness or disinterest (the strongest sign of heat in some shy mares), or less enthusiasm about efforts to kick him. Certain mares require palpation by a veterinarian to determine the true status of their heat cycles.

When teasing or breeding, the stallion should always be under control and have excellent manners. He should be allowed to approach the mare and make her acquaintance by nibbling and making sweet talk ("chuckling"), but never allow him to bite. If the mare is in heat and ready to breed, she will usually show it and be willing to stand for him. If she isn't, she will reject him firmly.

tions, but the solid chute prevents her from kicking or striking out. Another method is to use adjacent stalls with a window open between them.

Preparing the Mare for Breeding

Tease the mare again just before breeding to make sure she is still ready to accept the stallion and to ensure her cooperation — and to encourage her to urinate and defecate. This is something she would be doing naturally if out at pasture moving around and socializing with the stallion before the act of breeding. Breeding is easier if she doesn't have a full bladder and rectum. Then wrap her tail in a clean bandage or enclose it in a clean nylon stocking or pantyhose. This keeps the tail hairs from cutting the stallion's penis and also keeps the penis from passing through the tail hair and picking up dirt.

Washing the Mare and Stallion

Wash the mare's genital area and buttocks with a mild soap and rinse them thoroughly, then wipe dry with paper towels. Chlorhexidine (Nolvasan) is a good disinfectant that cleans without irritating. The stallion's penis and sheath should be washed with clean water (no soap) once he drops it down — which he'll do if he is near enough to see the mare. Repeated washings with soap or disinfectant kill normal skin bacteria, which make more likely the chances of infection with resistant pathogens. Repeated washings also kill sperm, so don't go overboard with the washing — and use plain water. For good sanitation, use a plastic garbage bag to line a bucket for wash water (have a separate bucket for mare and stallion) to be thrown away after each use. The water should be body temperature. Both mare and stallion should be well rinsed and dried because any remaining water tends to kill sperm. The mare can be briefly teased again so that she and the stallion are both ready.

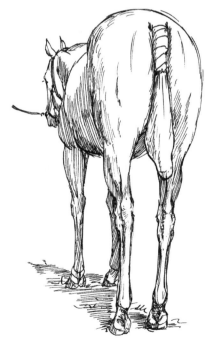

This tail is wrapped for breeding and enclosed in a nylon stocking.

Use of Breeding Hobbles

Some breeders use kicking hobbles and other restraints on the mare as a matter of course, but this is not always a good idea. Most mares will submit willingly to the stallion when they are ready to breed. A mare who kicks and fusses a lot is probably not ready. Hobbles are unnecessary on most mares and can be a hazard to both mare and stallion if he gets tangled up in them. A maiden mare (one that hasn't been bred before) may be nervous and need quite a bit of steadying to keep her from moving around too much, but hobbles and excessive restraint can do more harm than good, making the breeding a bad experience for her. Reserve the use of hobbles for the few mares who truly need them, mainly to protect the stallion.

Hobbles should be properly designed for the job, with a quick-release feature. With young mares, hobble them a few times before you expect to breed them. You may even need to twitch a mare or use a Stableizer before hobbling her.

Approaching the Mare

Lead the stallion up behind the mare, about three horse lengths behind her and to one side so she can see him. She may try to turn and face him if she is nervous; the person holding her must keep her calm and steady. If she's quite upset, lead the stallion around her at a safe distance for a few times so she can see him better and become less afraid. If she won't stand still, yet you are sure this is the right time to breed her, use a twitch or Stableizer to make her stand. (See chapter 5 for proper use of these restraints.)

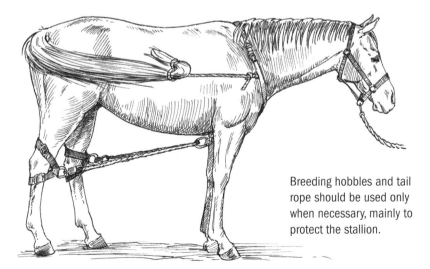

Breeding hobbles and tail rope should be used only when necessary, mainly to protect the stallion.

SAFETY TIP

Breeding should take place in a safe, open area with no hazards for the animals to run into, no dust that might introduce infection, and plenty of room for the horses and handlers to maneuver. Footing should be solid, dry ground — or a good floor in a breeding shed — with adequate traction for both mare and stallion. If the mare has a foal, he should be confined in a pen close by where the mare can see it, preferably in front of her.

Handling the Stallion

The stallion should be led up only when he is fully ready to breed her, with penis erect but not so large, or "belled," that he can't enter her. The approach should be a little to one side, at a 45 degree angle, in case the mare tries to kick. He should approach calmly, not at a run or on his hind legs. Let him mount from the mare's hip and work his way around to the proper position. Give him enough slack at the proper time so he won't be distracted by a pull on his head. Both handlers should be very alert at this time to try to prevent and be out of the way of any striking or kicking on the part of the mare, and to avoid flailing forelegs of the stallion.

If the stallion has been pasture bred, he may have developed his own pre-mating ritual; it's usually best to let him approach the mare in his own way as long as he does not rush and frighten her.

As he mounts, one person should hold the mare's tail to the side and make sure the stallion enters properly. If the stallion is an individual who likes to bite the mare while he is breeding, you should muzzle him or protect the mare's neck and withers with a pad.

Holding the Mare

The best method is to have a solid fence or panel with the mare's head over it and her chest up to the fence so she can't go forward much. The mare's handler stands on the opposite side of the fence (with the release rope for her hobbles — if hobbles are used — in his hands at all times, long enough for him to release it from his side). The mare will move forward a step or two as the stallion mounts, but the solid fence keeps her from lunging forward and gives her something to brace against so both she and the stallion can have solid footing. Moving forward a bit helps prevent injury to the mare and allows the stallion to find his proper position, but the mare should not rush forward. The mare's handler should keep her head high as the stallion comes up to her; this will inhibit kicking if she feels so inclined. If she starts to kick, the handler should pull her head around toward the stallion to get her hind feet away from him.

The Service

Copulation usually takes 30 seconds to a couple of minutes, followed by ejaculation. Most stallions show a rhythmic flagging of the tail while ejaculating. A few stallions do not ejaculate during the first mount and must do it again. After the stallion has finished his service let him dismount, or move the mare a step to the side to help him dismount. If you are collecting semen for checking, it is easy to obtain at this point because some usually runs out of the mare's vulva as the stallion dismounts. Catch it in a clean paper cup to check under a microscope.

The mare's handler should be alert as the stallion dismounts, turning her to keep her from kicking him. Lead her for a few moments to keep her from standing and straining, which would discharge much of the semen, then put her back in her stall or paddock.

Washing the Stallion After Breeding

After the stallion dismounts and is led, his penis should be rinsed. You don't need to disinfect, just rinse him off to make sure there is no dirt. It is more important to have the mare and her hindquarters clean before breeding so that no contamination is taken in with the breeding.

Check Mares Closely After Breeding

If a mare does not conceive after being bred, she will come into heat again about 15 to 17 days after going out. Because mares are often irregular, it is wise to watch every bred mare closely for several weeks after service to see whether she has settled. (Up to 10 percent of pregnant mares show one or more heat cycles after being bred and settled, and this can be deceptive.) Check her by teasing about 20 days after service and again at what would be her next heat cycle. If she shows no signs of heat for the next two possible heat periods, she is probably pregnant, but you should have her checked by a veterinarian to be sure. She can be checked by ultrasound as early as 11 to 12 days after ovulation, or by rectal examination any time after 20 days (by an experienced equine veterinarian). There is now a blood test that can accurately determine pregnancy any time after 70 days' gestation.

Artificial Insemination (AI)

Some breeders use AI to impregnate a mare who is difficult to breed or settle by natural service, or to extend the semen of a valuable or older stallion (one ejaculate can be divided and used to inseminate a number of mares). Some breeders use AI for all mares being bred to their stallions. The easiest way to collect semen is to let the stallion mount a mare in heat or a "phantom mare" (a breeding "dummy" designed for this purpose), and divert his penis into a specially designed artificial vagina to receive the ejaculate.

Cooled, Shipped Semen

In the breeds that allow use of AI, many breeders collect and ship semen so mares can be bred without having to go to the stallion. The semen must be shipped by overnight or same-day airfreight so the mare can be inseminated at the right time. This requires advance planning, letting the stallion owner know exactly when the mare will be ready to breed.

Fertility Problems in Mares

Most fertility problems can be detected ahead of breeding with a breeding-soundness exam, but sometimes mares do not settle for unknown reasons and return to heat. A fairly large number of embryos are lost in the first 45 days. The embryo dies for some reason and is reabsorbed by the uterus. Losses between 45 and 100 days have been estimated at up to 16 percent.

Infection

Uterine infection is a common cause of infertility. Antibiotics and suturing (Caslick repair) to prevent contamination can clear up some of these. Good breeding hygiene also prevents transmitting infection from one mare to another by the stallion. Almost all mares pick up infection at foaling time because the open genital tract inevitably comes in contact with some contamination, but most mares throw off this infection within 1 or 2 weeks without treatment. If the mare is in poor condition or for some other reason her reproductive tract tissues have lowered resistance at that time, she may not be able to overcome the infection.

Be aware that the genital tracts of some mares are more resistant to infection than others. The older the mare, the more susceptible she is to infection; the healthy young uterus is more able to ward off infection because it is not scarred and damaged.

It is unwise to breed a mare who has any infection; if she does settle, chances are she will abort later. It is better to spend a few weeks treating the infection and clear it up than to breed the mare and have her abort at 3 to 6 months' gestation. If a mare has an infection or any kind of fertility problem, work with your veterinarian to try to correct it before attempting to breed her.

Breeding at Foal Heat

Because most mares have minor infections for a short time after foaling, it often is unwise to rebreed a mare at foal heat, which occurs 3 to 22 days after foaling. Some mares (especially those who return to heat soon after foaling) are not fully recovered at this time; breeding may not be successful or may lead to further problems if the mare does conceive. Breeding before the infection has cleared up seals it in if the mare does become pregnant. During pregnancy, her cervix is tightly closed and there is no flushing activity to clear out the uterus. An infection thus sealed in may cause abortion. It's better to give her more time for drainage and fighting off the infection. Breeding at foal heat is not recommended unless the mare had a normal delivery with no complications, has an exceptionally healthy reproductive tract, and has her first heat late enough after foaling (2 weeks or more) to be fully recovered from the temporary infection.

Embryo Transfer (ET)

In recent years, a growing number of breeders have utilized embryo transfer (ET) for breeding. This entails taking a fertilized egg from a mare and inserting it into a recipient mare to carry the pregnancy, and give birth to and raise the foal. ET was originally used as a reproductive tool for older mares with fertility problems that hindered them from carrying a foal to term. An embryo could be collected from the older mare and put into a young, healthy surrogate mother. This is still done today, but most breeders now use ET to multiply the number of foals that can be produced by a good mare, collecting two or three embryos from her each year. Many people also use ET to raise a foal or two from a mare while she is still active in an athletic career.

21

Care of the Broodmare

THE ARRIVAL OF A NEW FOAL is a long-awaited event that requires careful planning and the best possible care of the mare. You wait approximately 11 months for the new arrival after the mare is bred, so it is important to make sure that all goes well; a lot of time and effort are invested in the coming foal. Proper care of the broodmare begins even before she is bred.

Care of the Mare Before Breeding

If a mare is to conceive, carry, and produce a healthy foal, she herself must be in excellent health — free of disease, parasites, and physical impairments that would jeopardize the pregnancy. She should be well fed, but not to the point of being fat. Mares who are obese and don't get enough exercise are often hard to settle.

Is She Pregnant?

After a mare is bred, you hope that she has settled. If she doesn't show signs of coming into heat again at the expected time, she may be with foal. Because many mares take a long time to show definite signs of pregnancy, it may be a good idea to have your veterinarian check her, using ultrasound, rectal palpation, or a blood test. In some mares, the pregnancy is fairly obvious by 5 or 6 months; the belly begins to drop and widen toward the lower part of the flanks. This drooping belly looks different from a round belly that is merely fat. When you look at the mare from the rear as she is standing still or traveling, if she is pregnant, one side of her belly toward the flank may appear a little larger than the other side, and as she moves, there may be a slight swinging motion.

Other mares, especially with first foals, do not look pregnant even right up to foaling time. In contrast, certain fat mares look deceptively pregnant even though they aren't. After about the eighth month, there may be signs of

NUTRITIONAL NEEDS VARY

Broodmares are categorized as "maiden mares" (those who have never been bred), "barren mares" (those that have had a foal but are not in foal presently), "mares in foal," and "lactating mares." Because these four groups have vastly different nutritional needs, they should not be fed as a single group. Maiden mares who are still young and growing need extra protein and calories. Mares in foal need extra protein and nutrients, especially during the last trimester of gestation. Lactating mares need the most nutrition because of the heavy demands of producing milk.

movement toward the mare's right flank as the fetus kicks and squirms, but you may not necessarily observe this. If you see movement, you know the mare is carrying a foal; however, not being able to see it does not mean she isn't.

Feeding the Pregnant Mare

While she is carrying a foal, she needs the best care possible. She must be kept in good body condition but not overfed. Nutritional requirements for protein and total calories increase significantly only during the final third of pregnancy. If you increase the mare's feed as soon as she becomes pregnant, she may be too fat when she foals. Proper amount and quality of feed are

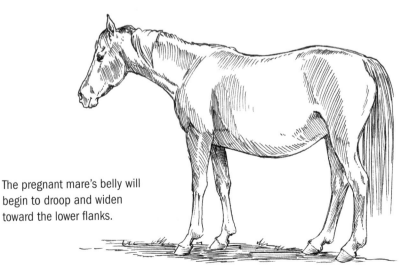

The pregnant mare's belly will begin to droop and widen toward the lower flanks.

important. The broodmare needs a well-balanced diet containing necessary nutrients for the growing fetus. Adequate trace minerals, for example, are essential to the foal's developing immune system. Also, the feed must be clean and free of mold and spoilage to eliminate the risk of abortion.

Exercise

Exercise is important for the optimum health of the mare. She should not be soft and flabby from inactivity. Don't limit her exercise for fear of hurting her or the unborn foal. She is not fragile.

If you were riding her before she was bred, continue riding her afterward. Just use good judgment in what you ask of her. A mare who participates in stressful athletic events before pregnancy can probably continue the same level of exercise during the first 6 months of pregnancy while the fetus is still small. Many pregnant mares have successfully completed endurance rides, for example, during the first half of pregnancy.

During the last month or two, many mares become quite heavy with foal, which cuts down on their agility and stamina. Usually after 300 days of gestation, it's best to slow down. The mare still needs regular exercise, so keep riding her if you wish, but make the rides shorter and slower. Stick to walking and jogging; avoid fast or strenuous workouts. Older mares should not be exercised as extensively as younger ones. Be aware that fatigue, strain, or overexertion are stresses that could lead to abortion.

Pasture Is Ideal

A good pasture with room to move around can provide needed exercise as well as proper nutrition. Pasture is an ideal environment for pregnant mares and those with foals. If it contains good types of grasses and fertile soil, it may provide all the nutrition a pregnant mare needs. During winter, however, and whenever pastures are dry, the mare will need supplemental feed.

When a mare is pregnant, keep her with a few other horses with whom she gets along — preferably other mares. Several pregnant individuals at pasture by themselves get along the best. Don't keep a pregnant mare in the same pasture with a group of young horses or geldings. You want to reduce the risk of injury to the mare from being kicked or slipping and falling while running. Geldings may tease and fuss with a mare and have been known to make them abort. As she gets ready to foal, put the mare in a separate area, by herself. If a mare foals at pasture while she is with other horses, a more dominant mare may steal the foal. Some geldings will even try to kill a newborn foal.

Preventing Edema

In late pregnancy, some mares develop edema, a swelling around the udder that extends forward along the flanks and is caused by fluid in the tissues. It results from disruption of the normal interchange of fluid among blood vessels, tissue spaces, and lymph vessels. Occasionally, the area along the mammary vein in the lower abdomen is swollen too, almost as far forward as the girth. A large fetus and the weight of its fluids

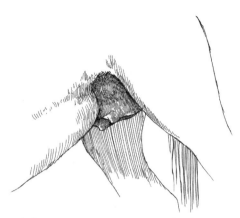

In late pregnancy, some mares develop swelling around the udder that extends forward along the flanks and belly.

distend the uterus, compressing the mammary veins and causing poor blood return. Inactivity leads to poor circulation and makes the problem worse.

As the mare gets closer to term and heavier with foal, she becomes more clumsy and sluggish and moves around less. It's a vicious cycle; the heavier she gets, the less she exercises, the more swelling she develops, and the less she feels like moving. She may have trouble getting up and down. Some mares get so clumsy, they rarely lie down during the last weeks or days of pregnancy because the effort of getting up is so great. The inactive mare's legs may also fill with fluid because of her sluggish circulation and capillary leakage.

A mare at pasture moves around during the normal activities of grazing and going for drinking water, so her edema may not become extreme. But a

DAILY EXERCISE

To prevent or alleviate edema in the heavily pregnant mare, make sure she gets mild exercise. This will help her circulation, clear away fluid from the tissues, and enable her to feel better (perhaps prompting her to exercise more on her own). She doesn't need vigorous exercise because she's clumsy and might stumble and fall, but brisk walking can be very helpful. If she is confined, give her a good walk at least twice a day at morning and evening (more often, if you can). Leading her at a brisk walk for at least 20 to 30 minutes at a time will help relieve the swelling and prevent more edema.

mare confined in a stall or paddock may not move around at all, especially as she gets uncomfortably heavy during late gestation.

Keep Her in a Hazard-Free Place

If the broodmare is on good pasture, she'll need very little extra care as long as she has good feed and water as well as shelter from flies and hot sun. Be sure to remove any obstacles or hazards from her pasture or paddock. Do not allow her in a pasture with ditches or gullies where she may get into trouble if she lies next to the ditch and rolls into it. Avoid narrow gateways and doorways when moving her out of the pasture or into a barn or stall. In the stall, remove projecting edges. During winter, make sure she has access to shelter if she is at pasture. If kept in a box stall, she should be turned out or exercised every day.

Abortion

Abortion, or failure to carry a foal to term, happens all too often with mares. It has been estimated that 20 to 30 percent of all equine conceptions end in abortion. It's a larger problem than many people realize because few aborted fetuses are ever found; the most common abortions occur while the fetus is small, and many times, it is simply reabsorbed rather than expelled. It is interesting to note that, before the use of ultrasound, it was difficult to accurately diagnose pregnancy until 30 days or more after conception, so pregnancies lost before 30 days were usually considered a failure of the mare to settle.

There are many causes of abortion. Among these are bacterial or viral infections such as equine viral arteritis, rhinopneumonitis, leptospirosis, bacterial contamination of the uterus, and — occasionally — mold or fungus. Noninfectious causes are not always as easy to detect and diagnose, but they are responsible for a large number of abortions.

Noninfectious Causes of Abortion

These include genetic defects of the embryo or fetus, lethal genes that cause a defective embryo to die early, twins, hormone problems (in particular, progesterone deficiency) that interfere with the continuation of pregnancy, and nutritional factors. In addition, some types of grasses and feeds, such as plants containing high levels of estrogen, can make a mare more susceptible to losing the pregnancy. Even a defect in the umbilical cord or placenta can interfere with the normal exchange of blood between mare and fetus; twisting of the cord is sometimes severe enough to cut off circulation to the fetus.

Placental Insufficiency

A common cause of abortion is placental insufficiency caused by degeneration of the uterine lining, especially in older mares. The uterine tissue becomes less efficient with age. Scarring from earlier pregnancies may not leave enough normal lining for adequate attachment of the placenta; therefore, blood supply to the lining is diminished. The developing embryo cannot make a good attachment and dies early, or the fetus dies later on because of inadequate nutrition from the poorly attached placenta. Less severe cases of uterine insufficiency may lead to the birth of small, weak foals, "windswept foals" with curved leg bones, and those with other abnormalities.

Twins

Twinning is another leading cause of lost pregnancy. Twins are often aborted because there is insufficient nutrition for both because of the limited surface area of the placental attachments. On the mare's placenta, there are no specialized areas that attach to the lining of the uterus (unlike the placenta of a cow or dog). It must have contact with the entire uterine wall. When twins share this space, neither one has quite enough contact. Occasionally, one dies and the other takes over its space and continues to develop. The surviving twin may still be small at birth as a result of nutrient deprivation during early gestation.

Loss of a twin is fairly common during the first 30 days of pregnancy, occurring in 3 to 5 percent of all pregnancies. Complications arise only when the loss

---

TWIN PREGNANCIES ARE FAIRLY COMMON

Mares often produce multiple eggs. When both fetuses die early and are reabsorbed or aborted, the owner usually doesn't know why the pregnancy ended unless the mare is checked by palpation or ultrasound and twin embryos are detected. If a veterinarian detects twins early (and one fetus doesn't reabsorb on its own), many breeders have the veterinarian terminate one to enhance the survival of the other. If the mare is less than 35 days pregnant, some breeders induce her to abort with prostaglandin and bring her back into heat again to be rebred.

occurs later; then the death of one twin causes the mare to abort both. Abortion of twins can take place at any stage of gestation but is most common from about the time the fetuses are 10 to 12 pounds (4.5–5.5 kg) up to full term.

Lack of Progesterone

The hormone progesterone stops the mare's heat cycles and keeps her pregnant. It also changes the uterine environment and enables it to support the fetus. Abortion partway through pregnancy sometimes occurs because a mare fails to make enough progesterone. Some mares habitually abort between 30 and 140 days of pregnancy because the level of progesterone falls too low. If you have a mare with this problem, your veterinarian can advise you on treatment with additional progesterone.

Stress

Actual physical trauma, such as being kicked, falling down, or hitting a doorway, rarely causes abortion. The uterus, placenta, and fluids in which the fetus floats usually provide adequate insulation against bumps from the outside world. A mare hardly ever aborts from the trauma of an injury itself, but the stress may be another matter.

Stress can trigger labor contractions and cause abortion, with the uninjured fetus dying because it is born before it can survive outside the uterus. The stress of any major problem — illness, severe colic, high fever — may lead to abortion. In fact, any drastic change in the mare's environment or handling can upset her, and in some cases, it causes the release of too much histamine and prostaglandins, which can lead to premature birth. If the fetus is almost at full term when this happens, it may survive the premature birth; otherwise, the fetus is lost.

Drugs

Certain drugs may cause abortion. Always read labels or consult a veterinarian before giving medications to pregnant mares. Most medications and drugs are safe to give during pregnancy, but it is always wise to err on the side of caution.

Feed Problems

Some feeds can cause a mare to abort, including fescue grass. The grass itself is blameless; the villain is a toxin-producing fungus in the bud of the plant. Not all fescue is infected with this fungus, but many fescue pastures are. The toxin can cause diminished blood supply to the placenta. It thickens, preventing nutrients from reaching the fetus. The affected mare either aborts or carries the foal to term, only to have problems during foaling. The placenta may become so thickened that the foal cannot break through the membranes at birth and suffocates.

Overfeeding of high-quality feeds on rare occasion leads to abortion. Rich alfalfa hay, high-protein supplements, or lush legume pasture sometimes cause a "flushing" effect that triggers estrogen production and premature opening of the cervix.

False Pregnancy

Sometimes a mare loses the embryo or fetus but the ovary or the uterus continues to secrete progesterone, the hormone that maintains pregnancy and keeps the mare from coming back into heat. Because she may not return to heat for several months, you might think she is still pregnant. Up to 15 percent of bred mares who have been checked and determined to be pregnant turn up empty before the hundredth day of gestation; some of these fail to return to heat for a while, giving the owner false hopes of a foal.

Infectious Causes of Abortion

Infections in the reproductive tract are a serious threat to pregnancy and should be cleared up before breeding. Certain infections are introduced at the time of breeding if good hygiene is lacking or if the stallion passes an infection to the mare. All mares bred by natural service are infected to some extent at breeding, but a healthy mare is able to overcome this infection. A natural flushing action of the uterus after breeding usually gets rid of any contamination before the tiny embryo comes down out of the fallopian tube and into the uterus. Any infection is quickly resolved, unless it is a particularly bad pathogen or the mare's uterus does not have normal immune defenses. Infection can also be introduced by unsterilized equipment during a breeding examination or AI.

If the mare becomes ill with equine viral arteritis (EVA), rhinopneumonitis, leptospirosis, or other diseases that affect the fetus, infection may invade the uterus through the bloodstream. Ask your veterinarian about vaccinating the mare for diseases that might be a problem in your locale and situation. She may need several vaccinations during her pregnancy to prevent certain diseases, such as rhinopneumonitis.

If a mare in late gestation starts bagging up too soon or acting as though she's in early labor, watch her closely and have your veterinarian examine her. Udder filling is the main sign of an impending abortion but, if caught in time, some of these cases can be saved with proper treatment. If ultrasound shows that the placenta is still intact, the mare can be treated and the abortion prevented. Mares who have prefoaling colic act as though they are in labor, so be sure to investigate any signs of abdominal pain. If a mare in late pregnancy is aborting, she may have a difficult labor. In any of these situations, you should not hesitate to call your veterinarian.

Udder Problems

Check the pregnant mare's udder periodically. This not only gives you a clue as to when her body is preparing for foaling (the udder will start to enlarge) but also alerts you to problems or abnormalities. Problems that occasionally occur in broodmares include inverted teats, small nipples, or even extra teats. Whereas extra teats are generally not a problem, inverted or small teats make it difficult or impossible for the newborn foal to nurse, especially if the udder is large and swollen. If a mare has this problem, watch closely when the foal is born to make sure it is able to nurse. If it can't, be prepared to bottle feed by milking out the mare by hand or using a milk pump such as the Udderly EZ mare milker until the swelling goes down and the foal can nurse more easily.

Dirty Udder

Sometimes dirt, oil, and sweat build up between the teats and cause discomfort and itching. The affected mare rubs her tail a lot (you might think she has pinworms) just because she's trying to get at the itch. If you look at her udder, you will see the flaky debris and accumulated material between the teats. Cleaning her udder periodically will resolve the problem.

Mastitis

Mastitis is infection and inflammation of the udder. Mares do not get mastitis as often as cows do, but the condition can be very serious nonetheless.

GET HER USED TO HAVING HER UDDER TOUCHED

The pregnant mare should have her udder touched regularly — especially if she's never had a foal before — so she'll be used to the sensation. Then, she won't be so upset when the foal tries to nurse. Also, if you have to milk her out to obtain colostrum for the foal, or because of a problem like blood type incompatibility (when the foal mustn't be allowed to drink her colostrum), or mastitis, or because you must try to help the foal nurse, she'll be cooperative and not resent your efforts. When handling the udder, be sure your hands are clean. Don't touch the ends of the teats. You can gently touch and massage the rest of the udder with your hands; or use a clean cloth, or a cloth and warm water, to wipe her udder and remove debris from between the teats.

A mare's udder is divided in halves with each half containing two quarters, like the udder of a cow. The two quarters on each side feed into a single teat with two openings. If you ever milk a mare, you'll notice two streams of milk from the nipple when you squeeze it. Mastitis usually affects just one or two quarters on the same side. Only rarely are quarters on opposite sides affected. Infection from bacterial invasion causes the affected quarter to become swollen, hard, and warm. The milk — if the mare is lactating — may become lumpy, thick, or watery.

Mastitis usually occurs in lactating mares, especially at weaning when the full udder is painful and slow to dry up. Infection may develop at any time up to 8 weeks after the foal is weaned. A mare doesn't have to be milking to develop mastitis. Even pregnant, barren, and maiden mares occasionally get udder infections from trauma, bruising, or insect bites. Mastitis can also occur in young fillies.

Many cases of mastitis occur in summer from insect bites. The inflammation is triggered by fly bites and other external irritations of the teat opening that introduce infection. Mastitis in a nonlactating mare may be caused by abnormally high estrogen levels from eating lush spring pasture or legume hay — causing mammary development and secretion of milk, even in mares who aren't pregnant and haven't foaled. Hypothyroidism can also cause false lactation and make a mare prone to mastitis. Moreover, udder injuries such as a kick, or contamination from bacteria getting into the teat openings from dirty bedding, can lead to mastitis.

Symptoms

Mastitis may be quite obvious or hard to detect, depending on the severity of infection and amount of pain involved. A mare with mastitis generally shows discomfort; she may stand off balance and rest one hind leg continually while trying to ease the swollen side of her udder. Or she may hold the stifle out from her body so that leg does not touch the swollen udder. She may show slight lameness in the hind leg next to the affected quarter, moving stiffly in an effort not to bump the udder. Because of her discomfort, a nursing mare may kick at her foal when he tries to nurse.

Signs of mastitis include swelling, heat, pain, and edema in surrounding areas. Usually, only one teat is affected. Some mares become depressed and feverish or go off feed. If she has a foal, he may avoid nursing the infected teat because the mare kicks or the milk tastes different; as a result, that side becomes even larger, adding to the mare's discomfort. In some cases of acute mastitis, abscesses form in the udder and nearby lymph glands, and these require surgical drainage. Occasionally, mastitis completely destroys the udder tissue and results in death of the mare. Highly pathogenic bacteria can lead to endotoxic shock and cause the death of both the mare and the nursing foal.

If you ever suspect mastitis, have your veterinarian examine and treat the mare. The veterinarian may take tissue smears or samples of milk from the affected quarter to identify the pathogen and determine which antibiotics will be most effective.

Treatment

A mare with mastitis needs antibiotics, some of which should be inserted directly into the affected quarter through the teat canal. Treatment includes frequent milking of that side of the udder. The mare may be very difficult to milk, however, because of the pain. She will resent any handling of the udder. Use of the Udderly EZ milk pump can make this job easier, quicker, and safer because it is not as painful to the mare as milking by hand. Hot packs and nonsteroidal anti-inflammatory medications (such as phenylbutazone or flunixin meglumine [Banamine]) help reduce the pain, swelling, and fever. Your veterinarian can instruct you on what to do. Some cases of mastitis occur simultaneously with a uterine infection; you may need to treat both to resolve the problem.

MASTITIS CAN CAUSE PERMANENT DAMAGE

If a mare has a serious case of mastitis, you won't know until her next foaling whether the affected quarter will produce milk again, even if the infection was treated immediately and cleared up. It depends on whether the mammary tissue was damaged. As she nears her next foaling date and the udder fills, you'll be able to tell. A damaged quarter that's not producing milk won't fill, and the udder will look a bit lopsided. If more than one quarter is damaged, you may have to find another source of milk for the foal.

Countdown to Foaling

When the mare is 6 to 7 months along in pregnancy, you may want to give her booster shots for flu and tetanus if she did not have them earlier. She will need rhinopneumonitis shots at 5, 7, and 9 months of gestation to make sure she does not become susceptible to this viral disease that could cause her to abort. Repeat the flu and tetanus shots at 2 to 4 weeks before foaling so she will have antibodies in her colostrum to give protection to the newborn foal. In some regions, it is a good idea to vaccinate against botulism as well. Discuss this with your veterinarian. As the mare gets close to foaling, also reduce or eliminate her grain ration if you've been feeding grain. Otherwise, she may have too much milk at first, which would give the foal diarrhea.

22

Foaling

THE BIRTH OF A FOAL IS ALWAYS EXCITING. If the mare is healthy and gives birth in a clean place, most of the time she encounters no problems. If something goes wrong, however, it's best if you're there to give assistance or to summon the veterinarian for help.

The Foaling Place

Well before the mare's foaling time approaches, decide where she will foal. If she must give birth in a small enclosure such as a shed, corral, or box stall, it must be cleaned thoroughly first. Take out old bedding and scrub the entire stall with disinfectant — floor, walls, and all. Sprinkle the clean floor with lime before putting in new bedding.

A foaling stall should be at least 14 × 16 feet (4.3 × 4.9 m), preferably larger, to minimize the risk of the mare lying too close to the wall during her labor. Identify a safe place in the stall where you could set up a heat lamp if the foal is born during cold weather.

Clean straw makes the best bedding for foaling.

SUPPLIES TO HAVE ON HAND AT FOALING

At least a month before the mare is due to foal, you should gather supplies and keep them in a clean, convenient place.

- A new, clean, 3- to 5-gallon (11.5–19 L) bucket
- A clean container to milk into if needed (an open pan works well because of the divergent flow from the teats — they don't "shoot straight"), or a trigger-operated, Udderly EZ milker with bottle attached (see Resources for info)
- Tail wraps or nylon stocking to bandage the mare's tail
- Pint or quart (0.5 or 1 L) of disinfectant such as chlorhexidine (Nolvasan) (for adding to wash water to clean the mare's hindquarters)
- Small bottle of tincture of iodine, betadine, or chlorhexadine (for disinfecting the foal's navel stump)
- Plastic shoulder-length obstetric gloves
- Clean bath towels, three or four
- Paper towels
- Roll of cotton
- Plastic garbage bag to put the placenta in
- Prepackaged enemas (human adult size in plastic squeeze bottles, fitted with a prelubricated rectal tube) or foal enema kit, or mineral oil, or mild dishwashing soap that can be mixed with warm water and administered from a large syringe
- Flashlight and new batteries
- Heat lamp
- Obstetric chains and handles, or nylon pull straps and handle
- Sterile syringes and needles
- Suction bulb, for clearing mucus and fluid out of the foal's nostrils
- Injectable antibiotic (refrigerated, if necessary)
- Obstetric lubricant
- 10 mL of oxytocin (refrigerated) to be given if advised by your veterinarian if the mare does not clean (shed her placenta)

Use Straw for Bedding

Shavings or sawdust make satisfactory bedding while the mare is pregnant, but when she foals, you need good clean straw instead. It won't stick to the foal or get into his nostrils as much as would other bedding; sawdust or shavings are sometimes sucked into a foal's air passages or swallowed by the mare when she tries to lick him. Also, wood products may carry *Klebsiella* bacteria, which cause uterine infection. The straw should be comfortably deep for the mare but not so deep that the foal will have a hard time getting up and moving around. Remove all obstacles from the stall — feed tubs, buckets, or anything else that might get in the mare's way during labor.

The Stall and Mare Must Be Clean

For foaling in a stall (especially if other mares have foaled there), wash and dry the mare's udder, belly, buttocks, and upper legs (any place the foal will nuzzle while trying to find the udder) with warm water and chlorhexidine. This will greatly reduce contamination to the foal when mouthing around trying to nurse. It is especially important if there have been any cases of *Escherichia coli* and other foalhood diseases on the farm. Also, pick up all manure in the foaling stall several times a day so the mare has no chance to get dirty. Washing her just before foaling reduces the risks of diarrhea and septicemia, which are the leading causes of death in newborns. Also, wash her udder and hind legs again as soon as she gets up after foaling to lessen disease risk.

If you think the foal may be at risk because other foals on the place have been sick, milk the mare and get a good amount of colostrum into the foal with a nursing bottle and nipple before he gets up and starts nosing around. It's a race between the bacteria and the colostral antibodies to be absorbed into the bloodstream before the gut lining closes. See chapter 23 for more information regarding the importance of having the foal nurse in a timely manner.

Consider Foaling at Pasture

The best place for a healthy mare to foal, especially if she has a history of easy deliveries, is a level grassy pasture where she can be by herself. It should be a clean, dry place with safe, smooth fencing and no brush, rocky ground, gullies, or other hazards — loose nails sticking out of fences, machinery or junk that might injure her or the foal, ditches, or puddles. Foals have been known to drown in tiny pools of water before having a chance to get to their feet. Even a dry ditch is dangerous if a foal rolls into it in his efforts to get up because he cannot stay on his back very long without suffocating. Foals have died because they were stuck on their backs under a fence. If a mare lies next to the fence when she foals, and the foal slides under the fence as he is born, you may have a problem.

Pasture Advantages

If the pasture is safe and weather is good, there are many advantages to foaling at pasture rather than in a confined box stall. A good, clean pasture offers less risk of infection than a stall or corral. Also, it's a more natural environment where the mare can seek her own privacy and be less upset and nervous. Other positive benefits are the natural feed source and the mare's ability to exercise at will.

If a mare is living at pasture, on green grass, she has self-initiated exercise and the resultant muscle tone, and looser, more natural bowel movements than a confined mare who is fed grass hay and grain. Green grass or alfalfa hay are more laxative than grass hay; the mare on pasture will have an easier time foaling than a hay-fed mare who is constipated.

Pasture Disadvantages

It's hard to keep constant watch on the mare if she's in a large pasture and not visible from your house. It helps to have a spotlight of some kind for checking on her at night. You may want the mare closer at hand so you can check on her often and be there in case she needs help, especially if it's her first foal. Foaling at pasture is not advisable in bad weather. Consider letting her out on nice days and keeping her in a stall at night, where you can more readily observe her.

When Will She Foal?

Gestation for mares can vary from 305 to 395 days. Mares are notoriously unpredictable. Veterinarians don't even agree as to what the "average" should be. Many factors influence length of gestation; these include the age of mare, time of year, genetics of the foal, stress, infection, and nutrition.

Normal gestation can vary between 320 and 360 days, with an average of 340 days. That leaves quite a period for watching and waiting up nights checking on a mare, so horse people try to predict the actual event more closely.

GENETICS OF THE FOAL AFFECTS GESTATION LENGTH

The individual foal's gestation length may or may not correspond with the due date. His genetic makeup, nutritional status, and other factors involving the mare influence his timetable for maturation. Some stallions sire foals who are carried longer or shorter than average. Also, light breeds tend to have longer gestation than draft breeds.

First Foals Sometimes Come Early

Certain factors affect the likely time of foaling. Keep in mind that these are generalities — averages or trends — and that mares don't always follow the rules.

Mares with first foals often have shorter gestations than individuals who have delivered before. Midlife mares generally carry foals longer than younger mares, but the first foal tends to have a shorter gestation regardless of the age of the mare. In other words, a 10- or 12-year-old mare bearing her first foal may have shorter gestation than a 7-year-old delivering her third. Because first-time mothers are likely to foal ahead of schedule, they need extra watching. They also may not show as much change in the udder or other areas as the veteran broodmares who often display signs of foaling several weeks ahead of the event. Older mares (age 15 and up) may have short gestations, especially if they have uterine insufficiencies.

Time of Year Makes a Difference

Mares foaling in January, February, or March usually have longer gestation length (10 days longer, on average) than those foaling during April, May, and June. Perhaps the lengthening days, warmer weather, green grass, and other beneficial changes in the environment help the mare bring a late spring or summer foal to term early. Indeed, some veterinarians believe the biggest factor affecting gestation length is the time of year the mares are due to foal.

Individual Patterns

Some mares consistently foal early, late, or on time. Often, a mare's own pattern is more reliable than her age, number of pregnancies, or foaling time of year. Always keep in mind that mares foal when they are ready (or more accurately, when the fetus is ready) rather than according to due date.

THE MARE CAN DELAY FOALING IF SHE'S UPSET

The foal triggers the onset of birth, but the mare determines when it actually happens. The final stages of labor and delivery depend on voluntary contractions of the mare's abdominal muscles as well as involuntary uterine contractions. If she is nervous or upset, she can delay foaling for a day or more. Many mares put off foaling when they are constantly watched. They are ready but prefer to do it in privacy. (Mares are notorious for putting off foaling, doing it suddenly when bleary-eyed owners head to the house for a cup of coffee!)

It is a mare's instinct to leave the herd and go off by herself to foal. This is nature's way of protecting the newborn from other horses who might try to steal him, injure him, or interfere with the bonding of foal and dam. Some motherly mares or dominant individuals are notorious for trying to steal another mare's baby.

Sometimes, a low-grade uterine infection causes the mare to have a longer or shorter gestation than normal if the fetus has been nutritionally deprived because of poor placental attachment. If a mare goes more than 3 or 4 weeks past her anticipated date, have your veterinarian check her to make sure nothing is wrong.

The Foal Triggers the Birth Process

When the unborn foal is full-term and ready, his hormones stimulate a change in the mare's hormones. Throughout the pregnancy, the mare's hormones have been inhibiting uterine contractions, but toward the end of gestation, this delicate balance shifts. The foal's hormones start triggering changes that cause her cervix to dilate, uterus to contract, and milk to let down.

Pinpointing Foaling

You can purchase foaling-prediction kits with test strips for checking the mare's milk before foaling (the strip has four zones that change color as the concentration of calcium increases in the milk being tested). Only a small amount of milk (about 0.03 fluid ounces [1 mL]) is needed for the test. It is recommended that milk be tested daily once the mare starts to develop an udder and twice daily after a color change occurs in two or more zones of the strip, with one of those tests taken late at night. The test is helpful when interpreted in conjunction with a mare's expected foaling date and signs of imminent foaling, but it is not 100 percent reliable.

Other methods involve installing television (TV) cameras in barn stalls (viewed through a TV screen or special monitor in the house) and acquiring alarm systems that signal the start of foaling. For example, the Birth Alarm fits around the broodmare's girth like a surcingle, sounding off in the house or sending a signal to your telephone or cell phone if she lies down prone. The Foal Alert is a transmitter that is sutured across the vaginal lips when the mare's Caslick repair is opened before foaling. It comes apart and gives a signal when pressure from the amniotic sac causes the vulva to open. The Breeder Alert attaches to the mare's halter and sends a signal if the mare lies down flat. The EquiFone is a similar device, placed on the mare's halter.

Signs of Approaching Labor

About a month before the mare is to foal, her udder begins to look fuller and larger, especially at night when she is at rest. Initially, this enlargement may recede during daytime. About 2 weeks before foaling, the udder will remain larger, fill with milk, and look shiny. A few days before foaling, the muscles on each side of the tail around the pelvic bones become relaxed and droop away from the root of the tail. The mare's vulva may appear relaxed and swollen. There may also be a secretion from the teats.

Waxing (and Foaling) Is Unpredictable

Secretion from the teats is one of the most common signals that foaling is imminent. The "wax" is formed by the congealing of secretions forced out of

---

COLLECT LEAKING MILK

If milk streams from the udder for several hours and the broodmare has not yet foaled, collect as much of it as you can and freeze it. This is the colostrum that is crucial to the foal, giving him important antibodies for temporary protection from disease. Don't milk her, however. That would stimulate her to let down her milk even more and lose more precious colostrum. If a mare leaks milk for more than 24 hours, have your veterinarian check her. Also check the foal after his first few hours of life to make sure he has sufficient antibodies from the milk.

HEALTH TIP

Keep the mare clean and wel groomed. The cleaner she is, the less chance there is of infection when the genital tract is exposed during foaling, and the fewer bacteria to be picked up by the foal as he noses around to begin nursing. Wrap the mare's tail in a clean bandage when you wash her for foaling to prevent dirt from getting into the open tract as she foals and to keep her tail free of blood and mucus.

---

the end of the teat. Most mares wax within 24 to 36 hours of labor, but there are always exceptions. Some don't wax at all. Others wax for as many as 10 days before foaling. Still others may make up a large udder and leak streams of milk before they foal, with the milk dripping down their hind legs. The mare may leak milk for just a few hours before foaling or for several days. A few mares foal without waxing or making much udder at all. Some have no milk when they foal; they come to their milk within a few hours after delivery. A mare may bag up, wax, then stop waxing. The most predictable thing about mares is that they are unpredictable. If you are worried about a mare who is not "going by the book," have your veterinarian check her.

Caslick Repair Check

If the mare has been sutured, her vaginal opening (vulva) will be very small; she'll have to be opened before foaling to prevent tearing and damage to these tissues. If you're not sure whether she's had a Caslick repair (if she's a mare you bought, for example), you can check by gently spreading apart the lips of the vulva, beginning at the bottom. Most mares do not object to gentle lifting of the tail to check. If you're in doubt as to her reaction, have someone hold her for you next to a gate or stable door that you can stand behind while you check.

In a mare who hasn't had a repair, the opening of the vulva will extend up to within about 2 inches (5 cm) of the anus.

In a mare who has been repaired, the top part of the vulva will be sealed shut with no opening and there will be just a small opening at the bottom — about 5 or 6 inches below the anus.

If the mare has a Caslick repair, it should be opened a few days in advance of the anticipated foaling so she will not tear herself if she foals unexpectedly and unobserved. Have the veterinarian do this for you, and watch how it is done so that you can do it yourself in the future if you wish. The scar tissue has no feeling and the mare will generally not protest if you carefully spread that tissue and snip it apart with surgical scissors — or any clean, sharp scissors, if you are careful not to poke her.

Give Her Privacy

As the mare gets ready to foal, give her privacy. If she's at pasture, she should be by herself, with no other horses who might pester her or injure the new foal. If she's in a stall, observe her quietly from a place where she can't see or hear you. If people are constantly hovering around, she may put off foaling. Remember that mares often foal at night, in the quiet seclusion of darkness.

Early Labor

The mare may be in early labor for just a few hours or for as many as 72 hours. Mild uterine contractions prepare her for delivery. During this first stage of labor, the uterus contracts and the foal is positioned for birth so that his head and front legs are aimed toward the birth canal.

The mare may not show much outward sign, but if you are observant, you'll notice subtle changes in her attitude or behavior. She may go to the far corner of the pasture or stand with a faraway look in her eye. She may become restless, eating a few bites, then pacing a bit before eating again. If you suspect she is in early labor, check her every 10 or 15 minutes.

ALERT YOUR VETERINARIAN

If you made arrangements for your veterinarian to be present at the birth, the time to call is when the mare is in early labor — when she is merely restless and periodically uncomfortable as the foal is shifting position inside her. Keep the veterinarian posted on progress so he or she can be there when the mare goes into second-stage (active) labor. Usually, everything goes according to plan and the mare foals with no problems at all; however, if anything does go wrong, there may not be much time to correct it because a mare usually delivers within 15 to 30 minutes of beginning second-stage labor.

Signs of Early Labor

During early labor, the mare is restless and nervous. Her pelvic muscles relax more fully, her tail rises, and her sides appear caved in just in front of the hip bones. The foal is shifting position, heading into the birth canal. As contractions occur, the mare shows distress, switching her tail and nervously pacing, pawing, or nosing at her flank. This is the time to bandage her tail, supply fresh bedding if she's in a stall, and then leave her alone (checking on her without her seeing you). If there is time, you can gently wash her teats and udder, and legs and buttocks if they are dirty, with soap and water or a mild disinfectant such as chlorhexidine, and rinse thoroughly.

> **Early Labor, Not Colic**
>
> The mare in early labor may sweat and paw, getting up and down and showing other symptoms similar to those of colic. She will also keep grabbing bites to eat, whereas the colicky horse generally is not interested in food.

Second-Stage (Active) Labor

Active labor occurs when the foal is expelled from the uterus in a series of hard contractions. This stage does not last long. During early labor, the mare moves around a lot and is fairly comfortable between contractions, but after she starts active labor, she is more uncomfortable because the contractions are coming faster. When she shows constant signs of distress and gets up and down, she is starting the second stage and the foal will come soon. She may go down only once and start to strain, delivering the foal in a very short time, or she may get up and down several times.

Be sure she has soft bedding because she may throw herself down rather violently as the pains hit. Mares have been known to rupture themselves internally by going down hard on a firm stall floor.

Rupture of Water Bag

During active labor, the water sac ruptures. You may see the dark-colored sac protrude from the vulva or just the stream of amber-colored fluid rushing out. When this happens, you'll know the foal should come within 5 to 30 minutes. Cows can safely be in active labor for several hours and still deliver a live calf, but if a mare takes longer than 20 to 30 minutes, she needs help — immediately. She is so strong that if she keeps straining with the foal in

After the water sac breaks, the clear white amnion sac should soon appear with the foal's feet within it.

an abnormal position, she will kill it or push part of it through her uterus. A first-time mother may take a little longer than other mares, but if nothing has happened within 30 minutes after the water breaks, the mare is in trouble. Either the foal is being presented wrong and cannot be born, or it is too large and not coming through the birth canal readily.

Normal Position

The normal position for the foal is head first with front feet extended (see drawing). A clear whitish membrane filled with fluid will appear, then, in a moment, the feet should be visible within it. As the feet protrude farther, the foal's nose should appear tucked between them. Usually, the feet are not even; one front foot may advance a bit before the other as the foal moves through the birth canal one shoulder at a time. Any deviation from front-feet-first position is abnormal; professional help will be needed at once, the sooner the better. A foal in an abnormal position can't be born, and both it and the mare may die.

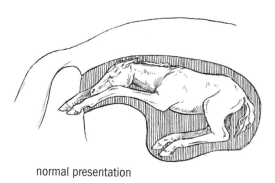

normal presentation

Passing the Shoulders

The foal's shoulders are the widest part to pass through the birth canal, and they usually come through one at a time. This is why one front foot may be extended farther than the other. If both are protruding evenly and the mare seems to be having difficulty moving him out, you can pull on one of the foal's

Pull on the foal's legs, one at a time, to bring the shoulders through the birth canal one at a time; the head is now appearing.

front legs as she strains to advance that shoulder so he can move along more easily. In most cases, however, you should not try to assist in the birth because the mare will manage fine on her own and too much pulling and tugging at the foal might injure the mare.

If you need to pull on a leg, it will be slippery and hard to get a grip with your bare hands. For a hard pull, you should use obstetric chains or straps, but for a little pull, you can do fine with just a small towel around the leg to give you more traction to hang on to it.

The mare will alternately strain and rest, often grunting as she strains, moving the foal gradually but steadily through the birth canal. After his shoulders pass through, he should slide out easily and quickly, with perhaps another hard strain or two from the mare while passing the hips. Then, he should lie there with hind feet still in the birth canal while the mare rests for a few moments after her labor.

Large Foal or Hard Birth

Occasionally, the birth may be hard even if the foal is presented normally. This is rare, but sometimes a mare takes longer than usual and needs some help. Once in active labor, she must deliver the foal quickly or his blood supply will be compromised and he will die. If the foal is partway out, the umbilical cord may be pinched off; if there is pressure on the foal's ribs and body, he will not be able to survive long.

If, for any reason, the mare is taking too long, grasp the foal's forelegs (using chains if you have them) and pull with each contraction. If it is a difficult pull, you may need one person on each leg. Pull the foal out and down toward the mare's hocks, directly in line with the birth canal. Pull strongly

If the mare is having trouble pushing the foal out, pull on the foal's legs each time she strains.

with each contraction, but do not pull when the mare is not straining. If you pull too hard when she is not pushing, you may damage the birth canal. If you tear the mare too much, she may hemorrhage.

Malpresentations

If the foal is not coming properly, his position must be corrected immediately. As soon as you see there is a problem, call your veterinarian or have someone else call while you get the mare up (if possible) to keep her from straining until the veterinarian arrives. You may have to put a halter on her and pull on her head or tap her hindquarters to encourage her to get up. If it is a simple problem, you may be able to correct it, and even if it's serious, you may have to try to help the mare while you wait. Unless the veterinarian is able to get there quickly, most serious malpresentations will result in a dead foal.

BE CLEAN WHEN HELPING A MARE

Don't reach into the birth canal without first taking precautions to prevent contamination. Bandage the mare's tail and wash her buttocks and genital area with a mild disinfectant such as chlorhexidine mixed with warm water. Be sure your hands are clean and scrubbed, and use a plastic obstetric glove if one is available. Lubricate the glove with plenty of obstetric lubricant or K-Y jelly (available at any drugstore). Carefully reach into the birth canal to feel how the foal is positioned.

One Foot Turned Back

If only one front foot appears, reach in and feel to determine whether the other foot is close enough that you can pull it out. If it's close but you cannot pull it out even with the first foot, get the mare up and push the first leg back in a little, while pulling on the

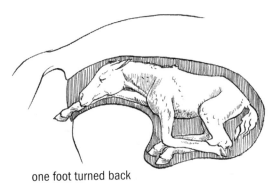

one foot turned back

missing one at the same time. If you can't find the missing leg — just the foal's nose instead — it is probably turned back at the shoulder. This will require veterinary assistance.

If the veterinarian can't get there soon enough, wash up again and apply plenty of lubricant to your gloved arm. Push the foal's nose back into the uterus and try to reach farther in and find the missing leg. The mare's strains will make it difficult so you'll need her to be on her feet with someone holding her, trying to keep her from straining so much. Be careful when trying to manipulate the foal; the uterus can be easily torn by a foot or even a nose pushing hard in the wrong place. If you can get hold of the missing foot, cup it in your hand so it won't tear the uterus, then bring it inward and upward — under the foal's neck and over the pelvic brim. Once both legs are in the birth canal and the head is in proper position, you can start pulling on the foal's legs. Make sure the foal's nose is in the right place and coming through the birth canal, and double check to make sure you haven't pushed the head back too far when bringing the missing leg up.

Nose But No Feet

Sometimes only the nose of the foal appears, with both legs turned back. This often means the foal is already dead; a live foal usually wiggles and squirms and gets his legs aimed into the birth canal during early contractions, but a dead one is limp and may not get positioned properly. The head-first, legs-back position is often typical of a dead twin. The second one might still be

---

SAFETY TIP

If you pull on a foal with chains or straps, place them above the fetlock joints. Otherwise, you may injure the joints or feet.

alive, so get help immediately. If the veterinarian can't come right away, try to get help from an experienced horse breeder, if possible. If you can't get help, you must try to push the dead foal far enough back into the uterus to make room to maneuver the legs into the birth canal.

Head Turned Back

If both feet protrude but no head is coming, it is serious trouble. Attempts to correct this malpresentation should be made only if your veterinarian is unable to get there within the next 20 minutes. If you must try to correct it yourself, scrub up and lubricate both arms to your shoulders. The mare must be on her feet, so have someone hold her. Push the foal back into the uterus with one hand and try to find the head with the other.

The head may be down between the legs or off to one side. If you can get hold of his lower jaw, you may be able to pull the head up into position. Even if you are eventually able to get the head up and the foal out, it may be dead by the time it's delivered. Any birth that takes much more than 30 minutes can result in a dead foal because the placenta begins to detach.

Unlike cattle, sheep, and goats, which have "button" attachments between the placenta and uterus that detach gradually, the mare's placenta comes loose from the uterine wall quickly. If the placenta begins detaching, the foal may die before you can get him out, even if you correct the malpresentation. This is why it's so urgent to have professional help immediately when there's a problem. Even if the foal dies, you must get him out to save the mare. If you try to stay as clean as possible and are careful not to damage the mare, she will probably be able to have another foal, even if you lose this one.

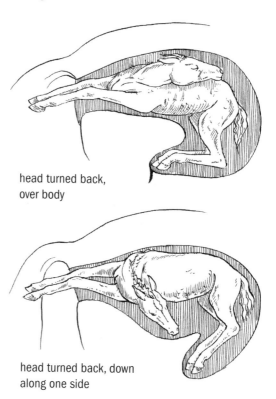

head turned back, over body

head turned back, down along one side

Three or Four Feet at Once

A rare but extremely difficult situation is when a foal comes front feet first with one or both hind feet entering into the birth canal as well. The hind foot or feet may not be discovered until you've already started to pull on the foal (the foal does not progress through the birth canal, and you may think he is just a large foal who needs help). It may be impossible to push the extra leg or legs back in. If the foal is still alive, your veterinarian may try to do a caesarean, but if he is already dead, the vet will probably choose to cut the foal into pieces with a special tool so he can be brought out of the mare through the birth canal, a procedure that is not without risk to the mare but safer than a caesarean.

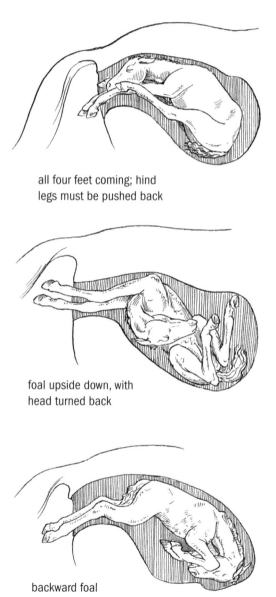

all four feet coming; hind
legs must be pushed back

Rotated Transverse Foal

A rare but frustrating situation is when a foal presents crossways, with part of his body in each horn of the uterus. This problem is usually seen only in large draft mares. The uterus is tipped so the foal cannot be reached from the birth canal at all. A caesarean is the only way to get the foal out.

foal upside down, with
head turned back

Backward Foal

If the hind feet are presented, the foal can still be born, but the birth must be accomplished quickly so the foal will not suffocate or breathe in some of the surrounding

backward foal

MAKE SURE YOU HAVE
FRONT FEET, NOT HINDS

Even if there are two feet coming, if you do not soon see the nose, you should always check to make sure they are front legs and not hind. Front feet have the soles down, whereas hinds point upward.

Even if they are pointed upward and you think they are hinds, it is wise to check. Reach in alongside the leg to see whether there are knees or hocks. The foal may be coming out frontward but upside down. If the foal is upside down and the head is turned back, you have a serious problem; but if the head is there, the foal just needs to be turned over. This may happen naturally if the mare gets up and down a few times; often as she finishes early labor she gets up and down a lot, helping the foal find the proper position. A small foal can be turned over without too much difficulty, but for a large one, you'll need help. To correct this position, push the foal back into the uterus, then take hold of each leg at the cannon bone and try to twist him into normal position. Once you get him rotated, check his head to make sure it's coming into the birth canal.

fluid before he is completely out of the birth canal. When you reach in, if you discover hind feet instead of fronts, and the veterinarian is not yet there, keep the mare on her feet and find someone to help you pull. The mare could deliver the backward foal by herself, but it will die without extra help because she will stop to rest during the birth (as when the hips are coming through). Because the umbilical cord is broken off or compressed by the foal's rib cage, the foal will have to start breathing — and if his head is still inside the mare, he will suffocate or drown in the amniotic fluid.

If there are at least two people to help, you can probably get the foal out alive. Put obstetric chains or straps above the fetlock joints, and when the mare starts to strain, pull with all your strength, one person on each leg. As the hips emerge, keep pulling, even when the mare stops to rest. The foal must come out immediately or he will die. Next, you must get him breathing.

The cord will break as the foal comes out, so there is no point in letting the foal lie quietly by his mother as you would with a normal birth — you must try to get him breathing. He may be limp and unconscious from being short of oxygen. Clear away any mucus from his nostrils with a suction bulb. If there is a lot of fluid in his air passages, have someone help you hold him upside down just briefly to let it drain out. If he doesn't start breathing, tickle his nostril with a piece of clean straw to make him cough and sneeze. Slap his rib cage if that doesn't work. If he still won't start breathing, put him back down on the ground and give artificial respiration. (See Giving Artificial Respiration on page 411.)

Breech

A breech presentation is much worse than a backward foal. Because the hind legs are forward in a sitting position, nothing appears. The mare is in labor but nothing protrudes. When you reach in to check, all you can feel is the tail and rump. *Seek veterinary help immediately.*

If you absolutely cannot get help, scrub up and have someone hold the mare (she should be standing) while you reach in with both hands, pushing the foal's buttocks forward with one and reaching for a hock with the other. If it is a long-legged foal, it may be impossible to reach the hock. Still, if you can get hold of the hock and bend it, you may be able to reach the cannon bone. If you do, slide your hand to the foot. Cup the hoof in your hand to keep it from tearing the uterus, and bring it up over the pelvis and into the birth canal. If you can get one foot, you can usually get the other. Then, follow the procedure above for delivering a backward foal.

If the feet can't be reached (the veterinarian may not be able to reach them either), the foal may have to be delivered by caesarean. If the foal is still alive after a breech position is corrected and pulled — a rare occurrence — he will need artificial respiration. (See page 411.)

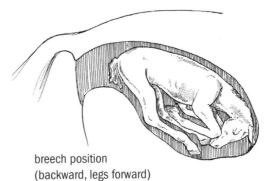

breech position
(backward, legs forward)

Placenta Previa

One situation that requires immediate action to save the foal is when the placenta detaches too soon and part of it precedes the foal through the birth canal. If you see a mass of dark-red membrane coming out when the mare starts labor, instead of the clear white amnion sac and feet of the foal, you should immediately scrub up and break through the thick membrane with your hands, reaching into the birth canal to get hold of the foal's legs and pull-

---

FECES

If a birth is difficult or the foal is severely stressed from shortage of oxygen, it's not unusual for him to pass feces during birth. A foal covered with yellow excretia is in stress and may be in worse trouble if any of this material is drawn into his lungs when he starts to breathe. Wipe off his nose immediately, and clear his nostrils with a suction bulb.

---

ing him out swiftly. Placenta previa signals that his lifeline is detaching from the uterus, and the sooner you can get him out, the better.

Uterine Torsion

Occasionally, a mare has torsion of the uterus in which the heavy uterus and its contents flip over, making a twist in the cervix. The foal cannot be delivered if the uterus stays that way. A mild torsion that occurs at foaling may correct itself when the mare rolls. Indeed, it is quite common for a mare in early labor to roll. If the foal is coming out a little crooked or the uterus has a mild twist, rolling may help.

In contrast, a complete twist prevents the foal from coming through. If this is the case, the veterinarian must try to correct the twist. Unless this is done promptly, the blood supply will be compromised, causing swelling and shock, death of the foal, and possible rupture of the uterus and death of the mare. Fortunately, this situation is rare.

The Importance of Breathing Soon After Birth

While the foal is developing in the uterus, he gets oxygen from the mother's bloodstream through the placenta and umbilical cord. At birth, he must begin to breathe for himself. Once the umbilical cord pinches off or breaks, he's on his own. At this point, the drop in oxygen level in his blood stimulates him to try to breathe (this is why the backward foal tries to breathe as soon as the umbilical cord crosses the pelvic brim and is compressed — and why you must get him out quickly). In a normal delivery, the birth process begins the transition. As the foal comes through the birth canal, his rib cage is compressed as he squeezes through the mare's pelvic opening. This forces fluid out of his air passages. As the foal's body emerges from the mare, the rib cage expands again, which creates a vacuum in the lungs, drawing in the foal's first breath.

Usually, he makes a smooth transition from dependence on oxygen in his mother's blood supply to his own breathing, but this process can be hindered by his position immediately after birth. To breathe effectively, he must be able to expand his rib cage and inflate his lungs, and his circulatory system must be working properly so blood reaches the lungs to pick up oxygen and carry it to the body tissues.

After a normal birth, the foal positions himself upright, resting on his chest with forelegs extended in front as he begins the process of trying to get to his feet. Because of the way his chest and rib cage are constructed, he breathes best standing up or lying in an upright posture (head up). In these positions, his lungs can fully expand, increasing the levels of oxygen in his blood.

A foal lying flat cannot fully inflate his lungs. If he is limp or weak after birth, gasping for breath, or lying flat instead of trying to raise his head and roll up onto his chest, you must help him get upright. It is crucial that he begin to breathe properly.

Clearing the Foal's Airways

Make sure his nostrils are clear of mucus. A squeeze-bulb suctioning device works well; keep one handy in your coat pocket when attending a foaling mare. If you don't have one and it's obvious the foal's airways are obstructed, position him briefly with head downward to help the fluid drain out, then gently squeeze out excess fluid by pressing thumb and forefinger along the top of the nostrils toward the muzzle — like squeezing a toothpaste tube.

Checking the Foal's Heartbeat

Sometimes a foal is limp and lifeless (as after a difficult delivery and prolonged birth, leaving him short on oxygen), but if he's still alive, there's a chance to save him. Check for a heartbeat with your hand on the lower left side of the

---

STIMULATING THE BLOOD CIRCULATION

After making sure his airways are not obstructed, position the foal upright on his chest. You can also stimulate his circulation — which will aid his breathing reflex — by lightly rubbing and drying him with a rough towel and by gently moving his legs.

rib cage, just behind and above the elbow. A young foal's heartbeat is easy to feel because there is very little tissue between it and the outside chest wall. If there's a heartbeat, try to get him breathing immediately.

Measuring heart rate is a good way to tell whether the foal is in trouble; it drops as body tissues are depleted of oxygen. If the foal fails to start breathing within 20 to 30 seconds of birth and has a heart rate lower than normal, he is in trouble (normal is about 60 to 100 beats per minute for the newborn foal in the first few minutes, then rising as high as 200 per minute in the first hour, before tapering off at about 70 to 130 beats per minute). If the heart rate has dropped to as low as 30, the foal's condition is critical. Also check his gums. They should be pink. If he's short on oxygen, they will be gray and colorless or blue. He needs to start breathing.

Giving Artificial Respiration

If the foal does not respond to a piece of straw tickled up his nose, clear his airways with suction, then roll him up onto his breastbone with head and chin resting on the ground, nose as low as possible, to allow more fluid to drain out the nose. Then lie him on his side, head and neck extended. Cover one nostril tightly with your hand, holding the mouth shut. Gently blow into the other nostril, with your mouth completely covering it. Don't blow rapidly or forcefully or you might rupture a lung.

Blow until you see the chest wall rise. Then let the air come back out on its own. Blow in another breath until the chest rises again as the lungs fill. Continue filling the lungs and letting them empty until the foal starts breathing on his own. Once his tissues become less oxygen-starved, his heart rate should rise and he will regain consciousness and start breathing. His breathing may be erratic at first, but if everything else is normal, he will develop a more regular pattern within a few minutes.

Normal Foaling — After the Birth

Let's look at what happens after a normal, unassisted birth. When the foal is born, he may be still wrapped in membrane but should quickly free himself by shaking his head. If he doesn't, and the sac has not broken, you should break it and make sure he begins breathing. Be certain that your hands are clean.

Don't Break the Cord Too Soon

He'll be lying with hind legs still in the birth canal. Don't try to pull him out or get the mare up. As soon as you are sure the sac is off his head and he's

---

KEEP THE CORD FROM BREAKING

If a mare foals standing up, as some do when nervous or when having help with a difficult birth, it's best if two people catch the foal as it is delivered and hold him up for a few minutes so the cord doesn't break immediately.

---

breathing, leave the mare and foal alone. The mare will generally lie there for 10 to 20 minutes before she gets up, regaining her strength after the effort of labor. Also, this is nature's way of making sure the foal gets his full blood supply. The placenta is full of blood that is pumped into the foal through the umbilical cord as the uterus starts to contract.

If the cord is broken too soon by the mare jumping up or you pulling the foal out, he'll be deprived of part of this blood and be somewhat weak and anemic for a few days. As much as 25 to 35 percent of the foal's blood is in the placenta at birth, but it drains into the foal if the mare stays down for a few minutes. If she gets up too soon, not only will the cord break prematurely (which can cause it to hemorrhage), but the placenta that is detaching and starting to work its way through the birth canal may fall back into the uterus. Because the uterus and cervix are contracting, it becomes harder for the mare to expel the placenta if it goes back into the uterus as she gets to her feet. Try not to disturb her as she lies there resting.

IF THE CORD BLEEDS

The cord rarely bleeds when broken naturally. The blood vessels of the cord have a greater elasticity than the outer covering, and if the cord is broken rather than cut, these vessels draw back into the stump and seal off the opening. If the cord does bleed, or if, for some reason, you have to cut it and it bleeds, sit with the foal and pinch the navel stump firmly with your fingers for a few minutes or use a clean clamp (such as a hemostat or the clamp used for umbilical cords on human babies) temporarily. Then apply disinfectant to the stump. Never tie it, as there will be a certain amount of natural drainage; tying could cause swelling and increase the risk of infection. Also, if you were to tie it, the tie would act as a wick for infection to enter the stump.

Disinfecting the Navel Stump

After the cord breaks — which usually happens as the foal struggles while trying to get to his feet, or when the mare gets up — thoroughly soak the entire stump in tincture of iodine or Nolvasan. This disinfects it, prevents the entrance of bacteria, and the iodine acts as an astringent to help it dry. A good way to iodine the navel stump is to immerse it completely in a small jar, making sure it is thoroughly dipped, without spilling any on the foal. One disadvantage to using iodine is that it is harsh and can burn the surrounding tissues. These caustic burns may cause irritation that can open the way for infection. Many veterinarians now recommend using Nolvasan, instead, but it must be repeated frequently until the stump has dried up.

If the cord does not break naturally, you can pull it apart with clean, well-washed hands (while you hold the cord close to the foal's body with one hand so there's no pull on his abdomen, which could injure him internally). It should be broken about 2 inches (5 cm) from his body; there is a natural stricture there, where it would normally break when the mare gets up.

If your veterinarian has suggested any medications for the foal (such as selenium, if you live in a deficient area), give these when you disinfect the navel stump, while the foal is still down.

Final Stage of Labor

The third and final stage of labor occurs when the mare expels the placenta. This should happen soon after foaling; the placenta can usually be seen hanging down after the mare gets up.

The placental membranes surrounded the foal in the uterus and were attached directly to the uterine lining except at the cervix. During labor, the cervix relaxed and uterine contractions put pressure on the fluid around the foal in his placental envelope. The only direction it could go was through the cervix. The placenta ruptured at that spot and allowed the inner envelope — the amnion — and fetus inside it to start through the birth canal. The fluids between the inner and outer sacs preceded the foal; this was the breaking of the water.

The placenta, the thick outer envelope, begins detaching from the uterine lining during labor, finishes detaching soon after the birth, and follows the foal out the birth canal 15 to 45 minutes later, except in abnormal situations when the placenta is retained. As soon as the placenta is shed (usually at about the same time the mare gets up, or soon thereafter), remove it from the stall and spread it out on a clean surface to see whether it is intact. A fresh, healthy placenta is bright red to mahogany, with a velvety appearance. An off-color placenta may be infected.

During third-stage labor when the uterus is shedding the placenta and begins shrinking, the contractions may be so severe that they cause the mare discomfort and colic. In severe cases of abdominal pain, she will need medication to halt the cramping and ease the discomfort. Consult your veterinarian if this is the case.

Checking the Afterbirth

If any part of the placenta is retained inside the mare, it will cause serious uterine infection. When the membranes are spread out for checking, they will look like a pair of trousers (one leg smaller than the other) with the ends of the legs closed. There should be only one tear in the membrane — at the "waist" of the trousers where the foal came through. If any part of the placenta is missing, it will usually be the tip of one of the closed trouser legs. If you think there might be a problem, save the placenta in a clean garbage bag for your veterinarian to examine. A placenta that weighs more than 15 pounds (6.5 kg) is abnormal and may be diseased.

Retained Placenta

The placenta attaches to the uterine wall during gestation with tiny finger-like projections that plug in to glandular openings in the uterine lining. This hookup connects the mare's bloodstream with the placental membranes, which send nutrients and oxygen to the fetus through the umbilical cord. Once labor starts, the placenta starts to detach, especially during second-stage labor, when the foal is being delivered. The placenta comes completely free after the birth when the uterus goes through more contractions and begins shrinking back to normal size. If, for some reason, it does not completely detach, it will hang out of the vulva, dragging on the ground.

If it takes the mare more than 2 hours to shed the placenta, this is abnormal. After 3 hours, you should suspect problems and consult your veterinarian.

IS THE MARE'S VULVA TORN?

Soon after the mare "cleans" (sheds her placenta), check to see whether her vulva ripped during foaling. If she tore very much, your veterinarian should repair her as soon as the swelling goes down. If she needs a Caslick repair, she should be sutured up again soon after foaling.

RETAINED PLACENTA

Retained placenta is fairly common after a difficult birth, twins, a premature foal, or induced labor. Sometimes, retained placenta is caused by dietary imbalance, such as a deficiency of phosphorus or selenium, or too much calcium from a diet too rich in legumes. Hypothyroid mares are also prone to retained placenta.

Anything longer than 8 hours means the mare may be in grave danger. If the afterbirth is hanging down and dragging, tie it into a ball above the level of the mare's hocks so neither she nor the foal will step on it and pull it out forcibly. Use very clean hands. The weight of the balled-up placenta will keep a gentle traction on the membranes to help ease them out. Never try to remove the afterbirth; it must come away by itself. If you pull on it, you may cause hemorrhage or tear it, leaving part of it inside the mare.

Risk of Infection
The placenta starts to decompose rapidly after the birth of the foal. The birth process opens the way for contamination, and the membranes hanging out of the mare serve as a wick by which bacteria can gain entrance. Most mares who retain the placenta longer than a few hours develop uterine infection. It may be temporary, local, and easy to eradicate, or it may become systemic and life threatening. Even a localized infection has serious consequences if it leads to pus in the uterus. The infection may reduce the mare's ability to have another foal.

Even though the infection may eventually clear up, the resulting fibrous scar tissue that develops during the healing process most likely will damage part of the uterine wall to such an extent that it cannot support another pregnancy. Extensive scar tissue in the uterus may interfere with conception or cause the embryo or fetus to die early in pregnancy.

Treatment
Your veterinarian may give the mare an IV or intramuscular injection of oxytocin to stimulate contraction of the uterus and hasten expulsion of the placental membranes. Anti-inflammatory drugs such as flunixin meglumine (Banamine) are usually given also to combat inflammation and prevent founder. The mare may need systemic antibiotics to treat infection. The veterinarian may also irrigate her uterus to remove any bits of afterbirth that might still be inside and to flush away debris that could foster infection.

Indeed, retained placenta can cause acute and fatal infection if bacteria get into the bloodstream. The mare becomes obviously sick with high fever, dullness and depression, lack of appetite, and a drop in milk production. The infection may also lead to founder.

Watch the mare closely after birth, especially if the placenta was retained. Check her temperature twice daily. Any fever, lameness, colic, loss of interest in feed, or drop in milk flow should be considered a medical emergency. Sometimes, in early stages of infection following a retained placenta, the mare seems normal, only to colic and founder a few days later. If this happens, only swift medical attention can save her life.

Foaling Founder

If a mare develops infection in the uterus after foaling, changes occur in the bacterial flora and endotoxins may be produced. These toxins set the stage for production of lactic acidosis, an excess of lactic acid in the bloodstream. The lactic acid and endotoxins in the blood affect the whole body, causing shock and damaging liver cells. The latter condition interferes with the normal filtering of poisons from the body. The toxins also damage tiny blood vessels, resulting in laminitis. The mare's extremities suffer from impaired circulation, and the horn-producing laminae in the feet are deprived of oxygen and nutrients. Mares with acute laminitis caused by uterine infection after foaling may be difficult to treat because all four feet are affected. Laminitis can also result just from inflammation (which causes the release of histamine and prostaglandins, substances that damage body tissues) without any infection being involved.

FOUNDER CAN BECOME CHRONIC

Some mares suffer foaling founder a day or two after foaling even if the birth was normal and the placenta shed promptly. An examination or a uterine flush, completed by your veterinarian to check the fluids or culture them, reveals an infected uterus. Most cases of severe founder follow a difficult birth or retained placenta and sometimes become chronic, giving the mare a low-grade laminitis that flares up periodically. Hypothyroid mares are also extremely susceptible to foaling founder.

Whether a mare will founder again when foaling in the future depends on the severity of the initial laminitis and the potential lingering presence of laminitis when she foals again. A mare whose founder episode is successfully treated, with full recovery, may not have further problems, but laminitis from bacterial endotoxins may recur if she is ever exposed to a sufficient quantity of endotoxins again. Once a horse suffers from founder, she's much more susceptible to founder in the future; therefore, a mare who founders because of uterine infection, may readily founder again if conditions in the uterus promote another infection.

Uterine Prolapse

Prolapse of the uterus is a rare problem. When this occurs, the uterus comes out of the birth canal after the foal and hangs down from the vulva all the way to the mare's hocks. This serious condition is fatal unless you get prompt help from a veterinarian. While waiting for him or her to arrive, saturate a clean blanket with a solution of disinfectant and warm water and use it to support the uterus. Tie the mare so she can't move around, and try to keep the inverted uterus supported so it can't come out any farther and put strain on its attachments. Too much trauma or strain might cause a blood vessel to break, resulting in fatal hemorrhage. It may take two people to support the prolapsed uterus, holding the blanket between them. If possible, it should be held up almost as high as the vulva.

Prolapses usually result from the mare's continual straining after the foal is born. The problem is most likely to occur after a hard foaling but can happen even under good conditions.

Hemorrhage After Foaling

Another serious complication of foaling is internal bleeding. This rare but life-threatening problem usually occurs within 24 to 48 hours of the birth. It is not easily recognized if you haven't witnessed it before, but is a condition you should be aware of. If the person attending the foaling realizes early on that something is wrong, there is a good chance of saving the mare and the foal. Some mares collapse suddenly onto the young foal, killing or injuring it.

Bleeding results from rupture of an artery in the uterine wall or of one of the major vessels that run along the uterine ligaments. The mare's chances for survival depend on the size and location of the damaged artery. Bleeding from the uterine wall is not likely to be fatal because blood seeping into the uterine tissues creates a large hematoma that tends to block and slow the flow. If the damaged vessels are in the uterine ligaments, however, they bleed right into the abdominal cavity, causing rapid and fatal blood loss.

Internal bleeding is usually caused by degeneration of artery walls rather than by a difficult foaling, and it is more common in mares older than 15 years. Continual stretching and contracting of the uterine arteries during pregnancy results in less elastic artery walls, and the force of foaling may break one. The mare shows no evidence of bleeding at the vulva and may die before the problem is recognized.

Hemorrhage sometimes occurs before foaling from uterine torsion or trauma during violent exertion, but most cases happen soon after the mare foals or as late as the next day.

Signs of Internal Bleeding

The discomfort of foaling and colicky cramps from placental expulsion may mask signs of hemorrhage, but certain subtle signs can lead you to suspect the colic is from bleeding. A mare with a ruptured artery is weak and may be reluctant to get up after foaling. If she doesn't get up, don't push her; the less exertion, the better. She may get up and go down again, being anxious, agitated, uncomfortable, and colicky. If she is standing, she may sweat and shift from one foot to another, sometimes curling her upper lip. She may paw and roll; if this is the case you should keep the foal away from her so he won't be injured. If bleeding continues, the mare becomes dull, her mucous membranes become pale, and her temperature drops and pulse weakens. Sweating is profuse and her skin is clammy as she goes into shock; she starts to quiver, ready to collapse.

If you suspect hemorrhage, keep the mare as quiet as possible and get veterinary help immediately. Blanket her to conserve body heat. Protect the foal (in a cage of hay bales just outside her stall door, so she can see him and not be too upset by his absence). The veterinarian will use tranquilizers to calm her and lower her blood pressure, as well as IV fluids or transfusions to keep up her blood volume. He may give oxytocin to hasten uterine contractions and seal broken blood vessels if the bleeding is in the uterine wall. Coagulants and painkillers also help. If the mare recovers, she will reabsorb the hematoma and may foal normally the next time.

Most veterinarians recommend that mares who survive a bleeding episode take a year off from breeding. If the mare cannot be saved, she should be milked out into a clean container or with the Udderly EZ pump immediately after death to provide colostrum for her foal.

MOST BIRTHS ARE NORMAL

There are many, many things that can go wrong at foaling, but do not be discouraged: most mares foal with no problems. It is important, however, to know about possible problems because many can be prevented with good care — such as cleanliness to avoid infection — or corrected with prompt assistance. If you know what to do in an emergency, you have a much greater chance of being able to save the foal — and the mare.

Care of the Mare After Foaling

After foaling, the mare will be tired, dehydrated, and somewhat constipated. Offer her plenty of clean, lukewarm water to drink and palatable food. Green grass or alfalfa hay is ideal. She might like a bran mash; it not only contains needed moisture but is also quite laxative. If she has been on a diet of grass hay, she may be constipated. (For a bran mash recipe, see chapter 2.)

Don't be alarmed if she passes no bowel movements for the first 12 hours after foaling. She "cleaned out" rather completely when she was nervous, restless, and pacing during early labor, and it may be a while before any more feces work through the digestive tract. Also, the mare may be sore from foaling and reluctant to pass a bowel movement because passing manure causes pain. The best remedy for constipation is a good, palatable, laxative feed such as green grass, alfalfa, or a bran mash. If the mare continues to be constipated for the first 24 hours (passing no manure or just a few firm, dry balls), consult your veterinarian. In serious cases of post-foaling constipation, the veterinarian may decide to give the mare mineral oil.

Deworm the mare with ivermectin during that first day after foaling to prevent parasitism in the foal by Strongyloides, a primary cause of so-called foal heat scours. (See the section on parasite control for the broodmare and her foal in chapter 24.) If the foal does not pick up worms from the mare's milk (tiny threadworms are passed to the foal from the mare's udder during the first days of life), he will scour less. He will also have less initial parasite load of small strongyles from eating his dam's manure — something foals do to obtain the intestinal flora necessary for digestion of roughage — since her manure will be clean.

23

Care of the Newborn Foal

YOUR MARE HAS JUST FOALED a beautiful long-legged baby. The foal will need immediate attention and then careful monitoring for the first few days to make sure all goes well. As described in the last chapter, your first order of business may be to get the foal breathing if he's having trouble or there have been problems at birth. Then, you must disinfect his naval stump, check that he nurses soon after birth to gain disease protection through absorption of antibodies in the mare's first milk, and be sure that he starts passing bowel contents.

Aside from the breathing, and disinfecting his navel stump, however, all of these things can occur in due time. Nature has a clock that must be followed, too, and the mare and her needs as a mother must be carefully considered.

Care Immediately After the Birth

There are several tasks you should attend to when the foal is born. Be sure to do these calmly and quietly. Too much fuss, noise, or activity can cause problems.

Avoid Foal Rejection

If you rush to disinfect the navel, help the foal to his feet, and aim him toward the mare's udder — and your family and friends are there to admire him — you may upset the mare. Sometimes, all this attention can lead to problems. Occasionally, a mare will ignore, reject, or even viciously attack her new foal, or she may seem unsure or confused about motherhood — especially if she's just had her first baby. Some mares balk at mothering the foal because of a hormone imbalance or physical problem (for example, pain from a sore, swollen udder when the foal tries to nurse), but often, the problem is too much human interference during the critical time when mare and foal should be alone for bonding, establishing that all-important mother–baby relationship.

Wait to Begin Imprint Training

A popular procedure is "imprint training"—handling the foal soon after birth and getting him used to the feel of your hands all over his body, impressing on him that human handling is part of his world and nothing to fear. Done properly, imprint training can be beneficial, making the foal more at ease later when you work with him. For example, he won't refuse to let you handle his ears, put something into his mouth, or pick up his foot. His subconscious will "remember" all of this has been done before.

Sometimes, however, the person who rushes to handle the new foal is focused on one thing only—imprinting the foal—and forgets how this might disrupt nature's program for the mare and her newborn. If you're going to imprint the foal, keep the mare in mind and stay tuned to her reactions. If she's an old, dependable broodmare and your handling of the foal is no problem for her, there is nothing to worry about. If she's a first-time mother or a highly nervous individual, go easy. Wait until that very important bonding process is fully accomplished before you do anything with the foal.

A detailed discussion of imprinting is given in *Storey's Guide to Training Horses* (Storey Publishing, Second Edition, 2010).

A first-time mother may be nervous and uneasy but needs to bond with her new baby.

Minimize Handling of the Foal

A nervous or insecure mare, or any first-time mother, may have her maternal instincts disrupted by too much human handling at foaling time. Even having people observe her during the birth may upset her and make her postpone labor for several hours or become less apt to mother the baby after she foals. Once the foal is born, don't rush right in unless his head is still encased in the amnion sac and he needs help to start breathing (see chapter 22). Let the mare and foal lie there until the cord is broken and it's time to disinfect the navel; then move in quietly, take care of it, and slip away again.

Rushing to disinfect the navel, clean up soiled bedding, dry the foal, or help him to his feet and to nurse can be too much disruption for some mares, causing them to reject the foal. If the weather is cold, you will have to dry the foal; otherwise, leave mare and foal alone once you have started the foal breathing if there's a problem, disinfected the navel, and given selenium injection if recommended by your vet. By moving slowly and quietly, you won't upset the mare. If everything is normal, leave the pair in peace for at least an hour. Watch from a distance. If the mare isn't worried or uneasy about hovering observers (who, in her mind, pose a threat to her baby), she'll be more likely to concentrate on the foal and begin her role as mother. Foaling time reactivates deep-seated instincts; even though your mare trusts you, too much human presence and fussing at this time will upset her.

Bonding Takes Place During the First Hour

During this time you should stay away, quietly observing from afar. Through-out the first hour, the mare smells, nuzzles, and licks her foal, permanently identifying it as hers and establishing the bond of scent that locks her new-born's identity into her brain.

Human interference can also confuse the foal because his instinct is to follow any large moving object. Under natural conditions, the only nearby large moving object is the mare. With people in the stall, he may be attracted to them instead and not seek the mare to nurse.

Don't Help Him Stand Up Unless He Really Needs Help

The foal's first attempts to get up are clumsy; he falls down several times before gaining his feet. It's tempting to help him, but usually unnecessary and occasionally harmful. Sometimes a foal has a fractured or cracked rib from the pressure of coming through the birth canal. These usually heal just fine (and you probably won't even know they occurred), but picking him up

> **NEW MOTHER SAFETY TIP**
>
> Be prepared for your friendly, trusting mare to have a sudden change of personality when she foals. Many mothers become extremely protective of their babies — it's their instinct to defend the newborn against predators, and they become aggressive toward intruders for the first few hours after birth. Keep an eye on the mare the first time you go into the stall or pen after foaling in case she decides to attack you as you tend to the foal to disinfect the navel. Keep all children, pets, and extra people away from the mare and foal the first day to avoid upsetting her.

or helping him stand may displace the fractured pieces and cause them to puncture a lung or other vital organs.

Out in a pasture, the foal generally has less trouble getting to his feet because the footing is better and he's not encumbered by the straw in a foaling stall or hindered by hitting walls or bumping against a manger or water tub. If he's born in a stall, make sure there is nothing there that might hurt him. Before the mare foals, remove tubs and other obstacles, as well as eye screws or protruding nails at foal height. Monitor his progress in getting up and blundering toward the udder, helping him only if he really needs it.

Nursing

By the time he's an hour old, the foal should be on his feet and looking for the udder. Make sure he finds it and gets his first drink of colostrum, the nutritious first milk that contains needed antibodies that keep a foal healthy. It's best if he can accomplish the first nursing by the time he is two hours old.

Waiting and Watching

You don't need to panic and assist if the foal is a little slow to find the udder. He'll get more strength and coordination with each unsuccessful try, unless he was born weak, oxygen deprived, or with some other abnormality. Even if he doesn't accomplish the first nursing within an hour, he is still not in immediate danger because his ability to absorb the colostrum's antibodies remains high for 2 to 3 hours following birth. Then, it begins to decline as the pores in the intestinal lining begin to close.

HER UDDER MAY BE SORE

The mare's udder is a little swollen and tender at first, and she may be nervous about the first nursing because it hurts her. Sometimes, the mare moves around a lot or squeals and kicks at the foal, but her kicks are not meant to hurt him; she's just upset because her udder hurts. Rarely does a mare kick viciously at a foal, but if this happens you'll have to intervene. Often, restraining the mare is more helpful than trying to guide the foal.

If you wait at least an hour before helping him, you'll be less apt to hinder the bonding process and the mare will be less confused by your attempts at intervention. As long as the foal gets a good nursing by 2 to 3 hours of age, and an adequate amount of total colostrum within 24 hours in subsequent nursings, he will obtain the necessary immunities. Let him try to do it on his own, and allow the mare to mother him.

Some mares are confused at first, especially first-time mothers or mares who had earlier bad experiences. Some habitually reject their foals or are indifferent, but most will mother the foal if given a chance. Usually, once he nurses, she accepts him. This is partly because nursing stimulates the production of hormones that prompt the hormonally triggered "maternal instinct." Also, the mare's udder won't be so sore and tight after some of the milk is out of it. Most mares who reject their foals eventually accept them if time, care, and patience are used, but you can often prevent potential problems if you keep a low profile at foaling time and give nature a chance.

Eventually, most foals find the right place to nurse, although, to an observer, it may seem like an eternity as the newborn noses about here and there. You only need to help if he is weak and tired from a long or hard birth and gives up too easily, or if the mare kicks at him. If she kicks, you'll have to restrain her until he has nursed.

Helping a Foal Nurse

If the foal has not succeeded in nursing by the time he's 2 hours old, help him before he becomes tired and discouraged to make sure he gets crucial colostrum. If the mare is moving around too much, halter her and stand her against the fence or stall wall. If she keeps moving her hindquarters to avoid the foal or insists on kicking at him, you may need to hold up her front leg or shoulder twitch her (grasp the skin over the shoulder with your hand) to keep her still and steady. Talk to her soothingly and try to keep her calm and still. If the foal is strong and vigorous, this may be all the help he needs. If he is clumsy and uncoordinated, or scared off by her kicking someone may need to guide him to the udder while you restrain the mare.

Use as little force as possible to keep an uncooperative mare calm and quiet, but use a twitch or Stableizer if necessary to hold her still (see chapter 5). Usually, by the second nursing, which should be tried 20 minutes after the first, the mare is more cooperative. In most cases, by the third or fourth nursing, she will accept the foal. You may merely need to hold her until he starts nursing, then you can release her gently and she'll let him continue to nurse.

Occasionally, a mare refuses her foal for 24 to 48 hours, but usually, if you are persistent in helping the foal at each nursing, she will come to accept it. Whenever a mare is aggressively vicious toward the foal, you should keep them separate except at nursing; it may even be necessary to find a substitute mother if she fails to accept the foal after several days.

When trying to help a foal nurse, hold him in place by hanging on to his tail. Don't handle his head much or you'll confuse him. If he can't get onto the udder himself, milk a little out of the mare into a clean container or use an Udderly EZ milker. Then put some of her colostrum on his nose and lips and all over the mare's udder (or use corn syrup, if necessary) to get him interested and help him find the right place. Never force his head down to the udder. He can't nurse in that position, and it will make him resist.

If a newborn foal is weak or confused, he may need help with his first nursing. If he has not nursed on his own by the time he is 2 hours old, you should assist.

If he simply will not nurse, milk more from the mare and use a sterile bottle (presterilize bottles in boiling water) and lamb nipple to get him sucking, or use the bottle and nipple that comes with the Udderly EZ milker. Usually, after he gets a taste, he'll be more eager and try harder, and you can guide him to the udder or lead him to it with bottle and nipple. If he's confused and tries to suck on you instead

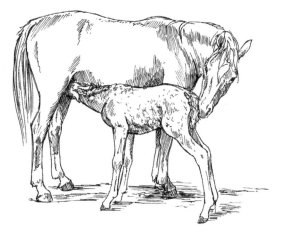

The foal's first nursing is extremely important: the mare's first milk (colostrum) contains nutrients crucial to his health.

of the mare, get him sucking the colostrum from your bottle and move it toward the udder until you can transfer his mouth from nipple to teat. This may be easiest if someone else is on the other side of the mare, holding the bottle underneath her and bringing it up next to her udder to "fool" the foal into getting into the proper position to nurse.

The Importance of Colostrum

The newborn's first nursings are very important. Foals who don't nurse soon after birth are more likely to become ill and die during the first weeks of life than foals who nurse promptly. This is because the mare's first milk contains ingredients crucial to the foal's health and survival. Colostrum has more energy, protein, vitamins, and minerals than regular milk, with twice the energy and five times the protein. It also serves as a gut stimulant to help the foal pass his first bowel movements. These first evacuations are often firm and hard-packed, and therefore difficult to pass. Sometimes an enema is necessary, but in most cases, the laxative effect of colostrum will get things moving.

Antibodies

Especially important are the antibodies in colostrum. The foal is born with no immunity, no resistance to disease. Unlike a human baby who picks up his or her mother's immunity while still in the uterus, the foal has to acquire his through colostrum when he nurses.

Disease Protection

During late pregnancy, when the mare's udder is filling, she makes this special milk, with a concentration of all the antibodies she's developed, to give the foal immunity for several weeks after birth. This passive transfer usually lasts until his own immune system has encountered bacteria and viruses in his environment and he starts producing his own antibodies. Passive transfer failure can occur, however, if there is an inadequate amount of antibodies in the mare's colostrum or if the foal is unable to get enough colostrum in time to receive proper levels.

If the mare has encountered diseases and developed antibodies, she can pass this immunity to the foal. Vaccination also stimulates her body to develop immunity against specific diseases. For example, if the mare has a tetanus, influenza, or botulism vaccine during gestation, she develops antibodies that will be present in her colostrum to protect the foal as soon as he nurses. Keeping the broodmare's vaccinations up to date is the best way to make sure the new foal will have immunity against most problem diseases.

A foal who fails to nurse soon enough runs a high risk of developing diarrhea, pneumonia, or other serious problems. Failure of passive immunity transfer is the greatest single factor resulting in disease and death in foals.

Timely Nursing

The foal must nurse within 2 or 3 hours to gain immune protection. At birth, his intestinal lining has openings directly into the bloodstream, and the antibodies can slip through. These pores, however, begin to close soon after he's born, so he must nurse promptly to get the full benefit of colostrum. Foals lose 75 percent of their ability to absorb antibodies within 4 hours after birth; the

---

BENEFITS OF COLOSTRUM

Colostrum contains a creamy fat, high in energy and easily digested. It's an ideal first meal, giving high-energy calories to generate body heat and strength. This is why a foal who nurses soon after birth is more lively, strong, and vigorous (and able to stay warmer if born on a cold night) than a foal who has not yet nursed. Once the foal gets some colostrum, he usually has the strength to keep looking for more, becoming stronger and more coordinated with each energy-laden meal. The newborn nurses in small amounts at frequent intervals, about every 10 to 20 minutes.

TESTING FOAL ANTIBODY LEVELS

If the foal is not able to nurse within the first few hours, or if the mare foals unexpectedly and, when you find the newborn, you don't know how old he is or whether he nursed soon enough, your veterinarian can check a blood sample after the foal is 18 hours old to test for immunoglobulin levels. The peak level of antibodies in the foal's blood is attained at 18 to 24 hours of age. The veterinarian can do a simple test at the clinic in just a few minutes. There are also test kits available to analyze blood samples; in fact, on many breeding farms, it is routine procedure to test foals. Test results are also required by some insurance companies as part of the criteria for determining foal insurance premiums.

large molecules can no longer slip through the gut lining into the lymph system and bloodstream. A foal who gets his first nursing after 4 hours receives only a fraction of the antibody protection he needs. The colostrum remains beneficial for protection against some gut infections because certain antibodies still have local activity within the gut. Even though he can no longer absorb antibodies after the first few hours, the foal needs colostrum in subsequent nursings throughout the first day because of its additional benefits. He not only needs the antibody protection, but also it's laxative effect and high energy content to keep him warm and give him strength.

If he's born during the night and you don't discover him until morning, you may not know whether he nursed soon enough. If he's up and about, strong and frisky, nursing vigorously, passing bowel movements and urinating (which most foals do after nursing), these are good signs that mean he's getting sufficient milk and probably nursed in a timely fashion. If he isn't passing urine or bowel movements, or you find him cold and shivering or slow and weak, make sure he nurses as soon as possible.

If there's failure of passive transfer for any reason — whether the foal failed to nurse in time or the mare had poor colostrum — the foal can be transfused with plasma from the mare or from another horse on your farm, or with commercial plasma preparations.

Lack of Colostrum

Sometimes, a mare does not have an adequate supply of colostrum in her first milk — either because the birth was premature or because most of the colostrum leaked out before foaling — or does not come to her milk until several hours after the birth. First milk should be thick, sticky, and more yellow-orange than regular milk. If it's thin and white, it's low in colostrum and may not give the foal much protection. Some mares have little or no milk—a common problem in mares grazing fungus-infested fescue pastures.

Usually, a mare with no milk comes to her milk if the foal nurses often, stimulating hormones that trigger milk production. Meanwhile, you must find an alternative source of colostrum (see Emergency Colostrum on page 430). If you feed him an alternative source, make sure he gets an adequate first nursing or two of colostrum, then keep him a bit hungry so he'll work on the mare's udder between feedings to stimulate the milk to come. Colostrum can be tested for antibody content; if you suspect a problem with the milk, or if the mare doesn't seem to have enough, consult your veterinarian.

Getting Colostrum into a Foal Who Hasn't Nursed

If he can nurse a bottle, that's the easiest way to get him fed if he can't nurse his mother. If he's weak and unable to suck, you must feed him by syringe or stomach tube. With weak foals, it is often beneficial to syringe milk into the mouth rather than use a stomach tube, just to keep the swallow reflex going. A stomach tube is generally used as a last resort on a foal who can't swallow, although some veterinarians prefer to use a tube for any foal who has trouble sucking.

If you must feed the foal his first colostrum, he needs about 6 to 24 ounces (0.2–0.7 L) each hour for the first 3 or 4 hours, depending on his size. A small foal (less than 80 pounds [36.5 kg]) will need just less than a cup (0.2 L) per hour, whereas a large foal (more than 100 pounds [45.5 kg]) needs as much as a pint and a half (0.7 L). It's best to divide that up and feed him every 30 minutes. Your veterinarian can advise you on the proper amount to feed.

Feeding by Tube

The stomach tube is a plastic tube approximately 3 feet (1 m) long that goes into the nostril to the back of the throat, then down the esophagus to the stomach. The veterinarian usually puts it in and leaves it, stitching it in place with a few strands of suture thread. Then, you can milk out the mare periodically (with clean hands, into a sterilized container, or with the

Avoid Stress

High stress at birth interferes with a foal's ability to absorb antibodies. A difficult birth or cold weather may reduce his ability to absorb antibodies. Stress seems to hasten closure of the intestinal pores.

Udderly EZ milker) and carefully funnel the milk down the tube at specific intervals and specified amounts as directed by your veterinarian.

This has advantages over bottle feeding if a foal is weak; you get the proper amount of milk into him without any effort on his part that might tire him, and without danger of getting milk into his windpipe, which would cause aspiration pneumonia. Also, the foal can still nurse the mare with the tube in place, and if you are not bottle feeding him, he won't be so apt to think of you as his mother; he may still try to nurse the mare. Once he starts nursing (leave him a little hungry now and then so he will try), the tube can be removed.

Emergency Colostrum

If the mare has no milk or poor colostrum, it may be possible to obtain colostrum from another mare or a breeder who has a supply for emergencies. Most large breeding farms keep frozen colostrum on hand, and if you have many mares, you should keep some too. Store it in small plastic containers or in snap-top freezer bags in 4- to 8-ounce (0.1–0.2 L) quantities — easy to thaw quickly when you need them — or in baby bottles with plastic liners. Milk the mare into a clean, nonglass container, pour the colostrum into the plastic liner or other container, seal it, label it with date and name of the mare, and put it in the freezer.

Taking Colostrum from a Mare

Colostrum is produced only during the first 24 hours; thereafter, the concentration of antibodies falls sharply. If a mare is a heavy milker and her foal has several good nursings, you can safely take extra colostrum from her during that first 24 hours to freeze and save. Before milking the mare, gently wash her udder with plain, warm water, or brush it clean with a dry, soft cloth to make sure you don't get any dirt in the milk you are collecting. Don't use anything but plain water for washing because the foal finds the udder by smell and you don't want to confuse him.

Using the Udderly EZ Mare Milker

In years past, the only way to obtain colostrum or milk from a mare was milking her by hand (a tedious process using thumb and forefinger on the short teats) or trying to use a human breast pump, or the vacuum suction of a large syringe with the end cut off to place over the teat. None of these methods are very efficient and all require both hands — and put you at risk for being kicked.

Now, there is an easier, safer way, using the trigger-operated Udderly EZ pump. First, make sure the udder and teat are clean and the teat is not sealed off; give it a squirt with your fingers before seating the soft flange over the teat. Once the flanged tube is well placed, give several trigger pulls to create a vacuum in the tube, and the milk will flow into the collection bottle. It takes only a few seconds to fill the bottle.

Because you merely reach under the mare with one hand to operate the pump, and don't have to bend down under the mare to milk into a container (a two-handed job), this is safer with less risk of being kicked. Also, it is easier on the mare and she is less apt to protest because the soft flange creates no friction on her tender skin. Milking by hand takes longer and can become painful to the mare (making her teats sore), and many mares don't have the patience for the time it takes. When using the Udderly EZ pump, follow directions for use and for washing.

Freezing and Thawing Colostrum

If colostrum is frozen quickly and kept frozen at low temperatures (approximately −5°F [−20.6°C]), it will keep well for about 2 years without losing the value of its antibodies. If it is older than that when you use it, check its potency by thawing a small amount and having it tested.

KEEPING A FOAL WARM

If it's cold when the mare foals, you may have to dry the newborn with towels and put a heat lamp in a corner of the stall where it can't be accidentally bumped. Another way to keep the foal warm is to use a zippered sweatshirt as an overcoat, putting his front legs in the sleeves and zipping the shirt under his belly. If the sweatshirt has a hood, use it to cover the top of the foal's neck. When most of the foal's body is covered, he won't get too chilled.

Dressing a foal in a zippered sweatshirt will keep him warm in cold weather.

Thaw colostrum in warm water, not a microwave; even a setting on low heat will destroy part of the antibodies if it gets much above body temperature (100°F [37.8°C]). Thaw it in water in which you can comfortably place your hand. Colostrum should never be thawed and refrozen.

Bowel Movements

The foal's first bowel movements are dark and sometimes hard. While in the uterus, he continually swallows some of the fluid in which he floats, but he never evacuates his intestinal tract until after he is born. Hard pellets accumulate in the tract, and the foal may have trouble getting rid of them. After he has nursed a few times, the colostrum stimulates his gut to move and his bowel movements become soft and yellow, but if he remains constipated, he will need an enema.

Constipation is a common cause of death in newborns. If a foal does not have a bowel movement during the first few hours of life, you should gently remove the pellets from his rectum with a well-lubricated (use vasoline or K-Y jelly) finger. To soften up pellets beyond finger's reach, give him an enema.

Giving an Enema

It is simplest to use an inexpensive preparation such as a human adult Fleet enema from a drugstore. If you don't have one of those, you can use a cup (0.2 L) of mineral oil mixed with a quart (1 L) of warm water or some slightly soapy water. A few drops of a mild dishwashing detergent can be added to a quart (1 L) of warm water. Gently put the water and oil, or soapy water, into the foal's rectum with a regular enema tube (inserted carefully into the rectum a few inches) and squeeze bulb, or use the nozzle of a large syringe to squirt it in. Have someone hold the foal for you with

Giving a foal an enema (mineral oil and warm water) with a large syringe can relieve constipation.

STALL MUST BE CLEAN

If the mare and foal spend any time in a stall, it must be very clean and safe. Remember that ammonia from urine and manure is highly damaging to young lungs. A stall might look clean and seem pleasant but still be noxious at floor level; the best way to check is to actually lie down in the straw to see whether you can smell ammonia. This is one of the leading causes of foal pneumonia in housed foals. If you can smell ammonia near the floor, where the foal spends most of his time when napping, you must clean the stall more thoroughly. The foal must have a clean stall and dust-free bedding.

---

one arm around his chest and the other around his buttocks to keep him still while you give the enema.

Stubborn Cases of Constipation

Some foals may need a small quantity (half cup to a cup [0.1–0.2 L]) of mineral oil orally, or magnesium hydroxide (Milk of Magnesia), to soften up a high impaction out of reach of an enema. If the foal continues to have trouble passing bowel movements or shows discomfort (restlessness, straining, rolling and colic, or nosing at his flank) even after several enemas, consult your vet. If constipation is severe, the vet may give the foal a sedative or mild tranquilizer to keep him from injuring himself until the laxative works.

Problems of a Newborn Foal

The first few days and weeks of life are a critical time for the new foal. Closely observe the pair so you can quickly detect and deal with a problem before it becomes serious or life threatening. Some problems are obvious right after birth — external congenital defects, limb deformities, and prematurity for example — and require professional care. If anything seems abnormal about the foal, don't hesitate to have your veterinarian examine him.

Birth Defects

Some foals are born with cleft palate, crooked head (from lying against the mare in an abnormal position), or crooked legs (from being cramped for space in the womb). Others have incomplete digestive systems; the intestine is closed off before it gets to the rectum or anus and the foal cannot pass

If the foal is born in a stall, both he and his dam should be turned out into a paddock or pasture as soon as he has nursed a few times, the pair has bonded well, the mare has cleaned, and everything is normal. The foal needs exercise; moving around will help him become stronger and more coordinated, and it will stimulate his appetite and digestive tract.

Turn the pair out in a safe enclosure with no other horses who might pester the mare or try to steal the foal. It's not unusual for mares to swap foals if several are in the same pasture. Keep each pair by themselves for the first few days until bonding is fully established. The paddock should be away from the influence of other horses because an overly protective mare may charge at an interested onlooker peering over the fence, possibly injuring her foal in the process. If the ground is wet or the weather is cold, leave them out only a few hours the first time, bringing them back in the barn when the foal gets tired and wants to lie down. If the weather is good, the pair are better off left outside.

bowel movements. Still other foals are born with a leaky navel (pervious urachus). In this condition, the canal in the fetus that connects the bladder with part of the umbilical cord and placenta does not seal off at birth, causing urine to leak from the bladder and out through the navel. Some foals may be weak, with severe heart defects. Among all these problems, some can be corrected or repaired with surgery whereas others can't. If a foal is born with turned-in eyelids, for example, a veterinarian can correct this condition.

Some abnormalities correct themselves with time, such as crooked legs or weak tendons. Occasionally, a male foal has trouble urinating because his sheath and penis have not come down yet; he crouches in urinating position for a long time and strains and dribbles. The first time you see this condition, you may think the foal is abnormal, but after a day or two, the situation usually resolves and the foal is perfectly normal.

Ruptured Bladder

Sometimes this condition is present from birth, either as a congenital defect, a result of trauma during birth (such as when a big foal is squeezed through the birth canal with a full bladder), or a weakness in the bladder wall that gives way later. The most common cause is an accident during birth, but a sudden jerk on the navel cord when the mare foals standing up or jumps to her feet immediately after foaling can also rupture the bladder.

A tiny hole or tear in the bladder will not keep the foal from passing some urine normally, but small amounts continue to leak into the abdominal cavity. A large tear will make him unable to pass urine because it goes out the leak faster than it can collect in the bladder — although he may often squat and strain from the uncomfortable full feeling. A large leak will make his condition obvious before he is a day old, but a small, slow leak may take 2 or 3 days to accumulate enough pressure for symptoms to appear.

DELICATE NERVOUS SYSTEM

Keep close watch on the foal and be aware of subtle signs of trouble. Time is critical when a foal is young because small problems can become serious in a hurry. For example, the foal's nervous system is immature and easily damaged by infection, oxygen deprivation, stress, or trauma. In mammals, the brain is protected from most insults (including drugs) by a system called the blood–brain barrier that keeps ordinary disease organisms and poisons in the bloodstream from damaging the brain. In the young foal (up to about a month of age), this barrier is not completely functional, so disease can have a devastating effect. A foal who has diarrhea or some other problem, or one who suffered stress and trauma during a hard birth and passed fecal material during birth, could experience neurologic consequences and collapse in a seizure because his brain has been adversely affected.

Symptoms

The foal may seem normal at first, but before the day is over, he becomes dull and uncomfortable, losing his desire to nurse. Periodically, he may raise his tail and try unsuccessfully to pass urine. His abdomen may become distended and tender as fluid seeps out of the ruptured bladder and into the abdominal and pelvic cavity. As the waste products of urine are absorbed into his body instead of being eliminated, the foal develops a toxic condition called "uremia." The fluid build-up puts pressure on his diaphragm and lungs; he pants shallowly, unable to get enough air. If this condition is not corrected quickly by surgery, he will go into shock and die.

Treatment

Your veterinarian can check for ruptured bladder by lying the foal on his side and inserting a needle into his abdominal cavity to drain off fluid and check it. If the abdomen is full of urine, it will flow freely out the needle and smell like urine. Blood tests can confirm the presence of urine products. Prompt surgical repair results in rapid and complete recovery. If the condition is neglected too long, the foal will die from uremic poisoning or suffocation.

Neonatal Isoerythrolysis

This problem may not show up for a day or two. The foal is born normal and is lively, vigorous, and nursing the mare. Then, he becomes weak. Sometimes called Rh foals, these newborns have a blood type different from the mare's, an incompatibility that causes antibodies in her colostrum to attack the foal's red blood cells. The condition can be fatal unless it is quickly corrected.

Cause

If the foal inherits an incompatible blood type from the sire and some of the foal's blood cells pass through the placenta into the mare's bloodstream, her immune system creates antibodies against the foal's red blood cells if these components are not already part of her normal blood makeup. The antibodies are usually produced in large amounts by the eighth to tenth month of pregnancy, but usually don't affect the unborn foal because they are too large to pass through the placenta from mare to foal; therefore, no reaction occurs until the foal is born and ingests the mare's colostrum, which contains a high concentration of the antibodies.

The antibodies are absorbed into the foal's blood after he nurses, where they attack his red blood cells and destroy them, inhibiting the flow of oxygen to body tissues. This results in anemia, bloody urine (from breakdown of red blood cells), and jaundice; his tissues look yellow because of the hemoglobin pigments that are scattered throughout his blood from the destroyed cells.

Symptoms

In acute cases, the foal becomes weak before you notice the jaundice. After the foal has nursed a few times, he becomes dull and lethargic and loses interest in nursing. There is no fever, but his pulse rate rises. Respiration remains normal until the anemia takes its toll, then he has trouble breathing and starts yawning; at this point his respiratory rate goes up to 80 breaths per minute. To save the foal, you must take him away from the mare immediately or muzzle him so he cannot nurse.

Some cases are not evident until the foal is two to four days old. Mild cases may not develop symptoms until day four after birth, and these may recover without treatment. If a foal shows symptoms soon after birth, you must take prompt action to save him. Your veterinarian can test his blood and the mare's colostrum for diagnosis.

Treatment

Treatment consists of feeding the foal an alternative source of milk and milking out the mare every 2 hours to keep up her milk flow (a task made much safer and easier with the Udderly EZ mare milker). This hastens the coming of her regular milk and gets rid of the last traces of colostrum that contain the deadly antibodies. Your veterinarian can check the foal's blood to see whether he needs a transfusion, which may save his life if he is weak and anemic. Of course, the blood used must be from a horse with a compatible blood type.

Prevention

Blood from both mare and stallion can be checked before breeding. If the mare is already pregnant, have your veterinarian test her blood serum against the sire's blood cells a week or two before she is due to foal. If the result is positive, meaning the foal will be endangered by her colostrum, you must not let the foal nurse his mother when he is born (until his own blood has been tested; it only takes a few minutes at the clinic). If the test shows the foal is at risk, do not let him nurse the mare until he is fully three days old. Usually after the second day of life, the foal's intestinal membranes can no longer absorb the antibodies and he is out of danger.

Be present when the foal is born, making sure he doesn't nurse. Have colostrum from another source available if you think there might be a problem. Muzzle the foal and test the milk; if it's harmful, feed a formula for the first 3 days.

Once a mare has produced an Rh foal, you need to prevent it from happening again. Breed her to a stallion with a compatible blood type.

Umbilical Hernia

Some foals are born with umbilical hernia, a soft tissue swelling or protrusion at the navel. While the fetus is growing in the uterus, the navel is a natural opening through which umbilical blood vessels enter. After he's born, these vessels begin a natural degeneration process; within the first month of life, the skin has healed over. In some foals, however, the underlying tissues of the belly wall do not close, leaving a hole.

Symptoms

Some hernias are obvious soon after birth, but others don't appear until a few weeks later. Trauma, infection, or even straining to have a bowel movement can cause more tissue to protrude through the opening, creating a visible bulge under the skin at the navel. As the foal grows, the increasing weight and fluid in the abdomen will cause the protrusion to enlarge if the hole in the abdomen wall is large.

INJECTIONS

In years past, most horse people gave the mare and foal injections at birth — vitamins, antibiotics, or iron shots. Usually, these are not necessary. In fact, they may even be harmful. For example, it used to be standard practice to give mares and foals tetanus antitoxin, but this sometimes leads to serum hepatitis. It's wiser to keep the mare's tetanus toxoid vaccinations up to date, with a booster a month before she foals. If she's on a well-balanced and adequate diet, her foal won't need supplemental vitamins. He won't need antibiotics either if you kept the foaling area clean and disinfected his navel. The only time the mare and foal need medication is when there is a specific problem, in which case your veterinarian will advise you.

Treatment

Large hernias require surgical repair, but a small one may disappear on its own as the foal grows. If you have any questions about the young foal's navel, ask your veterinarian to check it. Although umbilical hernias often heal on their own, if left untreated, a large one can have fatal complications if a loop of intestine comes through the hole and becomes pinched or strangulated.

Infections

Probably the most serious problems of young foals are infections. Not only is the new foal acutely susceptible to certain pathogens because of his immature immune system, but the viruses or bacteria often gain the upper hand swiftly, causing damage or death in just a few hours. Most infectious organisms that might harm the foal are common in the environment, especially on breeding farms or wherever horses have congregated. In large part, these pathogenic invaders are bacteria that enter the foal through contaminated material that is ingested or chewed on (soiled bedding, dirty stall walls), or through the respiratory system or navel stump if it's not quickly sealed off.

Bacterial infections in the foal can usually be treated with antibiotics early in the course of illness, but viral infections do not respond to antibiotics. The best treatment in these instances is good supportive care — giving extra fluids to prevent or combat dehydration and electrolyte imbalances, keeping the foal warm and dry, and reducing stress as much as possible — so his body is better able to fight the infection.

Navel Ill (Joint Ill)

Sometimes a foal is born with an infection already in his bloodstream — transmitted from a diseased uterus — and it settles in his joints. Or he may develop a similar infection soon after birth if bacteria enter the navel before it dries and seals off, or if he develops septicemia from ingested pathogens that are absorbed through the gut. Acute septicemia causes high fever and makes the foal weak, dull, and too listless to nurse. His condition will deteriorate rapidly and become fatal unless it's treated immediately.

In some cases, the infection localizes in the body, such as in the joints or organs, without as much general illness. There may be mild fever, lameness, and local swelling if the infection settles in the joints, in which case the disease runs a much longer course. It may or may not develop into fatal illness.

FOAL HEALTH TIP

When a foal is ill, it is often helpful to treat the symptoms, such as reducing fever and discomfort. If a foal has high fever, drugs such as flunixin meglumine (Banamine) or phenylbutazone (Butazolidin) can bring it down. Although these types of drugs can cause ulcers in foals, ulcers are the lesser of two evils when a foal is suffering from life-threatening high fever that can cause irreversible damage to his nervous and other systems. Once the fever and pain are reduced, the foal may feel well enough to start nursing again, taking in urgently needed nutrients and fluids.

If a foal has to be confined to a stall for treatment, your veterinarian may prescribe antiulcer medication such as ranitidine (Zantac), cimetidine (Tagamet), or omeprazole (Ulcergard, Gastrogard). Years ago, many foals were successfully treated for infectious disease or other serious conditions only to die later from gastric ulcer brought on by the stress of stall confinement and aggravated by medications for the primary illness.

If acute septicemia is halted in time to prevent death, infection may still have invaded the joints. Joint ill can be the aftermath of any serious infection that invades the circulatory system.

Joint Infection. Leg joints in the young foal are susceptible to irritation and subsequent infection because of the stress of weight bearing. If the condition is untreated, the joint surfaces become damaged, leaving the foal with enlarged and painful arthritic joints. The first sign of trouble is tenderness or lameness. The foal may have a fever and his joints may be hot, with swollen soft tissues. Streptococcus infection usually initiates a rapid destruction of the joint cartilage and bone, so it is essential to make an early diagnosis and begin treatment. The foal will need massive doses of antibiotics; the veterinarian may also flush the affected joint to remove the harmful enzymes.

Preventing Joint Ill. A high incidence of joint ill occurs in foals who suffer abnormalities at birth and in foals who don't nurse soon enough to get antibodies from colostrum. Attending all births and making sure all goes smoothly are important. So are making sure the foaling area is scrupulously clean and disinfecting the foal's navel stump immediately after the cord breaks. Uterine infections in the mare should be treated and cleared up before breeding to make sure no infection is transferred from mare to unborn foal. If subsequent uterine infection is suspected before foaling, your veterinarian may prescribe antibiotics for the mare.

Using IVs and Stomach Tubes

If a foal has a severe infection, the veterinarian may put him on a round-the-clock schedule of antibiotics administered by IV. For ease of treatment and to prevent the repeated damage and pain from putting needles into a vein, it's common practice to insert a catheter into the foal's jugular vein and secure it with a small suture so the medication can be given every few hours as needed, along with IV fluids. Dehydration is a serious problem in sick foals because once they stop nursing, the fever and diarrhea dehydrate them.

The veterinarian may choose to give fluids and medication through a stomach tube if the foal can absorb fluids through the gut, if the infection is in the digestive tract and will respond to oral medications, or if the diarrhea can be eased with magnesium hydroxide, mineral oil, or other preparations.

Crooked Legs and Weak Tendons

Sometimes, a foal is born with crooked legs or weak tendons that allow the limbs to sag or bend at abnormal angles. In some cases, the crookedness may result from nutritional imbalance or inadequate diet in the pregnant mare; in other cases, it may be caused from the way the foal was lying in the uterus or to uterine insufficiency. The leg deformity or lax tendons are not usually genetic defects that can be passed on to future offspring.

Treatment

Mild cases generally straighten on their own without treatment, leg braces, or corrective foot-trimming. If the foal can get up and around, nurse his mother, and travel, the weight bearing and exercise will strengthen the legs and they will gradually straighten. It may take a few days or weeks. If the foal has trouble getting around or is causing himself injury, consult your veterinarian or farrier for advice in bracing or splinting the leg until it becomes stronger.

Serious cases may need surgical correction. For example, periosteal stripping (removing a strip from the bone lining on one side of the cannon bone to stimulate one side to grow faster than the other) or epithelial stapling (putting metal staples on one side of the growth plate at the end of the long bone to hinder growth on that side) at the proper phase of early bone growth will often cause the leg to straighten to normal position.

This newborn foal has weak hind pasterns; notice how the fetlock joints appear to have dropped to the ground.

Premature Foal

If your mare gives birth to a premature foal or set of twins, the chances of survival depend on how fully developed the baby or twins are and what kind of care can be given. You'll need a good place for the mare and her fragile offspring — clean, warm, and dry, and well away from other horses. These tiny babies must be in isolation for a while until they gain strength and resistance to disease, and you must be careful not to introduce any disease germs yourself. Consult your veterinarian immediately after the birth for advice on trying to save and care for these newborns.

A premature foal or underdeveloped twin may need to be in an intensive care unit at the clinic. If he is strong enough, he may be able to stay in a roomy box stall with the mare. When weather is cold, you'll need a heat lamp in a dry corner where the foal can snuggle in but the mare can't. A pole across a corner, chest high to the mare, will keep her out, yet allow her to still see and smell the foal. If she has twins and they both live, she may or may not have enough milk for both. You might have to give supplementary bottles until they are big enough to eat more solid food.

Keep the stall absolutely clean, changing the bedding often. You can often save a premature foal or twins with diligent good care.

Orphan Foal

At some time, you may be confronted with an orphan. This happens when the mare dies at birth or shortly thereafter, if a mare won't mother her foal, or if she does not produce milk. Unless you can locate a nursemare, you'll need to feed a formula or milk replacer. The problem is more complicated if the foal is newborn and has had no colostrum; he will need at least a pint (0.5 L) of colostrum, preferably within 4 hours of birth.

Colostrum

If the dam died from foaling complications rather than disease, her colostrum can be milked out immediately after death to feed the foal. If the dam is fine but refuses to mother the foal, she can be restrained, twitched (or held still with a Stableizer), and milked of enough colostrum through that first day to feed the foal. When there is no way to obtain colostrum from the dam, you must find another source. You may use frozen colostrum, colostrum from another mare who has just foaled, or, as a last resort, give a commercial serum containing blood plasma and antibodies intravenously.

Milk Replacer

After the foal has had several feedings of colostrum, you can switch to a foal milk replacer such as Foal-Lac, Foal-Mate, Mares Match, or Mare's Milk Plus and mix it according to directions. All are composed of cow's milk, with additional ingredients. Prices vary, reflecting the quality and digestibility of the added ingredients. Match your foal's nutritional needs with a product's ingredients, asking your veterinarian which one would be best. Milk replacers containing a nonlactose type of carbohydrate such as maltose, a sugar found in grain, may not be tolerated well by sick or young foals. Foals under 2 weeks of age lack the enzymes to digest some types of carbohydrates. A foal who is premature or stressed by infection might do better on a high-fat milk replacer than on a high-carbohydrate product; he cannot digest the carbohydrates but can still get energy from the fats.

A milk replacer should not contain fats derived wholly from vegetable sources because they irritate the gut, causing diarrhea. Milk fats and animal fats are more easily digested. Milk replacer also contains vitamins and minerals, and some have probiotics (gut microbes) to aid digestion. Sick foals may do well with probiotics, but most foals don't need them.

Don't use calf milk replacer; most don't meet the foal's nutritional needs. If you don't have access to foal milk replacer, mix a home recipe.

Home Recipes

Don't use straight cow's milk. Mare's milk is lower in fat and protein, but higher in carbohydrates, than cow's milk. In making an emergency formula using cow's milk, you can modify it by lowering the fat and protein content and increasing the sugar. There are a number of workable formulas to try:

- 1 can (12 ounces [0.4 L]) evaporated cow's milk mixed with equal part water; 4 tablespoons (60 mL) limewater or tap water; 1 tablespoon (15 mL) sugar or corn syrup
- 1 pint (0.5 L) low-fat cow's milk (2 percent); 4 ounces (0.1 L) limewater or tap water; 1.5 teaspoons (7.5 mL) sugar (white sugar is harder to digest than corn syrup or lactose — it's best to use something other than white sugar)
- 1 pint (0.5 L) low-fat cow's milk; 2.5 ounces (74 mL) water; 2.5 ounces (74 mL) limewater or tap water; 2 teaspoons (10 mL) lactose
- For several feedings you can mix up 3 pints (1.5 L) low-fat cow's milk (2 percent); 1 pint (0.5 L) water; ¼ package Sure-Jell pectin or 1 ounces (20 mL) of 50 -percent medical glucose

Limewater (a solution of calcium hydroxide in water) has traditionally been used to provide extra calcium but isn't always available in an emergency. Most drugstores no longer sell it. Instead, you can use tap water unless it is soft (low in minerals) or high in fluoride. In these cases, use distilled water.

Two-percent cow's milk is the best basis for a homemade foal formula because it has the proper calcium-phosphorus ratio. For increasing the carbohydrates, add 0.5 cup (0.1 L) honey or corn syrup or ¼ package of Sure-Jell pectin to a gallon (3.8 L) of milk. Glucose (simple sugar), which is easy to digest, is an ideal source of extra carbohydrates if you can get it. Pectin (the thickening agent for homemade jelly or jam) is also a good carbohydrate that has a protective effect on the foal's stomach and intestines to prevent ulcers or diarrhea. If you use pectin as the carbohydrate source, add a few tablespoons of honey for palatability.

Amounts to Feed

A young foal will drink 2 gallons (8 L) in a 24-hour period. He should be fed every half hour for the first few hours, then once an hour for the first 5 days of life (about a half pint [0.2 L] at each feeding, then up to a pint [0.5 L]) and preferably for the first 10 days. Young foals nurse small amounts often. It helps if someone can share feeding duty so you can get some sleep. When the foal is

BOTTLE OR BUCKET?

If you are feeding with a bottle, get one with a nipple that doesn't run too fast or the foal may choke or get digestive upsets. A lamb nipple is a better size for the foal than a calf nipple. Be sure the bottle is clean and to warm the formula or milk replacer to body temperature (100°F [37.8°C]) before you feed him. A foal can also learn to drink from an easy-to-clean bucket. Start him on the bucket within the first few days or he may not want to switch from the bottle. To teach him to drink, get him sucking on your finger and submerge the finger in a pan of milk. If you get him drinking from a bucket, you can put in enough milk for 4 to 6 hours at a time.

10 days old, you can go to 2-hour intervals at night but continue hourly feedings during the day. As you increase the length of time between feedings, you'll need to increase the amount given at each feeding. By the time the foal is a month old, you can start cutting down to feeding every 4 to 5 hours once he is also eating solid food. If he is doing well, you can wean him by 6 to 8 weeks of age.

Signs of Illness

Because orphan foals easily become dehydrated, make sure the youngster is getting enough fluid, especially when you start him on dry feed and cut back on the number of fluid feedings. Also make sure fresh water is always available. Take his temperature daily. Remember that fever can be an early sign of sickness. Any abnormality, including reluctance to eat, weakness, fever, diarrhea or constipation, or distressed breathing, should be a signal to call your veterinarian. Many problems quickly prove fatal to a young foal. Modern drugs, transfusions, and IV feedings can often save him, but don't let sickness get a head start. Treatment should begin as soon as you suspect the foal is sick.

Diarrhea

If the foal gets diarrhea, temporarily substitute electrolyte formula for milk. A simple mix is 2 quarts (2 L) water, 1 package Sure-Jell pectin, 1 tablespoon (15 mL) low sodium salt (sodium chloride and potassium chloride), and 1 tablespoon (15 mL) flavored gelatin for palatability. Whereas

Goat's Milk

Goat's milk can be a substitute for mare's milk and is usually digested well by premature or sick foals.

milk irritates the gut, this formula gives it a chance to heal while providing fluid and nutrients, and replaces the electrolytes the foal is losing through diarrhea. If the problem is from an infection, he'll need an oral antibiotic; consult your veterinarian.

Growth and Weight Gain

Foals should gain 1 to 3 pounds (0.5–1.5 kg) per day. If your orphan is not growing well or has a rough coat, potbelly, or loose bowel movements, he needs a better milk replacer. If he feels and looks good but is always hungry, just feed him more of what you're using. Get him eating alfalfa hay, green grass, or foal pellets as soon as he shows interest. Milk replacers are not designed to be a foal's only source of nutrition during his first months of life.

Nursemare

An orphan can be raised on a substitute mare. The easiest time to persuade a mare to adopt an orphan is very soon after she herself has given birth. If her own foal dies at birth or soon thereafter, she is more inclined to accept a substitute immediately after her foal dies than she would be a few days or weeks later. It's harder to fool her after she's had her own baby a while. Also, the hormones in her body at foaling time heighten her maternal instincts and help in the bonding process; they make her more receptive to mothering. A mare who has lost her own foal generally makes the best nursemare. If you can't locate a nursemare, you can sometimes convince a mare to raise an orphan along with her own foal. This takes more work and careful monitoring.

Introduce the orphan as soon as possible after the mare loses her own foal, rubbing the mare's nose with Vicks VapoRub (to disguise other smells so she can't tell the orphan is not hers), and thoroughly rub the mare's fresh afterbirth over the orphan — especially rump and tail area, where the mare will be nuzzling and smelling as the youngster nurses. It also helps to put some Vicks on the foal, especially on his hind end.

Use Hobbles

Hobble the mare's hind legs so she can't kick. The hobbles should have sheepskin lining or some other soft material that won't chafe her pasterns in case you have to leave them on for several days. Keep her in a stall. She can still move around with hobbles but can't hurt the foal. For the first nursing, tie her to the side of the stall, with the rope not too short, allowing some freedom of head and neck. Alternatively, have someone hold her so she can still reach around and smell the foal but can't hurt him or knock him down. Introduce her to the foal and let her smell him.

If the orphan is critically in need of milk, accomplish the first nursing quickly with the mare tied, hobbled, and twitched (or restrained with a Stableizer); then allow bonding to take place afterward. After the foal has nursed a few times, the mare may mellow in her attitude. Many mares are more inclined to accept a foal after it has nursed. To make sure the mare doesn't hurt the orphan, tie her in the box stall with her hind feet hobbled and watch closely until you are sure bonding is complete and she accepts the foal. This may take 1 to 3 days, sometimes longer.

Raising Two

If a nursemare can't be located and you have a gentle mare who gives a lot of milk, you may be able to persuade her to raise the orphan along with her own. Complete success is most likely when you bring the extra foal to her immediately after she gives birth to her own. If you can be there at the birth and present her with both foals at once (using Vicks VapoRub on her nose, and rubbing birth fluids and membranes on the orphan), she may think both are hers. To make sure the adoption works, hobble and tie

Sometimes you can raise an orphan on a mare who already has a foal, and she will feed both babies.

the mare and monitor the nursings. Sometimes, a mare adopts both from the beginning, but usually, you must supervise the process for a while.

Keep Foals Penned Separate from the Mare

If the mare prefers her own foal and rejects the orphan, keep both foals in a pen by themselves between nursings, but in sight of the mare. Adjacent pens (with a safe fence such as diamond mesh wire, or boards or poles with very little space between them so the foals won't be injured while trying to stick their heads through) work best. The bonding process takes longer with two newborns, but after a time, most mares will accept both foals. Having the foals separate from the mare keeps her from bonding with only her own.

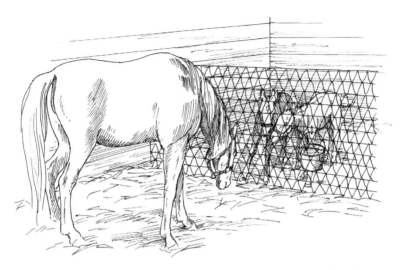

Keep both foals penned separate from the mare at first, except at nursing time.

Apply Vicks to her nose several times a day at first. Also put Vicks on the foals — especially their hindquarters — to make them smell the same, confusing the mare as to which is hers. At nursing time, tie the mare before bringing in the foals. Have them nurse together, one on each side. When they are finished, remove them to their pen. Don't leave them with the mare lest she mother her own and reject the other.

After the mare accepts the nursing procedure (it may take several days) she can be left free in the stall during nursing, with just the hobbles on so she won't try to kick the extra foal. You can remove the hobbles later, after she accepts her role as mother of "twins." She may never love the orphan as her own, but she will most likely come to tolerate him.

Feeding the "Sibling" Foals

The foals should have access to feed (such as Foal-Lac pellets) and water in their own pen. Some foals need encouragement to learn to eat pellets; you may have to put the feed into their mouths at first. They need good bedding and a heat lamp if weather is cold. Be sure the heat lamp is well away from anything flammable and out of reach of the foals.

Feeding the Nursemare

Because of the increased demand on her milk supply, the mare will produce more milk; therefore, you must increase her feed and make sure she always

For the first few days, the adoptive mare should wear hobbles when nursing so she can't kick the new foal.

has plenty of water so she can keep up with the demand. Most well-fed mares can easily raise two foals if the foals are given supplemental feed. Keep close watch on the foals to make sure they are doing well. After the mare is well bonded with the foals, they can be allowed to remain with her and no longer need supervision at nursing; however, you should continue to keep the mare and her babies separate from other horses.

The extra time it takes to graft an orphan onto a nursemare will pay off greatly. Not only will the mare feed the orphan, but the youngster growing up with a "mother" will do better physically and mentally. He'll be better adjusted than the hand-raised orphan who lacks contact with other horses. The mare can teach a foal a lot more about being a normal horse than you can. If you do have to raise an orphan yourself, let him live with a kindly old gelding or older mare, after he is about 1 month old, to teach him manners and "horse sense" while you provide his dinner by bottle or bucket.

SAFETY TIP

Mares often get upset when you are handling their foals. Always restrain the mare before putting the foals with her. She may also fuss when you take them back to their pen after nursing. Work quietly and firmly so as not to upset her and the foals.

24

Care of the Young Horse

RAISING A HORSE TO ADULTHOOD is a challenge and a pleasure. There's considerable satisfaction in watching the youngster grow and develop. It also takes good care and conscientious handling. The young horse is vulnerable to problems that are not a worry with adult horses, so you must be prepared to prevent or minimize them.

Because the young foal is most vulnerable to diseases and other problems, he needs more care and monitoring than the rest of your horses. If he obtained a good supply of colostrum from his mother, he'll be temporarily protected from the many diseases that would otherwise be quite life threatening during his first weeks of life. As temporary immunity wears off, however, he becomes susceptible to disease until his own immune system responds to pathogens in his environment. He will encounter all kinds of dangers — from parasites to respiratory illnesses — but with good care, you can reduce risk and stress and keep him strong and healthy as he makes the transition from baby to yearling.

Handling the Foal

The foal should be handled from the day he's born, not only to instill a trust of people but also to promote ease of management when it's time to give vaccinations, medications, or other health-care procedures. When he is very young, it's better to restrain him with your arms than to use a halter. He may resist restraint of his head by halter or rope, rear up, and flip over backward. Until he's more accustomed to people and to restraint, there is a risk of injury when using a halter and lead rope.

It's best to handle the foal with one arm around his front and one around the hindquarters; this is much safer and a lot less traumatic than a halter. If the foal must be treated, given an enema, or moved from stall to pasture, it's

If you must handle the foal, restrain him with one arm around his chest and the other around his hindquarters.

easiest with two people: one to hold the foal and one to treat him. If the foal is rambunctious, the most effective way to restrain him is the tail twitch—the person holding him places one arm around the front of the foal and with the other hand holds his tail straight up, over his back. This tends to make him stand still. When working by yourself, crowd the foal into a corner, use your hip to hold his shoulder and ribcage in a body block, and hold his tail straight up with one hand while you use your free hand to administer the enema, injection, or other treatment.

If you must move a mare and foal by yourself, use an adult horse halter as a foal restraint. By putting the big halter on the foal as a body harness, you can hang on to him with one hand and have the other hand free to lead the mare whenever you need her to go.

Using an Adult Halter as a Foal Harness

Using a full-size nylon web horse halter with round buckles, hold the halter upside down and slip the noseband over the foal's head like a collar, positioning it around the base of his neck above the shoulders. The halter's jaw strap will then be resting on the foal's withers, aligned with his backbone. The crown-piece (which would normally go behind the ears of the adult horse) will be at his girth. Buckle the crownpiece around the foal's belly like a cinch or girth, but not too tightly — just snug enough that he can't get a hind foot caught in it. This foal "harness" gives you a safe and sturdy handle for hanging on to or restraining the youngster.

Leading Lessons

Once the foal is several days old and familiar with handling — having been "led" a few times with your arms around his front and rear — you can progress to a rope around his chest and buttocks with a figure eight over his back. After he goes willingly with you in this type of restraint, start putting a foal halter on him; he can have his first leading lessons as you take him and the mare from stall to pasture or from pen to pasture. Be sure to keep using a rope around his buttocks during his early leading lessons. Then, if he pulls back or tries to get away, you have control over his movements and he can't pull back hard on the halter or be as likely to pull away or flip over backward. A few leading lessons, going along with the mare, will be invaluable later, even if he doesn't get much formal training until he's a weanling or older. Early handling will make your later work with him much easier.

An adult horse halter can become a foal harness that allows you to hang on to the foal with one hand.

In the foal's early leading lessons, use a rope around his buttocks so he can't hang back on the halter.

NURSING RATE

Foals are aggressive eaters, nursing more often than calves do. A young calf nurses about 4 times in a 24-hour period. A newborn foal nurses about 6 or 7 times per hour, during the first few days, and gradually increases the intervals to about every 30 minutes; when he's a few weeks old, about once an hour; then later, about 18 times per 24-hour period.

Feeding the Foal

If mare and foal are at pasture while he grows up, he will get all the nutrition he needs if the pasture is good and the mare milks well. If you live in a climate or region where pastures are dry or of poor quality, or if the mare and foal are kept in a paddock, they will need good hay and perhaps some grain. By the time the foal is a week old he'll be nibbling the mare's hay; make sure it's of good quality.

Mare's Milk

A good milking mare will provide her foal with proper and adequate nutrition without the foal needing any supplemental feed other than good pasture or good hay — as long as the mare is well fed on a well-balanced diet. Brood-mares usually produce about 3 percent of their body weight daily in milk during early lactation. A 1,200 pound (545 kg) mare gives about 36 pounds (16.5 kg) of milk daily (more than 4 gallons [15 L]). This is more milk than the average beef cow produces. Clearly, the lactating mare needs adequate food to meet her body's needs for milk production.

Milk production in the mare increases during the first 2 months and then begins to decline. The energy and protein content of the milk also decline. The foal's greatest need for energy and protein from milk is during those first 2 months; then, as he starts eating more solid foods, his nutritional needs are partly met by grass or hay and grain. By the time he is weaned, his need for mother's milk has passed, as long as he has an adequate and well-balanced diet.

Grain

By 3 weeks of age, the foal can be started on grain. If you're graining the mare, the foal will have already sampled hers and learned how to eat it. After all, foals learn best by mimicking their mothers; the easiest way to teach him to eat grain is to let him try some of his dam's. Most people feed mares and foals grain, but not all of them need it. You should never underfeed a lactating mare and growing foal, but don't overfeed them either. Some rapidly growing youngsters develop serious leg deformities and soundness problems if they are overfed on grain and rich feeds; this problem is addressed later in this chapter.

When feeding grain, make sure the mare doesn't eat the foal's. You can make a foal creep — putting his grain tub or manger in a corner of the pasture, with a pole he can go under but the mare can't (4 to 5 feet [1–1.5 m] high, depending on the size of your horses) — or make an enclosure the foal can go into. Tie the mare and foal while they eat in their separate end of a manger in a stall.

Whether you use grain or foal pellets, feed only the amount the foal can clean up within 20 minutes; do not overfeed. Don't let any uneaten grain accumulate in corners of a manger or in a feed tub. Grain that becomes moist from saliva can spoil. It can also be contaminated by birds. Always clean out leftover crumbs or kernels.

Health Care and Problems of the Foal

Pay special attention to cleanliness of the foal's environment so he is free of possible infection and parasites. You should regularly deworm the mare so she won't pass internal parasites to the foal, and he himself should be dewormed starting at 1 to 2 months of age. At 3 to 4 months he should be vaccinated with tetanus toxoid and started on other immunizations, including those for eastern and western encephalomyelitis and West Nile virus. He may need other vaccinations depending on your locale and situation; discuss this with your veterinarian. Foot care should begin at 2 or 3 months of age — earlier if the foal needs special trimming. Handle his feet regularly, and teach him to shift his weight and pick up his feet. (See chapters 4 and 5 for information on picking up a horse's foot.)

Diarrhea

Diarrhea is a serious problem; the younger the foal, the more critical it can be. Indeed, the young foal does not have much body reserve and can quickly dehydrate. He also has little resistance to infection. He may lose body fluid more rapidly than it can be replaced by nursing. If he is sick, he may not feel like nursing, which compounds the problem.

Diarrhea can best be treated if it is discovered early. It's important to try to determine the cause and treat accordingly. You need to know whether it's the result of something the foal has eaten, too much milk, infection, or the mare's first heat cycle.

Symptoms

The normal healthy foal is bright and alert, active and sassy. If he doesn't feel well, he becomes listless and dull, perhaps with drooping head or ears. The healthy young foal stays close to his dam. If you find him off by himself or the mare standing over him with a full udder and a worried attitude, something's wrong. In early stages of diarrhea, the foal may not have obvious evidence on his hindquarters. He may be squirting liquid, but he lifts his tail high and squirts it out in a stream. Even a close look at his pen, stall, or pasture may not reveal the liquid feces because they sink into the ground, grass, or straw. If you fail to find any fresh, normal foal manure, be suspicious.

Gut Pain. If diarrhea is severe, the foal will also have gut pain. His intestines are cramping as they work overtime; the gut is irritated and sore. If pain is mild, he may just stand listlessly, shifting his weight from one foot to the other. If it worsens, he may paw or look around at his flank. He may tread his feet or back into a corner in his attempt to get away from the pain. If it becomes severe, he may lie on the ground paddling with his feet or rolling. A foal's gut is so delicate and sensitive that a few cramps from diarrhea may put him in such misery he won't feel like nursing. The pain may be so debilitating that he seems weak, dragging his feet and staggering. You may think he is already severely dehydrated and weak, when in fact, the true culprit is pain from the cramping in his gut.

Merely easing the pain may make him feel so much better that he'll go ahead and nurse — if the diarrhea is still in the early stages and the foal is not yet too dehydrated and ill. An injection of flunixin meglumine (Banamine) may help him feel better. If he's dehydrated, he'll need additional fluids given orally by stomach tube or IV.

EFFECTS OF DIARRHEA

After the foal has had diarrhea for a few hours, it takes a toll on him. He feels so poorly that he no longer bothers to lift his tail when passing manure, soiling tail and buttocks. The high acid content of the feces burns and irritates the skin; if he's had diarrhea for any length of time, he'll lose hair from his buttocks. Each flick of the tail takes off more hair; the tail is matted and abrasive.

Dehydration

Dehydration, if not corrected, will ultimately kill the foal because his body can't function without fluid. At first, the affected foal is merely dull and his skin loses elasticity. If you pinch or pull out the skin on his neck, it doesn't spring right back into place but goes back slowly. As dehydration progresses, the skin becomes less and less elastic, and his eyes seem sunken back into their sockets; he has lost fluid in the fatty tissues around the eyes. As he goes into shock, his legs become cold and pulse weakens. If the condition is not reversed, his kidneys fail. Because there is no longer enough circulating blood volume (too much fluid has been lost into the gut), he slips into a coma and dies.

Too Much Milk

In the very young foal, the most common cause of diarrhea is too much milk. Many mares are heavy milkers, producing 3 to 5 gallons (11.5–19 L) a day. This is too much for a young foal. It will cause scouring. To avoid this, mares should not be overfed during the first weeks of lactation — no grain for the first 2 weeks or until after the first heat cycle — unless they are not giving enough milk. If that's the case, their rations can be increased.

Foal Heat Diarrhea

The most common and least dangerous diarrhea in young foals occurs around four to fourteen days of age, coinciding with the mare's first heat. Almost all foals scour a little at this time, with diarrhea lasting 1 to 3 days. Bowel movements may become soft, loose, or squirting, but the foal keeps nursing and

---

BISMUTH SUBSALICYLATE (PEPTO BISMOL) FOR DIARRHEA

If the foal has watery feces for more than a day, Pepto Bismol (about 2 tablespoons [30 mL], the adult human dose) can be given orally by syringe. Place the syringe (without needle) into the back of his mouth and gently squirt in the Pepto Bismol, a little at a time, as he swallows it. This will soothe his gut and help slow the diarrhea, making him more comfortable. Kaolin and pectin mix (Kaopectate) is another gut soother that can also be given by mouth. The usual dose for foals is 2 to 3 ounces (59–89 mL).

doesn't seem sick. This diarrhea usually requires no treatment other than soothing the foal's buttocks with petroleum jelly or saturating his tail and rear end with mineral oil to keep the acid feces from irritating his skin and taking all the hair off. Wash (and dry) his buttocks gently before applying the soothing coating. Keeping his buttocks clean and well-greased will make him more comfortable and prevent hair loss.

You can minimize foal heat diarrhea by deworming the mare with ivermectin immediately after foaling; there is a definite correlation between Strongyloides worms and foal heat scours.

A foal with diarrhea during the mare's first heat usually needs no drastic treatment but should be watched closely to make sure the diarrhea does not worsen. If the problem persists or he becomes dull, he may need fluid and electrolytes. Make sure he bounces back after the mare goes out of heat. If he seems at all sick, he will need medical attention. The stress of foal heat diarrhea may set him up for an infection and more serious problems.

Diarrhea Caused by Gut Infection

Diarrhea that develops after the third week of life is often caused by infection (as are diarrheas that occur before foal heat — and these are most deadly because the pathogens reached the gut before the protective antibodies from colostrum did). Most cases occur in foals who did not absorb enough antibodies from colostrum. The severely infected foal may have fever or reddened gums because the infection produces toxins that circulate throughout his body. In this case, the foal needs immediate antibiotic therapy. Some types of infectious diarrhea (including salmonella, rotavirus, and some caused by protozoa) kill a foal swiftly if not treated; consult your veterinarian at once for diagnosis and treatment.

Pneumonia

Foal pneumonia can be caused by many airborne organisms and is most common in youngsters 2 to 6 months of age, although cases can occur at any time after birth. Pneumonia is usually caused by a virus or bacteria, and sometimes by fungi or parasites. Bacterial pneumonias are more likely to be fatal. A virus may invade the lungs or damage the cilia and lining of the airways, allowing bacteria to gain entrance. Any disease that lowers the foal's resistance and is accompanied by stress can open the way for pneumonia; and it is often the pneumonia, rather than the primary disease, that kills the foal.

HEALTH TIP FOR RECOVERY

Pneumonia can be caused by foreign substances — contamination and subsequent infection — in the lungs. For example, giving a foal a large volume of liquid medication is dangerous if some gets into the windpipe and lungs.

Stress is also a factor. Wet or stormy weather, overcrowding, dusty conditions, and prolonged emotional or physical stress from injury or illness deplete the foal's energy reserves and make him more vulnerable.

If a foal develops pneumonia, stress should be avoided at all costs until he recovers fully. Foals with pneumonia should not be weaned, shipped, or even handled much. Total rest is crucial.

Symptoms

Typical symptoms are increased respiration (breathing rapidly and shallowly), fever, cough, increased pulse rate, dullness, and lack of appetite. Some foals continue to nurse and are fairly bright, maintaining body weight and nursing until just before they die. Others stop nursing and become progressively duller and weaker. The foal may stand with front feet wide apart to try to breathe more easily, or he may try to breathe with his mouth open. There may be nasal discharge — thin and runny or thick and pus-filled.

Treatment Requires Veterinary Help

With prompt and adequate treatment in the early stages, most bacterial pneumonias respond quickly and completely to antibiotics.; however, infections involving more than one organism are not uncommon, and treatment aimed at one may not affect the other. Sometimes, it's hard to tell whether the foal has pneumonia or an upper respiratory infection, such as a cold. If this is the case, have your veterinarian check the foal. Don't wait to see whether he gets better or worse. The foal may need supportive treatment — additional fluids and medications to help him breathe more easily.

Permanent Damage

If he's treated early and diligently, the affected foal has a chance for recovery. If the infection is neglected, treated too late, or not treated long enough, it may overwhelm the foal. Lung tissue is destroyed, the foal becomes weaker, and his chances of recovery diminish. Foals who do recover may never regain total lung capacity.

Preventing Pneumonia

Dust often carries bacteria that cause pneumonia. This is why foal pneumonia often hits in summer if foals are confined in dusty paddocks. The dust irritates their airways and lung lining, and the bacteria aboard the dust particles invade the tissues. If your foals spend time in a dusty paddock, reduce the dust by sprinkling regularly. Avoid confinement in stalls. The dusty, sometimes damp, ammonia-laden air in the lower part of the stall (and bacteria in the stall bedding) make a very unhealthy environment. Being kept in a moist barn (saturated bedding and high humidity) can make a foal susceptible to pneumonia. A clean, grassy pasture is the best place for him.

Prevention includes good sanitation and close observation. Where shelters have adequate ventilation and mares with foals have plenty of space, the problem is seldom encountered. Unnecessary "protection" of winter foals (keeping them in barns from January to March) and confinement in dusty paddocks are the prime causes of pneumonia. A winter-born foal should be allowed outside during mild weather or sunny days. He and his mother should be confined indoors only when weather is severe or temperatures plummet. The barn should always be clean and dry, with excellent ventilation (see chapter 1). Parasite control is also important; foals subjected to heavy parasite infestation are quite susceptible to pneumonia.

Gastric Ulcers

Overfeeding a foal or having the mare on a heavy grain ration can lead to foal ulcers. Feeds containing anabolic substances (such as rice oil, which is commonly used as a body builder for foals who will be shown at halter) can cause digestive problems. So can stress. Older horses have often learned to live with stress, but foals are prone to developing ulcers — raw spots in the digestive tract — from too much stress. If the problem persists, the ulcer may penetrate the gut lining and rupture the wall, causing peritonitis and death.

Stress

Temporary stress, such as fright during a thunderstorm, is not a problem; but prolonged, unrelieved tension and anxiety caused by an environmental condition or a medical problem can produce ulcers. For example, long-term confinement in a box stall increases the risk of ulcers. When a foal is confined in a stall, all his youthful energy is bottled up with no outlet and he becomes irritable, unhappy, and bored—which leads to tension and stress. Overuse of certain drugs such as phenylbutazone (Butazolidin) or other painkillers, as well as dietary deficiency of selenium, also cause ulcers.

Symptoms of Ulcers

A foal with an ulcer usually has gut pain. He may grind his teeth, show symptoms of colic, or lose weight. The pain tends to increase at night and between meals, as in human ulcers. The discomfort can be temporarily relieved by eating because food in the digestive tract tends to buffer the ulcer and keep digestive fluids from irritating the raw spot in the gut lining. Symptoms may be mild and subtle at first; the foal may just grind his teeth and slobber. He may try to eat wood or bedding, lie on his back, or prop himself in an odd position to try to relieve the discomfort.

Diagnosis and Treatment

A veterinarian can diagnose ulcers by means of physical examination (pressing deeply behind the edges of the rib cage to see whether this causes pain), lab tests, blood samples, looking into the stomach with an endoscope (a long, flexible tube), and evaluating abdominal fluid. If the foal has an ulcer, he can be

---

PARASITE HEALTH TIP

Several types of internal parasites cause problems for young horses from birth through the first year of life: intestinal threadworms (*Strongyloides westeri*), large roundworms (ascarids, *Parascaris equorum*), small strongyles (cyathostomes), and bloodworms (*Strongylus vulgaris*). The latter are a concern today only if there has not been a regular deworming program for horses on that particular farm. The mare is usually the primary source of infestation, so she should be kept wormed and her manure picked up on a regular basis.

Young foals are especially vulnerable to parasite damage in body tissues. Early damage in lungs, blood vessels, liver, and other tissues adversely affects the youngster's potential as a future athlete. A foal with serious parasite infestation may never reach his true potential.

treated with cimetidine (Tagamet) or omeprazole (Ulcergard or Gastrogard) and other medications to keep the gut lining from being damaged.

Parasite Control

Of utmost concern is minimizing a foal's exposure to internal parasites. The first step is to keep the mare dewormed during pregnancy. Some veterinarians recommend that she be dewormed about 1 month before she's due to foal, using ivermectin (approved for use on pregnant and lactating mares because it will not hurt the foal). Some also recommend deworming her again right after she foals, some time that same day, to protect the foal against threadworms. These are more damaging to the young foal than to older horses; foals develop immunity to them by 4 to 6 weeks of age. The parasites enter the foal through the mare's milk or even through the foal's skin if he has contact with infested ground, grass, or bedding. Threadworms can damage the lungs and liver of the young foal. They also cause diarrhea.

Even if a mare is regularly dewormed during pregnancy, she can still pass threadworms to her foal in large numbers right after she foals. This is because threadworm larvae remain dormant in the tissues around the mammary glands for several months and are out of reach of dewormers (even ivermectin). During labor, the larvae break their dormancy, perhaps because of hormonal changes in the mare's body. They migrate into the milk and enter the foal as he nurses. If the mare is not dewormed within 12 to 24 hours after the

---

WAYS TO MINIMIZE PARASITE PROBLEMS IN YOUNG HORSES

- Remove manure from stalls daily.
- Feed horses in mangers, troughs, or other feeders that hold hay off the ground so it can't be contaminated with manure and worm eggs.
- Never spread fresh manure over a pasture. Compost it first; heat from the compost pile will kill eggs and larvae.
- Rotate pastures. If you always use the same pasture for mares and foals or young horses, there will be lots of ascarid eggs there, waiting to infest foals.

For more information on parasites, see chapter 10.

birth, larvae will appear in her milk 4 to 20 days after foaling, with the highest number appearing at days 10 to 12. The larvae mature quickly, becoming adult worms approximately two weeks later in the foal's gut. They cause infection and diarrhea when he's 2 to 3 weeks old. The mare should be dewormed the day she foals, as soon as you have attended to the newborn.

A young foal eats his dam's fresh manure to obtain the microorganisms needed for breakdown and digestion of roughages. If the mare has not been dewormed, he picks up parasite eggs at the same time. The best way to protect the foal is to make sure the mare is dewormed and her udder cleaned at foaling time. During the first weeks of the foal's life, the mare is the primary source of parasites.

Weaning the Foal

Foals can be weaned at 2 to 6 months of age. It is detrimental to leave a fast-growing, heavy-bodied foal on a heavy milking dam past 2 or 3 months if he is developing leg problems because he has developmental orthopedic disease (see page 465) from overfeeding. Other foals do better physically and mentally if they are allowed to stay with their mothers until about 5 or 6 months of age. Foals can be left on their mothers longer than 6 months, but this is hard on the mare if she is pregnant again.

Weaning Methods

Horses are herd animals, happiest with other horses. The worst way to wean a foal is to separate him from the dam and leave him all by himself. He should be where he can at least see other horses and not feel alone and isolated. Left by himself in a stall or corral, he may become so frantic that he will literally climb the walls.

Weaning methods of yesteryear involved putting the foal in a stall and taking the mare away, or putting several foals together in a weaning pen. This is stressful for both mare and foal and can lead to injury and illness. The foal doesn't need his mother's milk anymore but still depends on her emotionally and is insecure without her. Foals who are put into weaning pens often run themselves to exhaustion and whinny themselves hoarse in frantic efforts to get out and find their mothers. All of them are desperate, pacing the fence, running back and forth, and stirring up dust that irritates lungs and opens the way for pneumonia. Any frantic activity by one sets off the others, and then they start pacing again, rarely stopping to eat or rest. After a few days, they resign themselves to life without their mothers, but this abrupt and traumatic separation is very hard on them. Well-fed foals may be physically ready to wean at 4 to 5 months of age but are not yet emotionally prepared for total separation from mother or from other horses.

Easy Weaning in Adjacent Pens

A better way to wean is to put the mare and foal into separate but adjacent pens where they can still see, hear, and smell each other. The pens should be built so the foal cannot reach through to try to nurse and high enough (6 feet [1.8 m] or higher) that neither mare nor foal will try to jump over. A safe mesh fence works well if it's tall and has a pole at the top so the mesh cannot be jumped or mashed down; diamond mesh wire works best. With this arrangement, the foal still has the emotional security of being near his mother.

Foals weaned in pens next to their dams usually spend most of their time near the fence that separates them, but they are not overly worried. There's little whinnying or pacing. They may sniff each other through the fence, but the foal is unable to nurse so the mare starts drying up. After a few days, the mares can be moved to a more distant pen or pasture with little fuss or protest, and the foals will not be frantic if they have each other or other horses for company. If you have only one foal to wean, let him live with a gentle,

Weaning the foal with mare and foal in separate but adjacent pens minimizes stress.

BE WORRY WISE AT WEANING

Minimize stress at weaning to avoid illness. Worry can also lead to injury if the foal spends a lot of time running or trying to get through a fence. If he works up a sweat, he may chill during cold nights. Cold, windy, or wet weather at weaning often leads to illness if foals are stressed.

easygoing gelding or some other tolerant "babysitter" horse after weaning or at least in a pen or pasture next to another horse; he'll be a lot happier than if he were all by himself.

Safe Fencing

Pens or pastures used for weaning should have safe and adequate fencing, and no hazardous obstacles for a foal to run into by accident. There should be no places where a foal could be tempted to put his head through or get a foot caught. If a group of foals is in a weaning pen, it should be large enough that the timid ones can get out of the way of more aggressive ones. The emotional trauma of weaning may turn the aggressive foals into bullies; they take out their frustrations on the timid ones. The pen should not be so large that a frantic foal could get up a lot of speed, trying to jump or crash the fence because he is running too fast to stop and turn in time. This kind of behavior is usually not a problem if foals are weaned next to their mothers.

Drying Up the Mare

The mare's udder fills and becomes swollen and uncomfortable after she is separated from her foal, and this fullness signals her body to stop producing milk. Any milk that remains is reabsorbed gradually. Watch the mare closely to make sure she dries up properly and does not get mastitis. If you try to milk her out, it will just prolong the drying-up process.

The best way to help her is to limit her feed and water for a few days. A day or two before weaning, cut down or eliminate her grain ration if she's been on grain; and on the day of weaning, stop feeding grain entirely. Keep her on plain grass hay (or cut down the amount of legume hay if that's all you have) and no grain until she has dried up (about 2 or 3 weeks). You can also limit her to approximately 10 gallons (38 L) of water daily, unless weather is hot, until her udder is less tight and she is drying up.

Feeding Weanlings and Yearlings

Once the foal has been weaned, he should be on a balanced diet that contains adequate protein and energy for his growing body. His needs can be met with a mix of good grass and alfalfa hay, and grain, if necessary. He needs more protein and total calories than a does mature horse, but he should not be overfed. This is a common mistake. More young horses are ruined by overfeeding than by underfeeding. The damage often begins even when they are foals.

Overfeeding and Leg Problems

Many breeders push young horses to grow rapidly with grain and high-protein feeds, but this can lead to contracted tendons, physitis (inflammation at the ends of growing leg bones), and osteochondritis dessicans (OCD), a developmental disorder in the growing skeleton that involves inflammation of cartilage and bone. The growth plates near the end of the young horse's long bones are soft, producing cartilage that turns to bone. As it produces more cartilage, the portion nearest the main bone shaft hardens and calcifies, causing the bone to grow in length. Growth plates are fragile; too much stress leads to permanent damage. Inflammation can result in bony enlargements, leg deformities, or contraction of the associated tendons.

Leg problems from overfeeding sometimes occur in foals under 2 months of age but may happen in foals up to 16 months of age. The youngster may appear to have swollen joints, but the problem is not in the joints themselves — it's caused from abnormal growth patterns at the ends of the long bones next to the joints. The degree of lameness may vary from stiffness and reluctance to move, to a definite limp.

A young horse with OCD has weak spots in the bone where the cartilage at the bone ends and joints does not turn into bone properly. These areas are vulnerable to inflammation, twisting, breaking, or tearing because they are weaker than normal. This can result in twisted bones, compressed vertebrae, shoulder cysts, swollen fetlock joints, and separated pieces of cartilage and bone within the stifle joints. It also results in youngsters developing spavin (hock problems) at an early age.

Dietary Imbalance

Too little or too much calcium, phosphorus, or protein disrupts normal cartilage-to-bone conversion; therefore, you should avoid excessive grain in the ration (high in phosphorus) and rich alfalfa (high in calcium). The younger the animal, the more sensitive he is to calcium excess because most of the growth of long bones takes place at an early age. The faster he grows, the greater the damage. The daily calcium requirement for a growing horse is approximately 1.1 ounces (30 g), but a typical alfalfa and sweet-feed diet gives a horse about 5.4 ounces (154 g) of calcium.

Most breeders feed high-energy diets to youngsters, with large amounts of grain and concentrates containing starches and nonfiber sugars in much higher levels than would be available in natural feeds. This overfeeding stimulates hormones that trigger cartilage growth. Immature cartilage at the bone ends is produced more rapidly than normal but matures more slowly, resulting in permanent damage to the equine skeleton. This leads to lameness and unsoundness when the horse grows up and is expected to work. Some researchers think there may be a genetic predisposition toward DOD problems (contracted tendons, physitis, OCD, and so on) because certain bloodlines that produce fast-growing foals seem more likely to develop skeletal problems.

Preventing Skeletal Problems

Allow young horses to grow up lean and active on natural feeds or on a diet carefully balanced in terms of minerals, energy, and protein. Natural feeds include pasture, grass hay (with only enough alfalfa added to give sufficient protein for normal growth, or alfalfa that is not too rich), and minimal grain, if any. Confinement, lack of exercise, and overfeeding, along with a genetic tendency toward rapid growth and early maturity (which many breeders select for) all work together to cause crippling bone disease. Adult horses are not as adversely affected by overfeeding or nutritional imbalance because these bone problems occur only in animals whose bones are still growing.

To protect the youngster from leg and joint damage, grow him a little slower. Give him room to run and exercise. Raising young horses in large

---

EXERCISE IS IMPORTANT

The less exercise a youngster gets, the less energy he can handle in feed. Stuffing a confined animal with grain is a sure way to make him unsound and lame by the time he's grown up. The horse who grows up at pasture without any grain may take a little longer to reach mature size and fill out than his pampered counterpart but, because of regular exercise, the pasture-raised youngster will stay more sound.

pastures at total liberty reduces the risk of OCD and physitis. The increased exercise level converts the sugars (obtained from grain feeding) to energy instead of increasing the hormone production that interferes with growing cartilage. Foals, weanlings, and yearlings should never be confined to box stalls or small paddocks. They need room to run to develop strong legs and lungs and burn off energy.

Care of Weanlings and Yearlings

Because growing horses needs more protein and energy than mature horses, feed weanlings and yearlings separately. If they are kept with older horses, they won't get their share because they don't eat as fast as the adults. Weanlings and yearlings (and sometimes 2-year-olds) eat smaller portions more often. They get full more quickly than the mature horse and stop eating. Then, when they go back to eat some more, the other horses have eaten it all up.

If you have several young horses, keep them with their various age groups and sexes. Don't have fillies and colts in the same pen or pasture, especially if the colts are not yet gelded, because young males tend to roughhouse more than fillies. Unless a colt is to be used later for breeding purposes, he should be castrated as a foal or during the spring of his yearling year.

Castrating (Gelding) Colts

The testicles are usually descended at birth or soon thereafter, so the first 3 weeks of life is a convenient time to geld. After a few weeks, the testicles may draw up against the abdominal wall and not descend again until the colt is a year old or older, but most stay down; therefore, most colts can be gelded at any time. It's best to do it at some point before weaning while he still has his mother for security.

Gelding at this age not only means the foal has his dam for comfort, but it also means she'll help exercise him. This is an advantage for you. There will be no aftercare (as with an older colt, who must be forced to exercise to prevent swelling and encourage drainage) — just a daily check to make sure the incision is draining properly and there is no swelling and infection. When gelding a colt this young, there is rarely much swelling or drainage, compared with gelding a larger animal. Also, if colts are gelded young, they don't have to be separated from the fillies at weaning time; they can all live together as weanlings and yearlings.

Years ago, most draft colts were castrated in the first weeks after birth, but in recent years, most colts have generally been left untouched until they

CASTRATING A COLT

A. After the sedative begins to take effect, steady the young colt as he goes down.

B. Ease the colt down on clean grass.

C. The veterinarian disinfects the scrotal area while one helper holds the colt's hind leg out of the way, and another helper protects the colt's eye from sunlight.

D. The testicles are removed.

E. As the sedative wears off, the colt wakes up.

F. Soon, the colt is back on his feet with his mama and ready to be turned loose again at pasture.

CASTRATE THE COLT ON CLEAN GRASS

Choose a clean, grassy area and hold the mare nearby as the veterinarian gives the colt an anesthetic, steadies him, and eases him to the ground. Grass is a practical surgical surface, being a clean place to work; the incision is not likely to be contaminated with dirt and dust. Moreover, the mare can graze near her colt, not likely to worry much about him. Because the surgery takes only a few moments, the pair can simply be turned loose (if it's done in their pasture) or taken through a gate to their own pasture or paddock after the anesthetic wears off.

became yearlings or 2-year-olds — the rationale being to let them develop more fully. Yet, most of the traits a colt acquires by age 2 are the ones people seek to eliminate by gelding.

The animal with a heavy, muscular, cresty neck is less balanced (less able to flex and collect) than the horse with a lighter front end. Also, the colt who is not castrated early develops the stallion's habits and attitudes and may become aggressive. Waiting until age 2 also means the veterinarian has to deal with a larger, stronger, and more aggressive patient.

Advantages to Gelding a Young Colt

Early castration does not adversely affect the animal's athletic ability and is less risky to his health than castration later. Also, colts gelded young grow taller (by as much as 1 or 2 inches [2.5–5 cm]) because the growth plates at the ends of the long bones stay active longer in a gelding; the hormones of puberty speed closure. Castration is minor surgery at this early age; the testicles are small, as are the blood vessels and supportive structures. There is much less trauma and stress to the colt (mentally and physically) when he is very young; the tissues removed and size of the incision are less extensive. The young castrated colt recovers from the surgery much more quickly than does the older colt, and with fewer complications.

Disadvantages to Gelding

The main disadvantage of early castration is a slightly greater chance after the surgery of scrotal hernia, a condition in which the intestines slip out through the incision if the colt has an inherited weakness. Looseness in the abdominal rings is a hereditary condition that can be corrected with a stitch or two during the surgery. The veterinarian doing the castration should check for this condition, regardless of the age of the horse. Advantages of early castration far outweigh potential risks. If a colt is to be gelded, there is no point in waiting.

Cryptorchid

Occasionally, one or both of the colt's testicles have not descended. In these cases, the missing testicle (usually only one) is partway up the passageway it came through during fetal development or it is clear up in the abdomen. If so, the colt cannot be castrated by ordinary means; major surgery is required to find and remove the missing testicle. If the testicle is partly descended, the veterinarian may wait until after the colt reaches puberty. The enlargement of the testicle at that time and its extra weight sometimes bring it down. If it's retained in the abdomen, there's not much chance it will descend. The colt will still be a stallion even if one descended testicle is removed in normal fashion and, although he appears to be a gelding (with no testicles in the scrotum), he will have a stallion's attitude, be able to mount mares (although he is infertile because the testicle is too warm for sperm viability), and most likely be dangerous to handle. Someone buying the horse may not realize he's a stallion and be unprepared for his actions and attitude. Also, the retained testicle may develop a tumor which alters the hormonal mix and may contribute to erratic behavior. The best solution is to have him gelded by means of abdominal surgery.

Epilogue

FOR THE HORSE LOVER, there's no greater pleasure than raising a foal and seeing him grow up to be the horse you've always wanted. Raising a foal to maturity may cost more in the long run than buying an adult animal, when you consider stallion fee, cost of feed for the mare and the young horse as it grows up, vaccinations and dewormings, veterinary services, visits from the farrier, and so on — but there are other factors that can't be measured in dollars.

The experience is priceless. It's fun, satisfying, and emotionally rewarding — providing hours of pleasure. Moreover, the young horse is special because he is uniquely yours, raised and handled the way you want him to be. Because you've handled him from birth, he trusts and respects you; the two of you can be a team from the beginning. If you handle him conscientiously, he will be a joy. If you spoil him, it's your own fault.

Yes, breeding a mare and raising a foal is time-consuming and costly, and there are risks involved. Things don't always go as planned or turn out with a happy ending; however, risk and challenge are part of what makes any undertaking worthwhile. The emotional rewards are all the greater if you succeed, and the accomplishment is special.

It is my hope that this book can help you along the way in your horse-raising venture. No matter how much you learn through reading, however, experience is the best teacher. Good luck.

Horse Anatomy

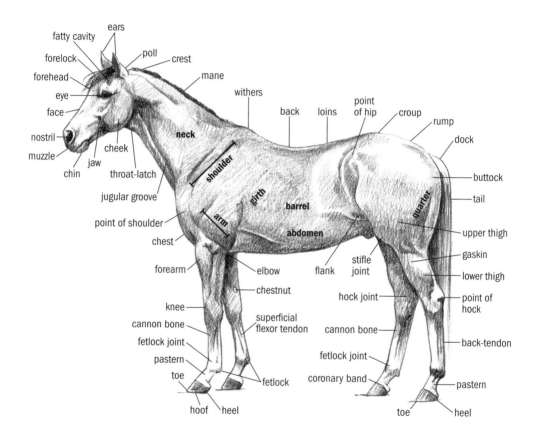

APPENDIX B

Resources

Breed Associations

American Connemara Pony Society
Middlebrook, Virginia
marynelleyles@hughes.net
www.acps.org

American Miniature Horse Association
Alvardo, Texas
817-783-5600
www.amha.com

American Morgan Horse Association
Shelburne, Vermont
802-985-4944
www.morganhorse.com

American Paint Horse Association
Fort Worth, Texas
817-834-2742
www.apha.com

American Quarter Horse Association
Amarillo, Texas
806-376-4811
www.aqha.com

American Saddlebred Horse Association
Lexington, Kentucky
859-259-2742
www.asha.net

American Shetland Pony Club
Morton, Illinois
309-263-4044
www.shetlandminiature.com

American Shire Horse Association
Effingham, South Carolina
843-629-0072
www.shirehorse.org

American Suffolk Horse Association
Ledbetter, Texas
979-249-5795
www.suffolkpunch.com

American Trakehner Association
Newark, Ohio
740-344-1111
www.americantrakehner.com

American Warmblood Society
Center Ridge, Arkansas
501-893-2777
www.americanwarmblood.org

Appaloosa Horse Club
Moscow, Idaho
208-882-5578
www.appaloosa.com

Arabian Horse Association
Aurora, Colorado
303-696-4500
www.arabianhorses.org

Belgian Draft Horse Corporation of America
Wabash, Indiana
260-563-3205
www.belgiancorp.com

Clydesdale Breeders of the USA
Pecatonica, Illinois
815-247-8780
www.clydesusa.com

Foundation Horse Registry
Apache Junction, Arizona
480-982-1551
www.foundationhorses.com

Foundation Quarter Horse Registry
Sterling, Colorado
970-522-7822
www.fqhrregistry.com

Friesian Horse Association of North America
Lexington, Kentucky
859-455-7457
www.fhana.com

International Andalusian and Lusitano Horse Association Registry
Birmingham, Alabama
205-995-8900
www.ialha.org

Jockey Club
Lexington, Kentucky
859-224-2700
www.jockeyclub.com

Lippizan Association of North America
Anderson, Indiana
sandy@lipizzan.org
www.lipizzan.org

Missouri Fox Trotting Horse Breed Association
Ava, Missouri
417-683-2468
www.mfthba.com

National Foundation Quarter Horse Association
Enterprise, Oregon
541-426-4403
www.nfqha.com

National Pinto Horse Registry
Thousand Oaks, California
805-241-5533
www.pintohorseregistry.com

National Show Horse Registry
Louisville, Kentucky
502-266-5100
www.nshregistry.org

North American Mustang Association and Registry
Mesquite, Texas
972-289-9344
namarmustangs@yahoo.com

North American Peruvian Horse Association
Burleson, Texas
817-447-7574
www.napha.net

Palomino Horse Association
Nelson, Missouri
660-859-2064
www.palominohorseassoc.com

Palomino Horse Breeders of America
Tulsa, Oklahoma
918-438-1234
www.palominohba.com

Paso Fino Horse Association
Plant City, Florida
813-719-7777
www.pfha.org

Percheron Horse Association of America
Fredericktown, Ohio
740-694-3602
www.percheronhorse.org

Performance Horse Registry
United States Equestrian Federation
Lexington, Kentucky
859-258-2472
www.phr.com

Pinto Horse Association of America
Bethany, Oklahoma
405-491-0111
www.pinto.org

Pony of the Americas Club
Indianapolis, Indiana
317-788-0107
www.poac.org

Spanish Barb Breeders Association
Silver City, New Mexico
info@spanishbarb.com
www.spanishbarb.com

Spanish Mustang Registry
Chilton, Texas
254-546-2177
www.spanishmustang.org

Tennessee Walking Horse Breeders' and Exhibitors' Association
Lewisburg, Tennessee
931-359-1574
www.twhbea.com

United States Lipizzan Registry
Salem, Oregon
503-589-3172
www.uslr.org

United States Trotting Association
Columbus, Ohio
877-800-8782
www.ustrotting.com

Welsh Pony and Cob Society of America
Stephens City, Virginia
540-868-7669
www.welshpony.org

More breed organizations can be found in the American Horse Council *Horse Industry Directory* (contact the American Horse Council, Washington, D.C., 202-296-4031, *www.horsecouncil.org,* for your copy). These include associations for Akhal Teke, American Sport Pony, American Warmblood Registry, Azteca Horse, Bashkir Curly, Belgian Warmblood, Buckskin, Caspian Horse, Cleveland Bay, Colorado Ranger Horse, Cracker Horse, Dales Pony, Dartmoor Pony, Donkey and Mule, Dutch Warmblood, Exmoor Pony, Fell Pony, Gypsy Cob, Hackney, Haflinger, Hanoverian, Holsteiner, Miniature Horse, Morab, New Forest Pony, Norwegian Fjord Horse, Oldenburg, Quarter

Pony, Racking Horse, Rocky Mountain Horse, Suffolk, Trakehner, Walking Pony, Welara, and others.

This directory offers listings for breed registries, educational organizations, equine veterinary schools, equine welfare organizations, federal information sources, organizations of interest to horse owners, guides to interstate health requirements, health and research organizations, equine publications and newsletters, show and sport organizations, state departments of agriculture, extension and horse specialists, state veterinarians, trail organizations, and more.

Note: There are many good books about all aspects of horse raising, horse care, breeding, training, and so on. Check with your local library or bookstore. Other good sources are the advertisements and book reviews in equine periodicals.

Suppliers
Foaling Alarm Systems
Allsman Enterprises
Rogue River, Oregon
541-582-2101
www.breederalert.com

Foalert, Inc.
Acworth, Georgia
800-237-8871
www.foalert.com

Kee-Port, Inc.
Arlington, Washington
425-345-1556
www.foalingalarm.com
Manufacturers of EquiFone and EquiPage

Foal Milk Replacers
Buckeye Nutrition
Dalton, Ohio
800-898-9467
www.buckeyenutrition.com

Calva Products, Inc.
Acampo, California
800-669-6455
www.calvaproducts.com

Manna Pro
Chesterfield, Missouri
800-690-9908
www.mannapro.com

Merrick's, Inc.
Middleton, Wisconsin
800-637-7425
www.merricks.com

PetAg
Hampshire, Illinois
800-323-6878
www.petag.com

Restraint Devices and Mare Milker
EZ Animal Products
Ellendale, Minnesota
800-287-4791
www.thestableizer.com
Makes the Stableizer and UdderlyEZ

APPENDIX C

Glossary

A

abortion. Loss of pregnancy; reabsorption or expulsion of embryo or fetus before it is viable

abscess. Localized collection of pus

acupuncture. Ancient healing art that involves pricking the skin or tissues with needles

alfalfa. A legume used for hay, high in protein and calcium

alternative therapies. Treatments such as acupuncture, massage, chiropractic, and herbal remedies used instead of or in conjunction with traditional Western medical remedies

amino acids. Group of organic compounds forming the basic structure of proteins

ammonia. Strong alkaline gas with pungent odor, formed by breakdown of urine and manure

amnion. Innermost membrane surrounding the fetus in the uterus

amniotic fluid. Fluid surrounding the fetus

anaphylactic shock. Hypersensitivity of tissues to an antigen; reaction against a previous dose, putting the animal into shock

anaphylaxis. Hypersensitivity in which exposure to a specific antigen results in life-threatening respiratory distress, circulatory collapse, and shock; see *anaphylactic shock*

anestrus. Period during which a mare does not come into heat

anhidrosis. Inability to sweat

antibody. A blood protein produced in response to an antigen for the purpose of counteracting it

antigen. A foreign protein that causes the body to produce antibodies to destroy or neutralize it

antihistamine. Substance that counteracts the effects of histamine, used in treatment of allergies and other conditions

antiseptic. Disinfectant used to prevent growth of disease-causing microbes

antiserum. A blood serum containing antibodies against specific antigens, injected to treat or protect against a disease

antitoxin. Antibody that counteracts a specific toxin

arthritis. Inflammation of a joint

artificial insemination (AI). Manually placing semen into the genital tract of the mare

ascarid. Large white intestinal parasitic worm; a type of roundworm

awns. Stiff bristles that protrude from a grain sheath or grass seed head

azoturia. Old term for cramping of the large muscles; "Monday morning disease"

B

Banamine. Flunixin meglumine; an anti-inflammatory, painkilling drug used for colic

bar. Support structure of the foot between the heel and the frog

barren mare. Mare that is not pregnant

base narrow. Feet too close together

base wide. Feet too wide apart

bean. Small, hard deposit of debris that collects near the end of the horse's penis

benzimidazoles. Group of deworming drugs that have similar action

betadine. Povidone iodine; tamed iodine

bighead. Condition characterized by enlargement of bones in the skull from calcium-phosphorus imbalance caused by eating too much bran

birth canal. Vagina; the passage from the uterus to the outside

bladder stone. Concretion of mineral salts in the bladder

blister beetle. Small flying insect containing a potent toxin, often baled in hay

bloodworm. Large strongyle; parasitic intestinal worm of which the larval stage damages the horse's arteries

body rope. Rope that goes around the horse's girth and through the halter ring, used for safely training a horse to tie

booster. Repeat immunization to increase immunity level

bot. Parasitic grub that lives in the horse's stomach; the adult fly lays eggs on the horse's hairs

botulism. Poisoning caused by a toxin produced by bacterium (*Clostridium*) sometimes found in decaying organic matter

bran. High-protein roughage supplement consisting of the outer husks of wheat kernels

breech. Foal positioned rump first (no legs in the birth canal)

"bute." Butazolidin, trade name for phenylbutazone, an anti-inflammatory, painkilling drug

buttock. Rounded back portion of the upper hind leg muscles

C

caesarean. Surgical removal of fetus from the uterus

calf knees. Back at the knees; a conformational flaw in which the knee bends backward

canine tooth. Small sharp tooth behind the incisors, present in most male horses and in some females

cannon bone. Bone between the fetlock joint and knee (or hock)

cap. Remains of a baby tooth covering the permanent tooth

carrier. An individual having a recessive gene that is not expressed but that could be passed to offspring; also, an animal that appears healthy but harbors pathogens or parasites that can be shed and passed to other animals

caslick repair. Surgical technique in which the lips of the mare's vulva are cut and sutured so they grow together, making the opening smaller

cast horse. Horse who is down in the stall (or next to a fence) on his back, unable to get up

cecum. Large pouch ("fermentation vat") between the small and large intestine where roughage is broken down by microbes

cellulose. Fibrous part of roughage

cervix. Constricture between uterus and vagina that opens during estrus and during labor to allow for birth of the foal

chiropractic therapy. Treatment involving manipulation of the horse's spine

chiroptic mange. Leg mange; skin irritation caused by mites

choke; choking. A serious condition in which there is an obstruction in the horse's esophagus and the possibility that he may inhale matter into his windpipe

chronic obstructive pulmonary disease (COPD). Heaves; respiratory ailment characterized by forced expiration and difficult breathing; caused by allergic sensitivity to dust and other irritants

club foot. Upright, stumpy foot with short toe and long heel

coffin bone. Bone inside the hoof (pedal bone, third phalanx)

coggins test. Laboratory test for equine infectious anemia

colic. Abdominal pain

colostrum. First milk, containing antibodies that give the newborn foal temporary disease resistance

colt. Male horse under 4 years of age

concentrates. Feeds low in fiber and high in food value, such as grains and oilmeals

concussion. Jarring impact on feet and legs when traveling

conformation. Body proportions and angles; how the horse is built

conjunctivitis. Inflammation of mucous membrane covering the front of the eyeball and inside of the eyelids

contracted feet. Condition in which there is narrowing of the hoof, especially at the heels

corn. Often located where the bars meet the heel, a bruised area on sole of the foot, caused by pressure from poorly fitted shoe

cornea. Transparent front of the eyeball

coronary band. Top of the hoof at the hairline, where hoof growth takes place

corticosteroids. Steroids produced by the body or synthetically (such as dexamethasone), used in treating allergy, pain, and inflammation

cow hocked. Having hocks too close together

cribbing. The act of grabbing a surface (such as fence or manger) with the teeth and sucking in air at the same time

crossbred. Horse with parents of different breeds

cross-tie. To tie a horse from two sides (from the halter to the walls or posts on either side)

croup. Highest point of the horse's rump

cryptorchid. Male horse with one or two retained testicles that did not descend into the scrotum

curb. Enlargement at the back of the hock

Cushing's syndrome. Metabolic disease that sometimes affects older horses (often caused by pituitary tumor); symptoms include failure to shed winter hair

D

dam. Mother of a horse

dermatitis. Inflammation of the skin

developmental orthopedic disease (DOD). Developmental disorder in the growing horse that adversely affects the health of cartilage, bones, or connective structures. This is a catch-all term that includes physitis, OCD, contracted tendons, and so on.

dexamethasone. Steroid used to reduce swelling, inflammation, pain, fever, shock, and allergic conditions

diestrus. Period between heat cycles when the mare is out of heat.

digital cushion. Spongy area inside the foot above the frog

dimethyl sulfoxide (DMSO). A solvent that has many medical uses, especially as an anti-inflammatory to reduce heat and swelling.

disposition. Temperament and attitude

distemper. Old term for strangles

DMSO. See dimethyl sulfoxide

dominant gene. One that produces the expressed trait in offspring when more than one gene for a certain trait are present

E

Eastern equine encephalomyelitis (EEE) "Sleeping sickness"; viral disease of horses affecting the brain, spread by mosquitoes

edema. Swelling caused by excessive fluid in the tissues

electrolytes. Important body salts that can be lost through sweating or diarrhea

embryo. The developing foal in earliest stage of pregnancy

encephalitis. Inflammation and infection in the brain

endorphins. Chemicals produced in the brain, having pain-relieving effects

endotoxemia. Presence of endotoxins in the blood, which may result in shock

endotoxic shock. Shock caused by body systems shutting down from reaction to bacterial poisons

endotoxin. Poison created when bacteria multiply in the body

enteritis. Inflammation of the intestines

enteroliths. Concretions of mineral salts ("stones") that build up in the intestinal tract around a foreign object

epinephrine. Adrenaline; a hormone affecting circulation and muscle action, used as a stimulant

epsom salts. Magnesium sulfate; used as a purgative or for soaking a foot to draw out infection

equine infectious anemia (EIA) "Swamp fever"; incurable viral disease spread by bloodsucking insects

equine protozoal myeloencephalitis (EPM). A parasitic infection spread by opossums, affecting the horse's spinal cord and the nerves that control his movements

equine viral arteritis (EVA). Viral disease that can cause respiratory disease, abortion, and infertility in horses

Escherichia coli (E. coli). Bacteria that sometimes cause gut infection

estrogen. Female hormone produced by developing follicles in the ovary; stimulates the mare to come into heat

estrous cycle. Recurring period of sexual receptivity

estrus. Heat period during which the mare ovulates and is receptive to the stallion

euthanasia. Act of ending a horse's life mercifully

F

farrier. Professional horseshoer

fecal count. Examination of manure under a microscope to count worm eggs so as to estimate degree of infestation

fetlock. Tuft of hair at the back of the fetlock joint

fetlock joint. Joint between pastern and cannon bone

fetus. Developing foal after 55 to 60 days of pregnancy

filaria. Tiny threadlike worms spread by biting flies

filly. Female horse under 4 years of age

flat bone. Term used to define depth of cannon, with the flexor tendon set well back of the bone

flatulent colic. Abdominal pain caused by excessive gas in the gut

flexor tendon. Large tendon behind the cannon bone

float. File used to rasp off sharp edges of a horse's teeth

foal. Newborn horse of either sex; the act of foaling

foal heat. Mare's first heat period after foaling

follicle. Fluid-filled sac on an ovary containing the egg

founder. Sequel to inflammation of laminae within the hoof (laminitis) in which there are changes that include dropped or rotated coffin bone and abnormal hoof growth.

Frog. Spongy V-shaped cushion on the bottom of the foot

G

gelding. Castrated male horse

gestation. Length of pregnancy

girth. Area behind the front legs; circumference of the horse's body at this location

girth itch. Skin disease caused by a fungus in the girth area

glycogen branching enzyme deficiency (GBED). Recessive genetic defect in certain family lines of Quarter Horses and related breeds, resulting in early death of affected foals.

granulation tissue. The tissue that grows to fill in a wound; scar tissue

granulosa cell tumor. Abnormal growth of certain cells in the ovary

grub. Fly larvae, such as the botfly or heel fly

guttural pouch. Sac between the ear and mouth that sometimes becomes infected

H

habronemiasis. "Summer sores" caused by larvae of a stomach worm trying to develop in the horse's skin instead of in the stomach, after being brought to the skin by flies

heat cycle. See estrus

heat stroke. Circulatory collapse and shock caused by exposure to high temperatures, especially while exerting

heave line. Ridge of muscle along the abdomen caused by the extra effort of forcing air out of the lungs; common in a horse with chronic obstructive pulmonary disease (heaves)

heaves. Breathing problem caused by respiratory irritation or allergy; see chronic obstructive pulmonary disease

hereditary equine regional dermal asthenia (HERDA). Inherited defect (in certain lines of Quarter Horses and related breeds) of skin and connective tissue, characterized by abnormal skin that tears readily and separates easily from the underlying tissues

heterosis. Hybrid vigor resulting from crossbreeding

heterozygous. Genes of a specific pair that are different

histamine. A chemical compound that dilates capillaries, constricts the smooth muscle of the lungs, raises the heart rate, and increases secretions of the stomach; released during a hypersensitivity allergic reaction

homeopathy. Treatment of disease by using tiny doses of substances that, in a healthy horse, would produce the same disease

homozygous. Genes of a specific pair that are the same

hoof tester. Tool used to put pressure on various areas of the bottom of the foot to check for pain or sensitivity

hook. Elongated portion of a tooth

hyperkalemic periodic paralysis (HYPP). Inherited muscle disease in certain lines of Quarter Horses and related breeds severely affecting the muscle metabolism.

hyperthermia. Use of heat to burn off a growth or tumor

I

immunotherapy. Prevention or treatment of disease with substances that stimulate an immune response

impaction. Blockage of digestive tract with food or other material

incisors. Front teeth

influenza. Respiratory disease caused by a virus

interdental space. Space between the incisors and molars where there are no teeth

isoerythrolysis. Condition in which antibodies in the mare's colostrum destroy the foal's red blood cells after he nurses; newborns with this condition are sometimes called Rh foals

ivermectin. Deworming drug that paralyzes the parasites; they cannot eat and eventually die

J-L

joint ill. Inflammation in the joints caused by bloodborne infection, often from bacteria entering the navel of the newborn foal

lactating. Producing milk

laminae. Attachments inside the hoof that bind the hoof wall to the inner tissues

laminitis. Inflammation of the sensitive laminae; may result in founder (changes within the foot and hoof wall)

lampas. Tenderness and inflammation of the soft tissue in the roof of the mouth, usually caused by eating hard feeds

larvae. Immature stage of a worm, tick, fly, or some other insect

legumes. Plants in the pea family, such as alfalfa and clover, that make their own nitrogen while growing; they have more protein than grasses

leptospirosis. Infectious disease caused by bacteria; may cause abortion in mares or eye problems (moonblindness)

lethal genes. Genes that if doubled up cause serious defects leading to death of the foal before or soon after birth

lice. Tiny insects that feed on the skin or suck blood

lip chain. Chain portion of lead shank placed over the gum above the top teeth (in the front of the mouth); pressure applied here sedates and immobilizes the horse as does a twitch

M

maiden mare. Mare that has never been bred

mammary glands. Milk-producing tissues

mange. Skin disease caused by mites

mare. Female horse after her fourth birthday

mastitis. Inflammation of the udder

meconium. Dark material passed in the foal's first bowel movements

melanin. Skin pigment

melanoma. Skin cancer; tumors in the pigment-producing cells

microbes. Microorganisms, germs; refers especially to bacteria, protozoa and fungi that cause disease or fermentation

mites. Tiny insects that feed on the skin

molars. Grinders; back teeth

"Monday morning disease." See *azoturia*

moonblindness. Iridocyclitis; periodic ophthalmia; inflammation of internal parts of the eye; recurrent attacks eventually cause blindness

Moxidectin. Deworming drug related to ivermectin, but having slightly different effects

mucous membrane. Mucus-secreting tissue that lines many body cavities

mucus. Secretion from a mucous membrane

N

navel ill. Disease caused by bacteria entering the foal's navel

navicular disease. Lameness from inflammation of the navicular bone that lies behind the coffin bone inside the foot

night blindness. Recessive genetic defect most commonly found in Appaloosas

nolvasan. Chlorhexidine; all-purpose disinfectant

O

oilseed meal. High-protein concentrate obtained after extraction of oil from flax, soybeans, or cottonseed

osteochondritis dessicans (OCD). Developmental disorder in the growing skeleton that involves inflammation of cartilage and bone

ovulation. Process by which a follicle ruptures on the ovary and releases an egg into the oviduct (the tube connecting the ovary and uterus)

oxytocin. Hormone that stimulates milk letdown and contraction of the uterus during foaling

P

paddling. Traveling with feet swinging outward

palpate. To feel, as when the veterinarian checks the uterus or ovaries through the rectal wall

parasite. Organism that lives on or within another animal

parrot mouth. Condition characterized by mismatched jawbones; upper teeth protrude over lower teeth

passive transfer. Temporary immunity transferred from mare to newborn foal through antibodies in colostrum

pastern. Area between the hoof and fetlock joint

patella. "Kneecap" on the stifle joint

pathogen. Invading organism (bacterium, virus, and so on) that causes disease

periodic ophthalmia. See moonblindness

phenylbutazone. See "bute"

photodynamic agents. Chemicals in a plant that cause a reaction in the horse's skin tissues when activated by sunlight

photosensitization. Serious skin disease caused by reaction to sunlight after a horse eats certain plants

physitis. Inflammation of the growing end of the long bones, usually caused by too much rich feed and dietary imbalance in the young horse

pigeon-toed. Having toes pointing toward one another instead of straight

pinworms. Parasites that live in the rectum and lay eggs at the anus, causing severe annoyance and itching

placenta. Afterbirth; membranes that attach the developing fetus to the uterus; it is shed after foaling

pneumonia. Infection in the lungs

poll. Top of the horse's head

polysaccharide storage myopathy (PSSM). Genetic defect in muscle metabolism that causes muscles to "tie up" and cramp during exercise; found in many breeds, especially heavily muscled horses such as Quarter Horses and draft horses

Potomac horse fever (PHF). Disease caused by a rickettsia spread by flukes that parasitize freshwater snails, causing severe diarrhea and laminitis in horses

pressure bandage. Bandage used to halt excessive bleeding

progesterone. Hormone that stops heat cycles and keeps the mare pregnant

prolapse. Protrusion of an inverted organ, such as the uterus

protozoa. One-celled animals; some cause disease

proud flesh. Excessive growth of granulation (scar) tissue in a wound

purebred. Animal with sire and dam of the same breed

Purpura hemorrhagica. Serious condition caused by internal bleeding from small blood vessels

Q-R

quicked nail. Nail driven into sensitive tissues when shoeing, causing pain and possible infection

quidding. Dropping pieces of partially chewed hay from the mouth

rabies. Fatal viral disease spread by bites of infected animals

rainrot. Fungal skin disease most common in wet conditions

recessive gene. One that must be received from both parents for its trait to be expressed in the offspring

rhabdomyelitis. Muscles tying up; see *azoturia*

rh foal. See isoerythrolysis

rhinopneumonitis. Viral herpes that causes respiratory problems and abortion

Rickettsia. A genus of bacteria that often live in the gut of ticks and insects and cause diseases such as Potomac horse fever, typhus, and Rocky Mountain spotted fever

ringbone. Arthritis of pastern joint or joint between pastern and coffin bone, causing excessive bone growth and lameness

ringworm. Fungal infection causing scaly patches of skin

roundworm (ascarid). A damaging parasite common in foals

S

salmonella. Bacteria that causes severe diarrhea

sand colic. Potentially fatal condition caused by eating sand with the feed

sarcoid. Skin tumor that sometimes develops after a wound

scours. Diarrhea

scratches. Skin disease of the lower legs caused by a fungus or bacteria

septicemia. Generalized infection in the bloodstream

settle. Become pregnant

severe combined immunodeficiency (SCID). Recessive genetic defect in some Arabians (affected foals cannot make antibodies and soon die)

shaker foal. Progressive weakness in a newborn foal caused by botulism

shear mouth. Condition in which teeth are worn at an angle so top and bottom teeth do not match but slide past one another

sidebone. Calcification of the cartilage just above the side of the hoof; bony enlargement above the coronary band

sidelining. Tying the hind leg to the front leg on the same side to prevent kicking

sire. Father of a horse

sole. Horny bottom of the foot

sound. Having no faults or injuries that impair usefulness and ability to travel freely or without pain

sow mouth. Mismatch of jawbones; lower jaw protrudes farther forward than upper jaw

spasmodic colic. Acute intermittent gut pain from spasms

spavin. Disease of the hock joint

splay footed. Having feet turned toe out instead of pointing straight ahead

splint. Bony growth on the splint bone

squamous cell carcinoma. Skin cancer usually affecting unpigmented areas of skin

Stableizer. Restraint device that goes over the poll and under the top lip; the loop is tightened to activate pressure points that stimulate release of endorphins, relaxing the horse and blocking out pain

stallion. Male horse 4 years old or older that is not castrated

stifle. Large joint high on the hind leg, next to the flank

stocks. Chute for restraining a horse

strangles. Respiratory disease caused by bacteria causing swollen and abscessed glands

strongyle. Bloodworm

"summer sore." Lesion that won't heal caused by worm larvae in the skin, spread by biting flies; see *habronemiasis*

surcingle. Belly band; strap that goes around the girth

"swamp fever." Equine infectious anemia (EIA)

T

tamed iodine. Povidone iodine (Betadine); type of iodine not as harsh on raw tissues as tincture of iodine

tapeworm. Segmented parasite that infests the intestine

teasing. Bringing a stallion near a mare to see whether she's in heat

thrush. Foul-smelling foot infection in tissues next to the frog, usually caused by the bacterium *Fusobacterium necrophorum*

tooth bumps. Enlargements under the lower jaw from impacted molars

total digestible nutrients (TDN). Digestible portion of diet

toxemia. Condition in which bacterial toxins invade the bloodstream and poison the body

toxin. A poison created by a living organism such as a bacterium, or a plant

toxoid. Vaccine that stimulates immunity to bacteria that produce toxins

twitch. Tool used to restrain a horse by putting pressure on the nose, which stimulates release of endorphins; to use such a tool

tying up. Muscle cramps in the large muscles

U

umbilical hernia. Protrusion of tissue by the navel

unsound. Having weakness or injury that makes a horse unable to perform without pain or disability

unthriftiness. Poor feed efficiency despite adequate nutrition; often caused by illness or parasites; characterized by thin condition and poor hair coat

urinary stones. Concretions of mineral salts that form in kidneys or bladder, sometimes blocking the urinary tract

urticaria. Hives; swellings in the skin caused by allergy

V

venezuelan equine encephalomyelitis (VEE); "Sleeping sickness"; form of encephalitis spread by a virus and transmitted by insects; more deadly than Eastern equine encephalomyelitis or Western equine encephalomyelitis, the Venezuelan variety is sometimes brought north from Central America

vital signs. Temperature, pulse, and respiration rates

W

warble. Lump in the skin of the back where the grub of a heel fly is developing

warmblood. Any European breed of heavy horse who has some Arabian blood in his ancestry; also used to describe the result of crossing cold blood (draft) with hot blood (Arabian or Thoroughbred)

water bag. Membrane filled with amber-colored fluid that precedes the foal through the birth canal

wax. Secretion that coagulates at the end of the teat just before foaling

weanling. Foal that has been weaned but is not yet a yearling

Western equine encephalomyelitis (WEE). "Sleeping sickness"; see *Eastern equine encephalomyelitis*

windpuffs. Puffy enlargements at the fetlock joints

winging. Having feet swing inward as the horse travels

withers. Top of the shoulders of a horse

wolf tooth. Small vestigial first premolar

worms. Helminth parasites that live in the body of the host animal

Y

yearling. A horse between 1 and 2 years old

INDEX

Page numbers in *italics* indicate illustrations or photos;
numbers in **bold** indicate charts or tables.

STOREY'S GUIDE TO RAISING SERIES

For decades, animal lovers around the world have been turning to Storey's classic guides for the best instruction on everything from hatching chickens, tending sheep, and caring for horses to starting and maintaining a full-fledged livestock business. Now we're pleased to offer revised editions of the Storey's Guide to Raising series — plus one much-requested new book.

Whether you have been raising animals for a few months or a few decades, each book in the series offers clear, in-depth information on new breeds, latest production methods, and updated health care advice. Each book has been completely updated for the twenty-first century and contains all the information you will need to raise healthy, content, productive animals.

Storey's Guide to Raising BEEF CATTLE (3rd edition)

Storey's Guide to Raising RABBITS (4th edition)

Storey's Guide to Raising SHEEP (4th edition)

Storey's Guide to Raising HORSES (2nd edition)

Storey's Guide to Raising PIGS (3rd edition)

Storey's Guide to Raising CHICKENS (3rd edition)

Storey's Guide to Raising MINIATURE LIVESTOCK (NEW!)

Storey's Guide to Raising DAIRY GOATS

Storey's Guide to Raising MEAT GOATS

Storey's Guide to Raising TURKEYS

Storey's Guide to Raising POULTRY

Storey's Guide to Raising LLAMAS

Storey's Guide to Raising DUCKS